Lecture Notes in Civil Engineering 638

exacerbate velocity pulses. They emphasize the high sensitivity of structural response when records are rotated to fault-normal or maximum-pulse directions.

In the **second topic Seismic Performance of Bridges and Similar Systems**, Huang *et al.* (2025) use OpenSees to simulate small-scale PLEXUS 1g tests on integral abutment bridges, demonstrating close agreement with experimental pressure distributions behind abutments. They emphasize the interface modeling strategy and the importance of nonlinear soil behavior. Shid *et al.* (2025) analyze friction-based isolated piers under vertical dynamic loads, showing that vertical force fluctuations influence the isolator's friction coefficient. They find that including vertical variability can better predict overall pier-isolator performance. Sadeghzadeh *et al.* (2025) develop a parametric 3D finite element model for the Colle Castino Bridge, enabling extensive statistical evaluations. They explore uncertainties in geometry and materials, highlighting how a modular approach streamlines reliability studies. Di Benedetto *et al.* (2025) focus on dapped-end beams with corrosion damage, validating an external steel retrofit technique. They demonstrate that refined modeling can capture complex stress redistribution in half-joints.

Under the **third topic Industrial Equipment and Storage Tanks**, Rapone *et al.* (2025) derive fragility curves for steel storage tanks using both far- and near-field records, emphasizing the choice of ground motion sets. They illustrate that near-fault pulses can intensify tank damage states, significantly affecting seismic reliability. Chalarca *et al.* (2025) investigate pressurized tank–pipeline configurations, evidencing that dynamic interaction between tanks and pipelines can amplify local stresses. They propose careful modeling of node constraints to capture true structural demand. Rabi *et al.* (2025) develop fragility curves for deformed, ground-supported liquid tanks, targeting failure modes such as elephant foot buckling. They highlight the importance of modeling geometric imperfections for conservative design. Shang *et al.* (2025) developed an OpenSees-based hysteresis model for piping systems, reproducing experimental moment-rotation curves with high accuracy. They validate their model's applicability to large-scale suspended piping systems under seismic loads.

In the **fourth topic Masonry Structures and Geotechnical**, Vailati *et al.* (2025) present a specialized OpenSeesPy tool for seismic risk analysis of masonry clusters, leveraging typological-mechanical fragility curves. They illustrate how this approach can expedite territorial-scale evaluations. Villar *et al.* (2025) explicitly model masonry aggregates with diverse interconnections to study global seismic response. They reveal that partial connectivity leads to more pronounced local failures, underscoring the need for thorough structural unit identification. Duzandeh *et al.* (2025) explore basin effects through 2D site response analyses and compare various boundary condition formulations. They conclude that boundary modeling strategies can significantly affect computed surface accelerations in complex sedimentary basins. Sadeghzadeh *et al.* (2025) also compare nonlinear and equivalent-linear site response analyses in OpenSees, PLAXIS, and DeepSoil. They show that accurate soil modeling and hazard-compatible records are essential to reliable seismic design spectra.

Finally, the **fifth topic Advanced Modeling, Materials, and Optimization** includes Rabi *et al.* (2025) who derive closed-form moment-curvature expressions for hollow and full rectangular sections under combined axial load and bending. They confirm that these equations closely match fiber-section analyses in OpenSees, facilitating large-scale fragility evaluations. Contiguglia *et al.* (2025) apply machine learning to generate equivalent-strut models for CLT infills in RC frames, calibrating strut parameters with high-fidelity finite element studies. They report improved prediction of global force-displacement responses under lateral loading. Wang *et al.* (2025) investigate 3D-printed fiber concrete's anisotropy and implement a customized constitutive model in OpenSees, coupling it with topology optimization for performance-based designs. They observe that print orientation strongly influences material strength, which must be factored into advanced simulations. Lastly, Zhou *et al.* (2025) propose an augmented Kalman filter for moving force identification, integrating OpenSees-based simulations and genetic algorithms to optimize sensor placement. They demonstrate robust and efficient force reconstruction in bridge structural health monitoring.

Collectively, these works provide a valuable reference for further investigations, demonstrating how OpenSees continues to inspire novel approaches in both research and professional practice, fueling progress in earthquake engineering and related fields. In closing, we would like to express our deepest appreciation to all authors, reviewers, session chairs, and participants, whose active engagement and dedication made the EOSD 2024 conference a resounding success. Their enthusiasm, professionalism, and willingness to share knowledge nurture the global growth and cohesion of the OpenSees community.

July 2024

Xinzheng Lu
Cristoforo Demartino
Fabio Di Trapani
Kaiqi Lin
Giorgio Monti

Organization

Scientific Committee

Filip Filippou, Asif Usmani, Giorgio Monti, Frank McKenna, Humberto Varum, Joel Conte, José Miguel Castro, Cristoforo Demartino, Zhongguo Guan, Liming Jiang, Lizhong Jiang, Kazuhiko Kasai, Dimitrios Lignos, Fabrizio Mollaioli, Massimo Petracca, Giuseppe Quaranta, Xavier Romão, Enrico Spacone, Fabio Di Trapani, Rui Wang, Longhe Xu, Zhaodong Xu, and Christos Zeris.

Organizing Committee

Xinzheng Lu, Fabio Di Trapani, Cristoforo Demartino, Giorgio Monti, Antonio Sberna, Linlin Xie, Kaiqi Lin, Yuan Tian, Donglian Gu, and Yongjia Xu

Official Partners

Asdea Disruptive Engineering & Technology, Via Tiburtina Valeria, 287/3, 65129, Pescara (PE), Italy

Eurasian OpenSees (EOS)

Eurasian OpenSees (EOS) is an international non-profit association comprising academics and engineers who share the vision of advancing efficient computational tools for complex engineering analysis and design. The organization's mission is to:

- **Encourage and promote** rigorous structural engineering principles based on international research, particularly—though not exclusively—within the OpenSees computational framework developed at the University of California, Berkeley.
- **Organize, promote, and disseminate** research results, training initiatives, design advancements, and publications aimed at fostering fruitful exchanges among members, scholars, and institutions.
- **Convene and support** courses, workshops, seminars, conferences, and similar events that enhance global collaboration among structural engineers.

EOS capitalizes on significant contributions from European and Asian researchers in computational mechanics and structural engineering, aspiring to disseminate scientific and professional achievements realized via OpenSees. The overarching goal is to bridge the continental gap between East and West, building an Eurasian OpenSees-based community where ideas and innovation circulate freely.

EOS Organization

- **President:** Prof. Filip C. Filippou
- **Vice-President:** Prof. Asif Usmani
- **Treasurer:** Prof. Giorgio Monti
- **Counselors:** Dr. Frank McKenna, Prof. Humberto Varum

The **2024 Eurasian OpenSees Days (EOSD 2024)** conference and the accompanying **OpenSees Course** in Beijing, hosted by Tsinghua University, will continue the tradition of fostering an open and dynamic environment for scholars and professionals worldwide. These events aim to promote the exchange of innovative ideas, strengthen collaborative research in structural and geotechnical engineering, and further expand the global OpenSees community.

EOS
EURASIAN OPENSEES

EOSD 2024

Current and Past EOS Events

Conferences

- *First OpenSees Days Europe.* 24–25 May, 2012, Sapienza University of Rome, Italy
- *Second OpenSees Days Europe.* 10–11 June, 2015, University of Salerno, Italy

- *Third OpenSees Days Europe.* 19–20 June, 2017, University of Porto, Portugal
- *First OpenSees Days Eurasia.* 20–21 June, 2019, The Hong Kong Polytechnic University, Hong Kong
- *Second OpenSees Days Eurasia.* 7–8 July, 2022, Politecnico di Torino, Italy
- ***Sixth Eurasian OpenSees Days (EOSD 2024).*** 24–25 July, 2024, Tsinghua University, Beijing, China

OpenSees Courses

- *International Short Course on Seismic Analysis of Structures using OpenSees.* Sapienza University of Rome, March 2019
- *International Short Course on Seismic Analysis of Structures using OpenSees (Winter School).* Politecnico di Torino, Turin, January 2020
- *International Short Course on Seismic Analysis of Structures using OpenSees (Summer School).* University of Palermo, Palermo, July 2021
- *International Short Course on Seismic Analysis of Structures using OpenSees (Summer School).* Politecnico di Torino, Turin, July 2022
- ***International Short Course on Seismic Analysis of Structures using OpenSees (Summer School).*** Tsinghua University, Beijing, July 2024

Keynote Speakers

1. Anastasios Sextos, University of Bristol, UK
   "Multi-hazard assessment of offshore wind turbines: modeling and conceptual challenges"
2. Sashi Kunnath, University of California, Davis, USA
   "Simulation of strength degradation in structural components: advances and challenges"
3. Zhaodong Xu, Southeast University, P. R. China
   "On multi-dimensional earthquake isolation and mitigation for structures"
4. Giuseppe Marano, Politecnico di Torino, Italy
   "Recent trends in structural optimization: state-of-art and new perspectives fostered by OpenSees"
5. Guido Camata, Università degli Studi "G. D'Annunzio" Chieti, Italy
   "Advanced analysis and digital twinning with STKO and OpenSees"
6. Liming Jiang, The Hong Kong Polytechnic University, Hong Kong
   "Recent development of OpenSees for fire modeling structures in fire and AI-FE hybrid simulation"

Contents

Advanced Modeling, Materials, and Optimization

Seismic Behavior of Buildings and Frames

Design Value of Global Capacity in NLA with Global Factor Integrating Concurrent Failure Mechanisms

Christina El Moussawi, Raihan Rahmat Rabi(✉), and Giorgio Monti

Sapienza University, Rome, Italy
{christina.elmoussawi,giorgio.monti}@uniroma1.it,
raihan.rahmatrabi@gmail.com

Abstract. This study introduces a method for evaluating the design capacity of a structural system at different limit states using nonlinear analysis (NLA). At the core of this approach lies the Global Factor Method (GFM), which simplifies complex structural behaviour into concurrent local failure mechanisms (LFMs). These LFMs are organized hierarchically, based on their series or parallel arrangement within the system, allowing for the derivation of a global factor used for estimating the design value of the global capacity, matching a given target probability. Notably, this method eliminates the need for alternative metrics like interstorey drift or post-peak capacity reduction, directly identifying global failure as driven by concurrent LFMs. Its practical application involves conducting two nonlinear analyses—one with mean values of the basic variables and another accounting for uncertainties in the LFMs. Correlation among LFMs is handled pragmatically, by considering extremes of zero and full correlation to enhance usability in practical design and assessment scenarios. The efficiency of the proposed method is demonstrated through its application to a reinforced concrete frame modelled using OpenSeesPy. The developed method is intended for practical use in NLA, specifically tailored for practitioners seeking an accessible approach to estimate design values within a simplified framework. In a sense, it provides an important step toward formalizing design capacity assessment within everyday engineering practice. While it is not a fully rigorous solution, it offers a reasonably accurate estimate of the design capacity, bridging the gap left by current codes, which have so far overlooked this critical aspect in NLA.

Keywords: Structural global capacity · Global Factor Method (GFM) · Local Failure mechanisms (LFMs) · Non-Linear Analysis (NLA)

1 Introduction

Nonlinear Analysis (NLA), including pushover and incremental dynamic analysis, is increasingly applied to evaluate the structural performance and safety of existing constructions. This evaluation involves comparing demand-related quantities, such as displacements or accelerations, against corresponding capacity values. These quantities are

X. Lu et al. (Eds.): EOS 2024, LNCE 638, pp. 3–14, 2025.
https://doi.org/10.1007/978-3-031-90690-9_1

affected by inherent and epistemic uncertainties, necessitating the comparison of their design values. Current codes specify seismic demand in terms of action with a defined exceedance probability but lack a robust framework for determining a design value for the global capacity through NLA.

Such framework should aim at two main objectives: (1) to objectively identify the attainment of distinct global Limit States (LSs), such as damage, severe damage, and collapse prevention, and (2) to determine the corresponding global capacities in terms of design values matching predefined target LS-exceedance probabilities.

Regarding (1), using interstorey drift limits as a metric has significant drawbacks, mainly due to the wide variability in reported drift thresholds across the literature. Furthermore, in three-dimensional models, the interaction between translational and rotational motions complicates the interpretation of drift ratios, particularly in complex geometries. Thus, interstorey drifts may not help identifying global LSs.

An alternative method examines the pushover curve and identifies LSs in correspondence of significant changes in the curve, such as yield plateau formation or slope reductions. However, limits placed on a pushover curve are of phenomenological nature and inherently subjective. Thus, while practical, this approach should be critically assessed in the context of global LS identification.

A more objective approach requires tracking the development of plastic hinges as localized mechanisms to determine the attainment of global LSs. This method examines their spatial distribution, structural role, and plasticization extent. Analysing the sequence and degree of plasticization provides a more robust basis for objectively identifying LSs.

Regarding (2), the *fib* Model Code 2020 [1] introduces a framework that emphasizes probabilistic and semi-probabilistic approaches. Specifically, for NLA, it provides multiple safety formats to estimate structural capacity under seismic actions, such as full probabilistic, Partial Factors (PFM), and Global Factor (GFM) methods. The probabilistic method is the most rigorous but is resource-intensive, often requiring extensive simulations. The PFM and GFM are more practical for typical engineering projects, allowing designers to estimate the design capacity with a balance between computational efficiency and reliability.

The PFM establishes design values for material strengths and load effects to achieve the desired safety level but does not explicitly define *how* to identify specific global LSs. Instead, it provides a structured approach for deriving the design capacity by applying partial safety factors, adjusted to account for model uncertainties and variability in material properties and loading conditions.

The GFM in Model Code 2020 provides a simplified, semi-probabilistic approach to evaluating the design capacity of structures using NLA. Unlike the PFM, the GFM applies a single global factor to the entire structural capacity—whether it is expressed as global force, displacement, or any other relevant parameter—, essentially providing an "all-in-one" adjustment to account for uncertainties. This single factor encompasses intrinsic and epistemic randomness, yielding a design capacity matching the target reliability, which vary based on the intended performance level (*e.g.*, for different LSs like damage, severe damage, or collapse).

To apply GFM, the design capacity is calculated by taking the NLA result, obtained using the mean values of the basic properties, and by scaling it using a global factor. This approach is advantageous for complex nonlinear models, particularly when detailed data on individual partial factors is unavailable. However, it requires a reliable calibration of the global factor to ensure that it meets the target exceedance probabilities. The aim of this work is to arrive at an analytical definition of the global factor through an equation expressed as function of the target probability, of the global capacity uncertainty, and of the structural system logic.

The main challenge lies in estimating the uncertainty of global capacity. The *fib* Model Code [1] introduces a single global factor to adjust NLA results, accounting for analysis uncertainties and calibrated to meet reliability targets. However, the provided values are limited and may not reflect the variety of cases in existing buildings. The ECOV method [2], in contrast, suggests deriving global response uncertainty directly from the NLA. This practical approach estimates the coefficient of variation (CoV) by running two analyses—one with mean values and another with characteristic values—to approximate the mean and characteristic global capacities. This method, however, relies on the strong assumption that the characteristic-value analysis yields the characteristic global capacity, which is not strictly valid for nonlinear systems.

Numerous studies have explored the effectiveness of various GFMs, leading to several comparative investigations. Among these, important contributions have come from researchers such as Schlune et al. [3], Cervenka [2], Allaix et al. [4], and Castaldo et al. [5].

Although structural system reliability has been extensively studied, current design codes still lack an accepted, standardized procedure for identifying global failure. System effects are usually addressed indirectly through parameters like interstorey drift limits or phenomenological indicators on the pushover curve, alongside local reliability criteria.

This study proposes an improved GFM approach, with the premise that global factors should be derived directly from NLA outcomes. Here, the failure mode observed in NLA—performed with mean values—is represented as an arrangement of local failure mechanisms organized into series and parallel subsystems. The resulting global factor is expressed through an equation as a function of the system's logic, allowing for an integrated calculation that accounts for uncertainties across subsystems. In this way, the global factor reflects the response and the uncertainties of the very system being analysed, without resorting to predefined typological values.

By basing the global design capacity on an analytical expression for the global factor that reflects observed failure mechanisms, this enhanced GFM overcomes the limitations of predetermined factors and oversimplified assumptions. Its flexibility makes it applicable across a variety of structural systems and limit states, supporting a more reliable structural assessment approach.

2 Proposed Method

The proposed method defines the design capacity of a structural frame at a specified limit state (LS)—for instance, displacement at yield or collapse—by matching it with a target LS-exceedance probability. Nonlinear analysis (NLA) is used to track the evolution of all local failure mechanisms (LFMs), classified as ductile (*e.g.*, plastic hinges) and brittle (*e.g.*, shear or joint failures). Brittle mechanisms are excluded if they are properly subordinated to ductile ones through proper capacity design or local strengthening.

As NLA progresses, ductile LFMs can sequentially exceed yield, ultimate, or collapse LSs, depending on the magnitude of the local engineering demand parameter, EDP_{loc}, which may be rotation or curvature for a plastic hinge. To account for uncertainty in the basic variables, each LFM's LS is represented by a fragility function, $F_{LFM}(EDP_{loc})$. During NLA, EDP_{loc} values are expressed in terms of a global EDP, such as the frame top displacement, thereby linking all LFM fragility functions to the same global EDP, as $F_{ijk}(EDP)$. The notation $F_{ijk}(EDP)$ distinguishes the different LFM fragility functions, with the suffix ijk denoting the k-th LFM in the j-th column on the i-th storey.

The global fragility curves for the frame are determined by aggregating the local fragility curves across all LFMs, considering their hierarchical arrangement and role within the structure. Specifically, the capacity of the entire structural system is expressed as a function of the capacities of its components, with frames modeled as an alternating hierarchy of series and parallel subsystems (see Fig. 1). Here, the frame is treated as a series system of i stories, each story as a parallel system of j columns, and each column as a series system of k LFMs.

This hierarchical structure allows for a systematic approach to reliability assessment, building from local to global levels by combining fragility functions $F_{ijk}(R)$ of individual LFMs into fragility functions $F_{ij}(R)$ for columns, $F_i(R)$ for storeys, and ultimately $F(R)$ for the frame. The resulting global fragility function reflects the system layered nature, using failure probability criteria for both series and parallel systems.

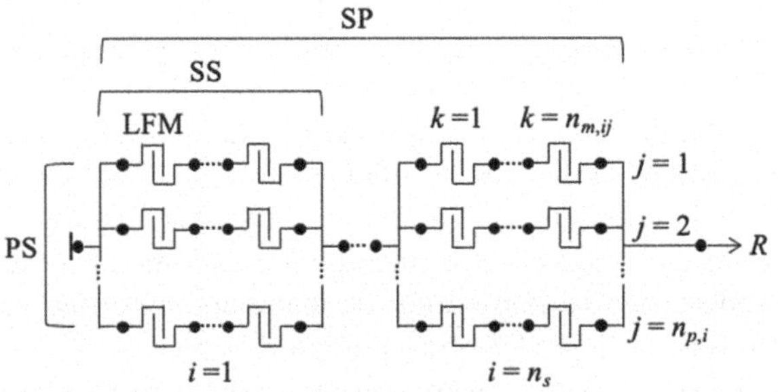

Fig. 1. Structural system logic: LFM = local failure mechanism, SS = series subsystem of LFM, PS = parallel subsystem of SSs, SP = series subsystem of PSs.

To account for the correlation among LFMs, the method evaluates fragility functions under both uncorrelated and fully correlated conditions, as in Table 1.

By combining the equations in Table 1, the fragility functions of the overall system for the two cases of uncorrelated and correlated LFMs can be written as:

$$F(R)_{unc} = 1 - \prod_i \left(1 - \prod_j \left(1 - \prod_k (1 - F_{ijk}(R))\right)\right) \tag{1}$$

Table 1. Fragility functions for columns, storeys, frame, as function of the fragility functions of the LFMs, in the two conditions of uncorrelation and correlation.

	Uncorrelated LFMs	Correlated LFMs
Column $F_{ij}(R)$	$1 - \prod_k (1 - F_{ijk}(R))$	$\max_k F_{ijk}(R)$
Storey $F_i(R)$	$\prod_j F_{ij}(R)$	$\min_j F_{ij}(R)$
Frame $F(R)$	$1 - \prod_i (1 - F_i(R))$	$\max_i F_i(R)$

$$F(R)_{cor} = \max_i \min_j \max_k F_{ijk}(R) \quad (2)$$

During the NLA, as R increases and more LFMs approach their respective LS, the probability of global LS-exceedance for the system increases according to the previous equations. When this probability reaches the predefined target for the LS in question, the system is considered to have reached the global LS. The corresponding value of the global $EDP = R$ at this point is defined as the design global capacity of the structural system, denoted as $R_{d,LS}$, satisfying:

$$F\left(R_{d,LS}\right) = \Phi\left(-\alpha_R \widehat{\beta_{LS}}\right) \quad (3)$$

where the target probability is expressed in terms of the sensitivity α_R attributed to the capacity and of the target safety index $\widehat{\beta_{LS}}$ attributed to the LS considered.

The values obtained under the assumptions of zero correlation and full correlation represent the bounds for the design value $R_{d,LS}$ of the global EDP for the given LS.

These can be obtained either by applying the full probabilistic approach shown above or by running a limited number of NLA to estimate mean and standard deviation of $F(R)$ in the two conditions of uncorrelation and correlation for all the LS of interest, as function of the fragility function parameters—mean and standard deviation—of each LFM, derived from the basic variables involved. Through these, a value for the appropriate global factor can be found using a purposely developed equation and then applied to the result of the NLFEA run with mean values of the basic variables to finally obtain the design value of the capacity.

3 Application and Results

The method's efficiency is demonstrated through its application to a reinforced concrete frame under lateral loads with column-driven failures. The structural system is modelled and simulated using OpenSeesPy, a Python-based framework for finite element modelling and analysis, particularly suitable for non-linear structural analysis. In this study, a three-storey, two-bay reinforced concrete (RC) frame structure is analysed (Fig. 2). The RC frame has been modelled as a two-dimensional structure consisting of and three storeys, labelled i, and arranged into series subsystem (SS), each having three columns, numbered with, j arranged into parallel subsystems (PS), each having two LFMs, denoted as k, representing the resisting mechanism at the bottom and the top of each element and arranged into series subsystems (SS).

The beams are assumed to exhibit linear elastic behaviour so are modelled using *elasticBeamColumn* element type. The columns are modelled using force-based beam-column elements, which allow for distributed plasticity along the length of the columns, eliminating the need to predetermine the location of plastic hinges, thus ensuring greater accuracy. The cross sections are represented using *FiberSection* object, enabling detailed modelling of the material behaviour at various locations across the cross-section. Fiber section allows each fiber to be assigned the constitutive relationships of its constituent materials. As for the materials, the *Concrete01* command, which simulates a uniaxial material with zero tensile stress, is used to model the unconfined concrete, while Mander's formulations has been used to model the confined one and finally, the reinforcement is modelled as uniaxial material with *Steel01*.

The material properties of concrete and steel are modelled as random variables following a normal distribution to account for inherent uncertainties as shown in Table 2. The non-linear analysis is conducted using Monte Carlo simulations, where 1000 random samples of these material properties are generated based on their respective mean values and standard deviations. These samples are then incorporated into the finite element model to analyse the structural response under lateral load. The analysis provides insights into the variability in the structure's behaviour, characterised by top displacement and lateral force.

The mean capacity $\mu_{R,ijk}$ of the ijk-th LFM is the value of the chosen EDP, corresponding to the attainment of the mean local capacity in the LFM at i, j, k. In Fig. 3 LFMs resulting from the nonlinear analysis of the structure are shown. At the local level, there are two resistance mechanisms that work in series, arranged at the ends of each element, for the two Limit States considered. In the nomenclature adopted, the first number represents the i-th storey, the second represents the j-th element; the blue and light blue colours represent respectively the mechanism at the base and at the top of the element for the damage LS (DLS), while the colours orange and yellow represent respectively the mechanism at the base and at the top of the element for the life safety LS (LLS).

The failure of each element, as shown in Fig. 4, occurs when one of the two local resistant mechanisms arranged in series fails, responsible for this failure is the weakest mechanism. The storey is considered as a parallel system, where the strongest elements are the considered, since the failure of one element does not necessarily lead to the failure of the storey (Fig. 5). Finally, the mean global capacity of the frame R_m, consisting of three storeys arranged in series, will be determined by the weakest storey with the minimum capacity among them. As final step of the proposed approach, the design capacity R_d is obtained as the ratio between the global mean capacity R_m and the global safety factor γ_R for each LS considered and in the extreme cases of zero and full correlation. The results are shown in Fig. 6, the mean global capacity R_m of the frame is represented with full circle markers, the design capacities R_d in the cases of full correlation represented

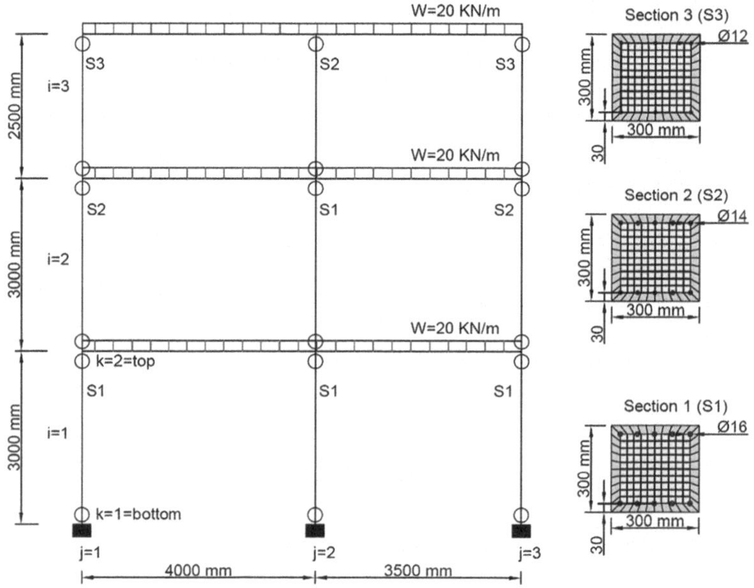

Fig. 2. 2D model of a 3-storey – 2-bay RC frame.

Table 2. Mean and coefficient of variation (CoV) of material properties.

Material Property	Mean μ []	CoV
Concrete compressive strength f_c [MPa]	20	0.2
Steel yield strength f_{sy} [MPa]	400	0.08
Steel modulus of elasticity E_s [GPa]	210	0.05

with triangle marker and zero correlation, represented with the square one, at DLS in blue and LLS in orange.

In a subsequent phase, the fragility curves of each LFM were numerically derived through Monte Carlo analysis. From these, progressing from the local to the global level, the fragility curve (FC) of the system is determined by combining the failure probabilities of series and parallel systems, considering the extreme cases of full and zero correlation, for the two limit states DLS and LLS. The results obtained were verified using the expressions by Nadarajah & Kotz [6], which provide the exact distribution of the maximum and minimum of two Gaussian random variables. The FCs derived from Nadarajah & Kotz's expressions are represented in the following figures with black dashed lines, showing how they overlap with those determined numerically. The design capacity is obtained using the mean and standard deviation values of the final fragility curve of the global system, and the inverse of the predefined target reliability probability.

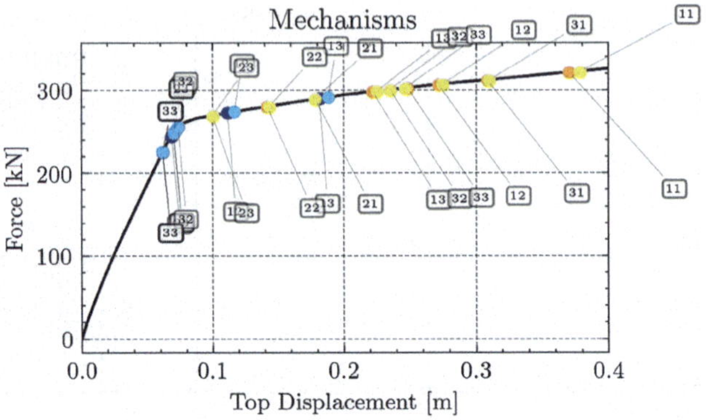

Fig. 3. Capacity of LFMs obtained directly from NLA through Monte Carlo analysis.

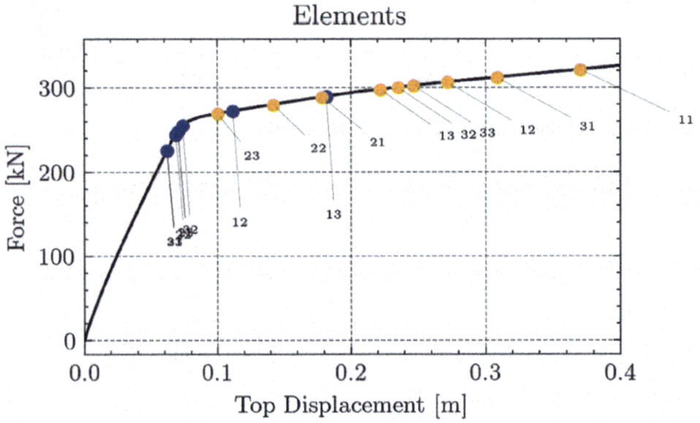

Fig. 4. Capacity of elements at DLS (blue) and LLS (orange)

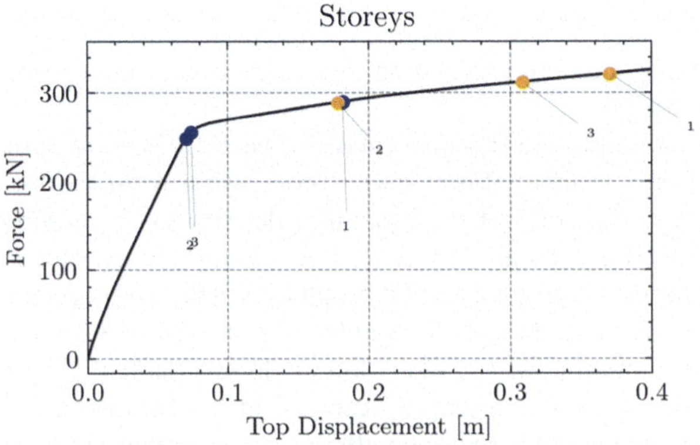

Fig. 5. Capacity of the storeys at DLS and LLS

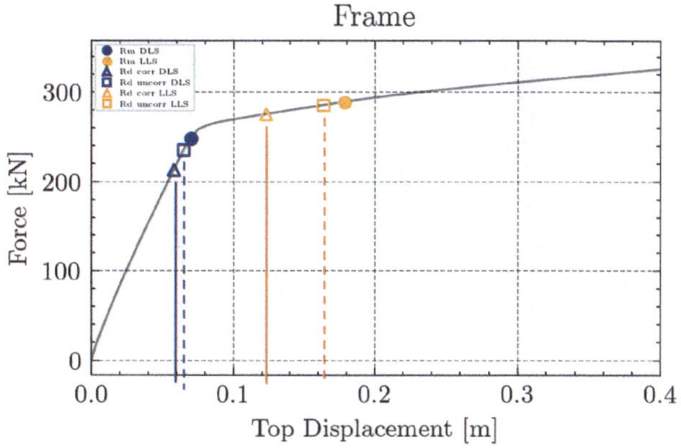

Fig. 6. Mean global capacity R_m (circle marker) of the frame identified by the weakest storeys in series, and design capacities R_d in the cases of full correlation (triangle marker) and zero correlation (square marker) at DLS (blue) and LLS (orange).

Figure 7 shows the fragility curves of the LFMs $F_{ijk}(R)$ resulting from the nonlinear analysis of the structure and obtained with Monte Carlo analysis for several iterations equal to 1000. At the local level, therefore, there are two resistance mechanisms that work in series, arranged at the ends of each element, for the two LLS considered. The probability of failure of the mechanisms arranged in series within each element determines the probability of failure of the elements $F_{ij}(R)$, Fig. 8. While the probabilities of failure of the $F_i(R)$ storeys, shown in Fig. 9, are determined by considering the two rightmost curves, which represent the strongest and the weakest elements that determine the failure of the system in parallel.

Finally, the probability of failure of the frame, $F(R)$, is obtained by considering the two leftmost curves representing the weaker storeys that determine the failure of the series storeys system, Fig. 10. It should be noted that, in the case of full correlation, the probability of failure of LFMs corresponds, from local to global, to the mechanism

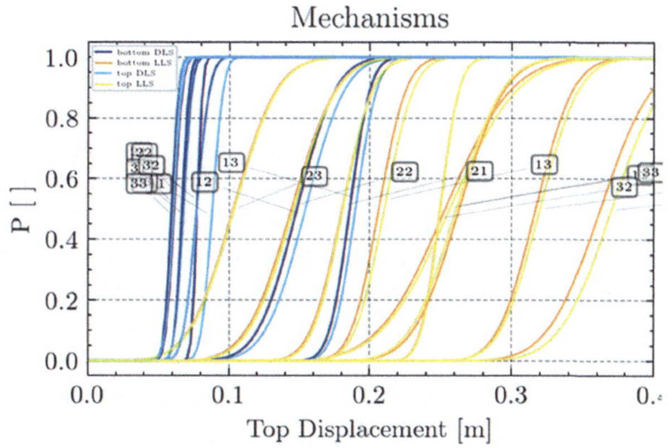

Fig. 7. Fragility curves of LFMs arranged in series for DLS (blue) and LLS (orange).

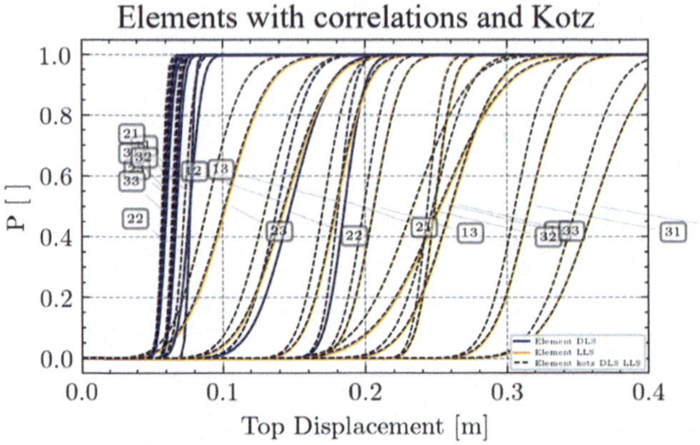

Fig. 8. Fragility curves of columns arranged in parallel, including those derived from the Nadarajah & Kotz expressions (black dashed line) for DLS (blue) and LLS (orange), in cases of full correlation (solid line) and zero correlation (dashed line).

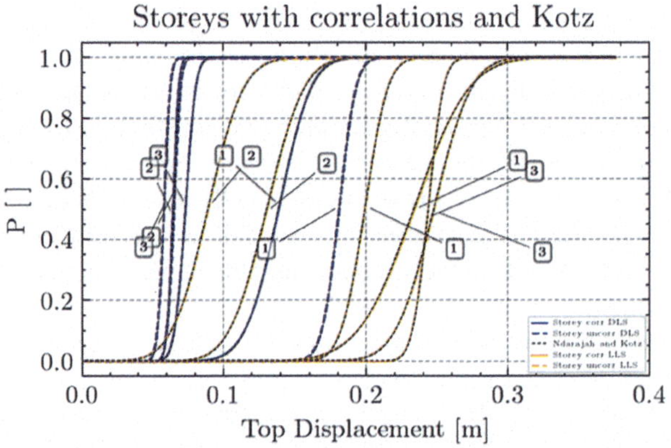

Fig. 9. Fragility curves of storeys arranged in series, including those derived from the Nadarajah & Kotz expressions (black dashed line) for DLS (blue) and LLS (orange), in cases of full correlation (solid line) and zero correlation (dashed line).

with the weakest capacity, the strongest structural elements, and the weakest storeys. Conversely, in the case of zero correlation, new fragility curves emerge, shifting to the left in series systems and to the right in PS compared to the initial curves, thereby altering the capacity of each system. Comparing these two cases within the same system, it is observed that in SS, non-correlation reduces the system's capacity, as the new curve is shifted leftward relative to the full correlation curve. In contrast, in PS, non-correlation enhances the system's capacity, as the new curve shifts rightward, albeit only in the initial section.

In the final comparison of results, the fully probabilistic approach yields the exact design capacity values, which have been shown to closely match those obtained through the proposed procedure.

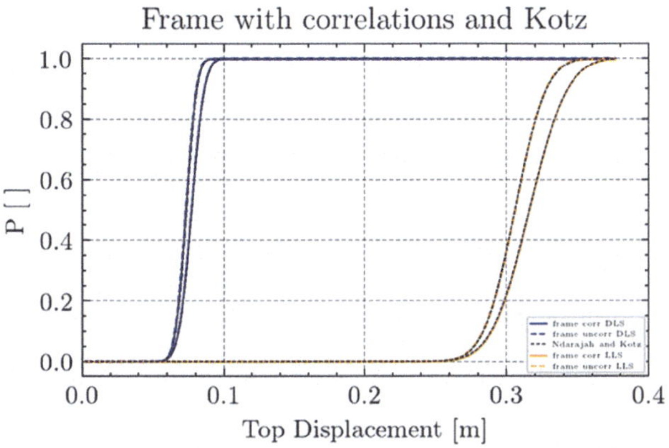

Fig. 10. Fragility curves of frame, including those derived from the Nadarajah & Kotz expressions (black dashed line) for DLS (blue) and LLS (orange), in cases of full correlation (solid line) and zero correlation (dashed line).

4 Conclusions

The proposed method offers a systematic approach to objectively determine the design capacity of frame systems subjected to lateral loads, aligned with specified safety targets. This study refines the Global Factor Method (GFM) by deriving global design capacities directly from the results of nonlinear analysis (NLA), considering the most likely failure mode and defining it as a series and parallel arrangement of local failure mechanisms (LFMs).

By integrating the structural system logic, this method establishes a clear link between LFMs and overall system failures, facilitating the definition of an objective metric for identifying global Limit State exceedance based on concurrent LFMs. Existing methods, such as the interstorey drift or phenomenological consideration on the pushover curve, do not adequately capture this relationship. Additionally, this approach delineates bounds for design capacity by considering scenarios of both uncorrelated and fully correlated LFMs, making it practical for application by practitioners in the field.

OpenSeesPy played a critical role in facilitating the NLA required for this study. Its fibre-based column models provided a detailed representation of material non-linearity. OpenSeesPy's support for Monte Carlo simulation further enhanced the study by allowing material variability to be incorporated into the analysis. This enabled a comprehensive assessment of the structural system's reliability under different correlation scenarios, contributing to the robustness of the method.

Future research will focus on extending the proposed framework to more complex structures. Studies and further investigations are currently underway on the partial correlation between basic random variables, which represents one of the most challenging aspects to address. These efforts aim to ensure the method's broad applicability, particularly for real-world engineering practices where variable geometries and detailing further complicate the assessment of global failure mechanisms.

References

1. Fédération internationale du béton. fib Model Code for Concrete Structures 2020. fib, Lausanne (2020)
2. Cervenka, V.: Reliability-Based Non-Linear Analysis According to fib Model Code 2010. Wiley-Blackwell (2013). https://doi.org/10.1002/suco.201200022
3. Schlune, H., Plos, M., Gylltoft, K.: Safety formats for non-linear analysis of concrete structures. Mag. Concr. Res. **64**(7), 563–574 (2012). https://doi.org/10.1680/macr.11.00046
4. Allaix, D.L., Carbone, V.I., Mancini, G.: Global safety format for non-linear analysis of reinforced concrete structures. Struct. Concr. **14**(1), 29–42 (2013). https://doi.org/10.1002/suco.201200017
5. Castaldo, P., Gino, D., Mancini, G.: Safety formats for non-linear finite element analysis of reinforced concrete structures: discussion, comparison and proposals. Eng. Struct. **193**, 136–153 (2019). https://doi.org/10.1016/j.engstruct.2019.05.029
6. Nadarajah, S., Kotz, S.: Exact distribution of the max/min of two Gaussian random variables. IEEE Trans. Very Large Scale Integr. VLSI Syst. **16**(2), 210–212 (2008). https://doi.org/10.1109/tvlsi.2007.912191

Numerical Modeling of Steel Frames with BRB-VED Hybrid Dampers

Mario Aguaguiña, Ying Zhou, and Yi Xiao(✉)

State Key Laboratory of Disaster Reduction in Civil Engineering, Tongji University, Shanghai 200092, China
yixiao@tongji.edu.cn

Abstract. BRB-VED hybrid dampers, resulting from the combination of a buckling-restrained brace (BRB) and viscoelastic dampers (VEDs), exhibit a multi-stage energy dissipation mechanism, with specific characteristics depending on the arrangement of their main components and the level of demand. This paper presents the numerical modeling schemes of three different configurations for BRB-VED hybrid dampers in OpenSees. The first two correspond to existing prototype devices, where the BRB and VEDs are combined either (1) in parallel or (2) in series. The third one corresponds to a new device setting using an (3) in-parallel damper combination with a gap system in series with the BRB. This work aims to establish practical modeling methods for BRB-VED hybrid dampers that can capture the main characteristics of their hysteretic response. First, BRB, VED, and gap system models are established and calibrated independently based on experimental data. Then, these components are combined according to the respective configuration. To validate the proposed models and observe their hysteretic behavior over a range of demand amplitudes, a damper model is subjected to displacement-controlled cyclic loading representing multiple levels of inter-story drift. The results show that the conceptual working mechanisms of the three configurations for BRB-VED hybrid dampers were effectively simulated. Therefore, the developed models can be used for seismic response analysis of frames incorporating BRB-VED hybrid dampers. A 4-story steel frame is employed as an example structure for this purpose.

Keywords: Energy Dissipation Device · Hybrid Damper · Buckling-restrained Brace · Viscoelastic Damper · Modeling Scheme · Seismic Response Analysis

1 Introduction

Hybrid dampers are special energy dissipation devices (EDDs) that combine different materials or mechanisms to control the structural response effectively across a wide range of vibration levels. To date, multiple device settings have been developed for hybrid EDDs, with brace-type configurations being the most prevalent [1]. Among these, the majority include a BRB in combination with VEDs, referred to as BRB-VED hybrid dampers. The combination of these EDDs is considered suitable because their advantages are complementary, and their limitations can be offset by the characteristics of one

X. Lu et al. (Eds.): EOS 2024, LNCE 638, pp. 15–27, 2025.
https://doi.org/10.1007/978-3-031-90690-9_2

another. Specifically, BRBs provide substantial hysteretic energy dissipation upon yielding, especially under strong excitations where large deformation demands are imposed, while VEDs can dissipate energy through shear deformation of their viscoelastic (VE) layers, without a yielding mechanism. This feature makes VEDs suitable for supplemental damping at all vibration levels and compensates for the inability of BRBs to dissipate energy under small deformations.

Existing BRB-VED hybrid dampers can be divided into two types, based on the arrangement of their main components: (1) in-parallel [2–5]; and (2) in series [6–8]. These configurations have distinct characteristics and exhibit a particular behavior across multiple demands levels [2–8]. Notably, the feasibility of a third device configuration with a new working mechanism is subject of study by one of the authors, whose characteristics, together with those of previous alternatives, are introduced in Sect. 2. In general, the broad energy dissipation capability and seismic response control effectiveness of previous BRB-VED hybrid dampers have been demonstrated through testing and seismic response analyses. Nevertheless, an optimal solution in terms of device configuration is still unclear due to the lack of a systematic comparative study of the seismic response control effectiveness of the different alternatives for BRB-VED hybrid dampers. In addition, despite the fact that numerical models for previous BRB-VED hybrid dampers have been provided, these models have some downsides limiting their application, i.e., they are either (a) particular of each device, and thus are unable to be generalized [2, 4] or (b) overly simplified, e.g., lumped mass shear model [6], hybrid damper not modeled as a brace [8]. Therefore, general and practical modeling strategies for these hybrid EDDs, which correctly represent their main components and accurately simulate their hysteretic behavior, are still needed.

This paper has three main objectives: (i) to introduce the working mechanism and main characteristics of three device configurations for BRB-VED hybrid dampers; (ii) to illustrate the development of the numerical modeling schemes of BRB-VED hybrid dampers using the built-in material and element libraries in OpenSees [9], including their validation through cyclic loading damper test simulations; and (iii) to show the results from the application of the developed models in the seismic response analysis of a 4-story steel frame with BRB-VED hybrid dampers.

2 BRB-VED Hybrid Dampers

2.1 Existing Device Configurations

Six prototype BRB-VED hybrid dampers have been proposed and investigated in the past, which can be categorized as (1) in-parallel and (2) in-series EDD combination.

In-parallel EDD Combination. Figure 1a shows the construction schemes and conceptual models of the four prototype devices using this configuration. In this arrangement type, the BRB and the VEDs are set side-by-side. As a result, both EDDs experience the same deformation across all excitation levels, and always contribute to resisting the demand. In practice, this configuration is achieved by attaching the VEDs to the BRB restrainer through bolting or welding. In such a setting, the restrainer not only provides continuous lateral support for the steel core, but also works as an elastic axial element in series with the VEDs to transfer their force along the brace.

In terms of energy dissipation, these BRB-VED hybrid dampers exhibit a VED-only behavior under small deformations (up to BRB yielding). In this stage, the damping capacity of VEDs is not fully exploited due to the contribution of the relatively large elastic stiffness of the steel core. In a second stage, once the demand triggers yielding of the BRB, the energy dissipation of the steel core starts and adds on to that of the VEDs. Here, as demand increases, e.g., strong earthquakes, a larger contribution in energy dissipation from the BRB, in relation to that of VEDs, is expected.

In-series EDD Combination. Figure 1b shows the construction schemes and conceptual models of the two prototype devices using this configuration, where the BRB and the VEDs are connected end-to-end. Yet, this combination is only feasible by the incorporation of a third element called 'gap system', set in parallel with the VED, which has two functions: (1) delay the activation of the BRB for a certain deformation; and (2) provide the VED with added strength and stiffness to yield the BRB once engaged. A gap system, typically formed by a slotted bolted connection, is characterized by a two-stage behavior: (a) for deformations smaller than the 'gap size', zero-force is provided, allowing VED to resist the force alone within the VED-gap ensemble; and (b) for deformations larger than the 'gap size', a sudden increase of stiffness occurs as the bolts come into contact with the edges of the slotted holes, causing the bolts to bear the entire force within the ensemble and creating a 'lock out' effect on the VED.

Here, BRB-VED hybrid dampers exhibit a VED-only or a BRB-only behavior for deformations within or beyond gap size, respectively. In this way, the damping capacity of the VED is exploited more effectively in the small deformation range (controlled by the gap size) as the VED dissipates energy on its own (without the BRB).

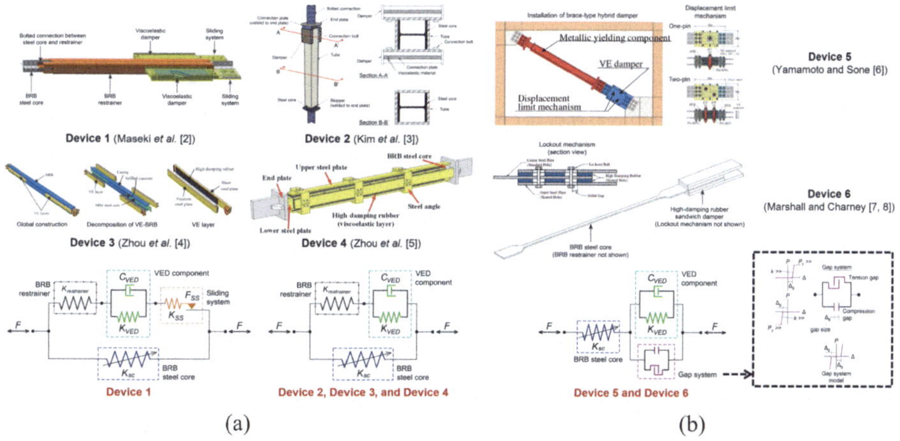

Fig. 1. Construction schemes and conceptual models of existing BRB-VED hybrid dampers: (a) in-parallel EDD combination; and (b) in-series EDD combination.

2.2 New Device Configuration

The new BRB-VED hybrid damper combines a BRB with a pair of VEDs in parallel and adds a gap system in series with the BRB steel core. Figure 2 shows the construction scheme and conceptual model of the new BRB-VED hybrid damper. A specially designed steel member, called 'bolted-to-pinned end connection adapter', has two functions: (1) creating the gap, using a set of plates with slotted holes arranged at the BRB end, and (2) enabling the in-parallel EDD action, through a thick end plate where the BRB-gap plates and additional top and bottom plates for VEDs fixation are welded.

The new configuration has a three-fold purpose: (1) to maximize the damping performance of the VEDs in the small deformation range, for deformations smaller than the gap size; (2) to control the deformation level at which the BRB is engaged, by adjusting the gap size of the gap system; and (3) to allow a combined action of the two EDDs once the BRB is engaged, for deformations larger than the gap size.

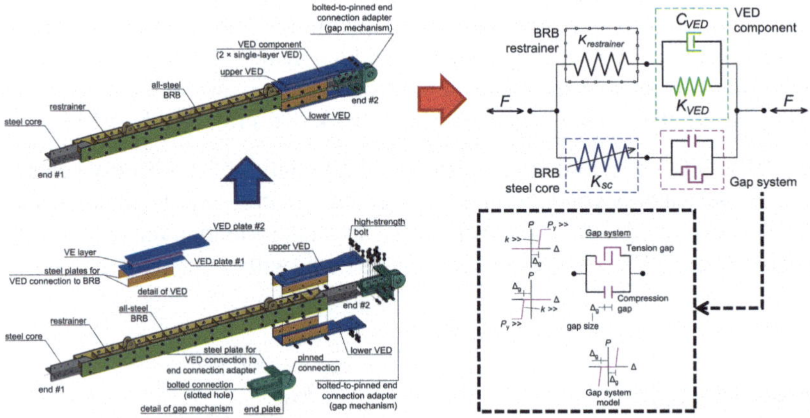

Fig. 2. Construction scheme and conceptual model of the new BRB-VED hybrid damper.

3 Numerical Modeling of BRB-VED Hybrid Dampers

3.1 Modeling of Individual Components

For the development of the models of BRB-VED hybrid dampers in OpenSees, it was necessary to first establish and calibrate the models of their individual components, i.e., BRB, VED, and gap system. A common single-story, single-bay frame, with a height of 6 m and a width of 6.4 m (i.e., work-point length $L_{wp} = 8.77$ m and brace inclination angle $\varphi = 43.15°$), was employed for the construction of all the models.

BRB Modeling. The BRB steel core (Q235 steel, $F_y = 293.75$ N/mm^2, $E_s = 206 \times 10^3$ N/mm^2) consists of three elements: a central yielding segment with cross-sectional area $A_{sc} = 5000$ mm^2 and a length $L_{ys} = 0.5L_{wp}$; and two non-yielding segments (one per end), with a length $L_{nys} = 0.1L_{wp}$ each, representing the transition regions and connection zones of the steel core. These segments were modeled as *elasticBeamColumn* elements with a high bending stiffness and a cross-sectional area $A_{nys} = 3.2A_{sc}$.

The yielding segment, where the inelastic response of the steel core occurs, was modeled as a corotational truss (*corotTruss*) element assigned with a composed material, i.e., *Steel02 uniaxialMaterial* wrapped with *Fatigue uniaxialMaterial* of OpenSees. *Steel02* corresponds to the Giuffre-Menegotto-Pinto constitutive model, which can simulate the hysteretic behavior of the steel core due to cyclic loading, considering kinematic and isotropic hardening. Calibration of *Steel02* parameters was done based on BRB tests [10]: $b = 0.005$, $R0 = 23.0$, $cR1 = 0.925$, $cR2 = 0.15$, $a1 = 0.05$, $a2 = 1.0$, $a3 = 0.05$, and $a4 = 1.0$. *Fatigue*, on the other hand, can capture the cumulative damage due to cyclic loading, based on the Coffin-Manson relationship for low-cycle fatigue. The parameters ε_0 and m for definition of the Coffin-Manson curve were 0.387 and 0.713, respectively, based on BRB test data [11]. Finally, a strain limit, ε_{max}, of 5% was assigned to simulate BRB fracture due to excessive deformation.

VED Modeling. The approach proposed by Zhou *et al.* [12] for the modeling of nonlinear VEDs was employed. Here, the VEDs are modeled using *twoNodeLink* elements assigned with a customized material. The so-called *NL-VED uniaxialMaterial* is the numerical implementation of a mechanical model for VEDs presenting multiple sources of nonlinearity, developed by Zhou *et al.* [13]. The VED used in this work has 4 VE layers, with a shear area $A_v = 200 \times 600$ mm^2 and a thickness $h_v = 25$ mm.

The *NL-VED uniaxialMaterial* developed in OpenSees consists of an in-parallel combination of three elements: (1) a nonlinear spring K_1, representing the softening phenomenon caused by the Mullins effect; (2) a nonlinear spring K_2, representing the relationship between the maximum force and deformation of the VED, especially the hardening under large shear strain; and (3) a nonlinear dashpot C-α, modeling the energy dissipation of nonlinear VEDs, which exhibit a low sensitivity to loading frequency. Finally, this model also includes force correction factors, λ_1 and λ_2, considering the effect of ambient temperature and softening due to heating and fatigue.

Gap System Modeling. The construction of the *Gap uniaxialMaterial* is illustrated in Fig. 3. It was built based on the *ElasticPPGap uniaxialMaterial* of OpenSees, which allows the definition of a unidirectional force-deformation relationship with an initial deformation gap, over which the restoring force is zero. As the gap system is bidirectional, tension and compression gaps are defined separately and combined in parallel.

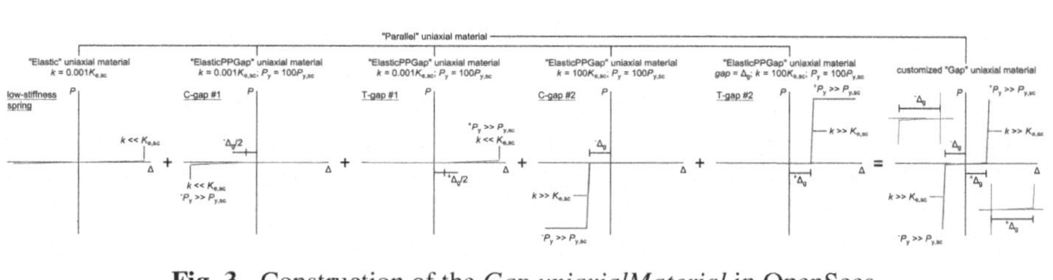

Fig. 3. Construction of the *Gap uniaxialMaterial* in OpenSees.

3.2 Modeling of BRB-VED Hybrid Dampers

Figure 4 shows the modeling schemes of the three configurations for BRB-VED hybrid dampers: (1) BRB-VED hybrid damper with in-parallel combination, referred to as Config. No. 1 – BRB-VED; (2) BRB-VED hybrid damper with in-series combination, named as Config. No. 2 – BRB-VEDg; and (3) new BRB-VED hybrid damper, designated as Config. No. 3 – BRBg-VED. The letter 'g' in Configs. No. 2 and No. 3 indicates the incorporation of a gap system – in parallel with the VEDs and in series with the BRB, respectively. Also, note that for Config. No. 1, only the conceptual model corresponding to Devices 2, 3, and 4 (see Fig. 1a) was considered in this study.

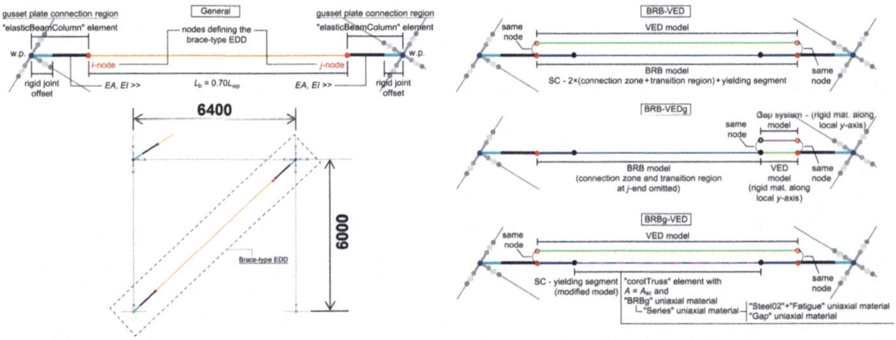

Fig. 4. Modeling schemes of three device configurations for BRB-VED hybrid dampers.

Config. No. 1: BRB-VED. The modeling of Config. No. 1 is straightforward because the BRB and VED models are directly applied as they were originally developed. Both EDDs are defined using the same end nodes. This ensures that the two dampers are subjected to the same deformation, thereby simulating an in-parallel action.

Config. No. 2: BRB-VEDg. The modeling of Config. No. 2 required adaptations of the BRB and VED models: (1) the non-yielding portion at the *j*-end of the steel core was omitted to facilitate the definition of the *twoNodeLink* elements corresponding to the VED and gap system; (2) the yielding segment of the steel core was modeled as a *twoNodeLink* element (instead of *corotTruss*) to achieve the in-series connection with the VED and gap system; and (3) an *Elastic uniaxialMaterial* with high stiffness was defined along the transverse direction of the VED element to prevent instability of the brace-type damper resulting from the in-series connection of axial elements.

Config. No. 3: BRBg-VED. Here, the in-series connection between the BRB steel core and gap system was accomplished by a combination of their respective materials through the OpenSees command *Series uniaxialMaterial*. As a result, a new material, called *BRBg uniaxialMaterial*, was created and applied only to the yielding segment of the steel core. The VED model was implemented as in Config. No. 1.

3.3 BRB-VED Hybrid Damper Test Simulation

To verify the modeling of BRB-VED hybrid dampers, cyclic loading test simulations were performed in OpenSees. Figure 5 shows the loading protocol used for these tests, which was formed based on several criteria: (i) displacement-controlled sine loading (f = 0.5 Hz) applied in the small (L1), middle (L2), and large deformation range (L3); (ii) the brace deformation $\Delta_{b,i}$ was set in terms of the inter-story drift ratio (IDR) $\theta_{s,i}$, based on Eq. (1); (iii) IDR limits specified by the Chinese seismic code [14] for minor and major earthquakes, i.e., $\Theta_{L1} = 0.4\%$ and $\Theta_{L3} = 2.0\%$, respectively, were used to compute the maximum amplitude for L1 and L3, respectively. The maximum amplitude for L2 was taken as half of that of L3; (iv) the applied values of $\theta_{s,i}$ were 0.1, 0.2, 0.4% for L1, 0.6, 0.8, 1.0% for L2, and 1.2, 1.5, 2.0% for L3, with three cycles each. In Configs. No. 2 and No. 3, a gap size $\Delta_g = 16$ mm was assumed ($\theta_s = 0.37\% \approx \Theta_{L1}$). Figure 6 presents the results of the numerical tests of BRB-VED hybrid dampers.

$$\Delta_{b,i} = \theta_{s,i} L_0 \sin\varphi \cos\varphi \tag{1}$$

where L_0 is the undeformed brace length (L_{wp}), and φ is the brace inclination angle.

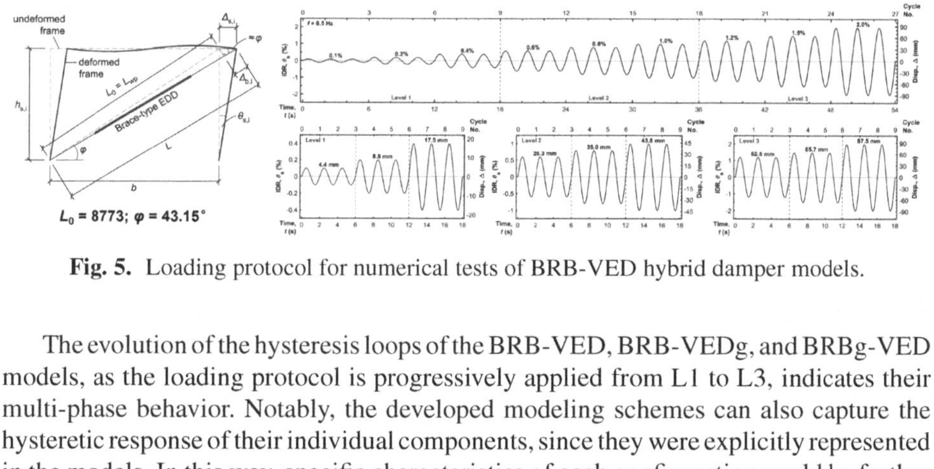

Fig. 5. Loading protocol for numerical tests of BRB-VED hybrid damper models.

The evolution of the hysteresis loops of the BRB-VED, BRB-VEDg, and BRBg-VED models, as the loading protocol is progressively applied from L1 to L3, indicates their multi-phase behavior. Notably, the developed modeling schemes can also capture the hysteretic response of their individual components, since they were explicitly represented in the models. In this way, specific characteristics of each configuration could be further verified. Figure 7 shows the hysteretic response of the three BRB-VED hybrid damper configurations and, directly beneath it, that of each component. In Config. No. 1, the BRB and VED are subjected to the same deformation due to their in-parallel combination. However, because of this, the VED is not able to resist the largest deformation amplitude of the loading sequence as its ultimate shear strain ($\gamma_u = 300\%$) is reached. In Config. No. 2, the 'lock-out' effect of the gap system on VED can be observed, i.e., VED deformation occurs only between limits defined by the gap size Δ_g in tension and compression and stops beyond Δ_g. In Config. No. 3, the VED is subjected to the same deformation as the BRB-gap ensemble (BRBg). The gap here, as in Config. No. 2, delays the activation of the BRB until a deformation of Δ_g is reached. A key difference between Configs. No. 2 and No. 3 is that the latter exhibits a larger stiffness post-Δ_g, due to the in-parallel combination between the two dampers. Finally, it is important to note that the gap

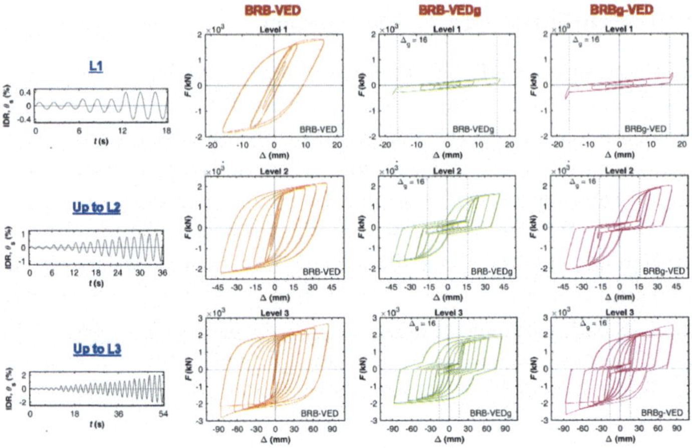

Fig. 6. Hysteretic response of models of Configs. No. 1, No. 2, and No. 3 to cyclic loading.

system in Configs. No. 2 and No. 3, reduces the deformation experienced by the BRB in comparison to Config. No. 1.

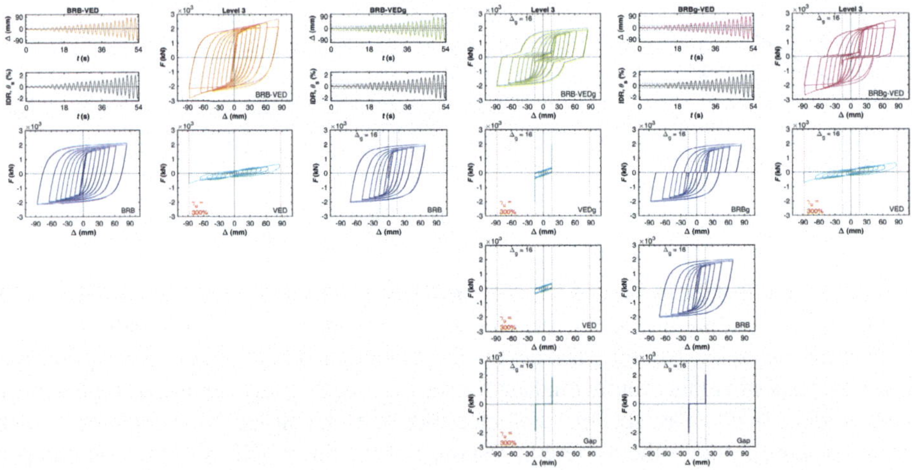

Fig. 7. Hysteretic response of the individual components of BRB-VED hybrid dampers.

4 Application Example

4.1 Prototype Structure

The developed models were applied in the seismic response analysis of a steel frame with BRB-VED hybrid dampers (three types). Prototype frames were designed per Chinese code provisions, and analyses were performed under three earthquake levels.

Building Description. The 4-story steel building from the JSSI Manual [15], shown in Fig. 8, was selected as the example structure. An interior frame in the X-direction was chosen as the analysis model. Figure 8 also provides information regarding the building's use, location, materials, and gravity loads assumed for design and analysis. Each prototype frame consists of a steel moment frame, primarily designed for gravity loads, with added BRB-VED hybrid dampers (three types), thus forming a dual system to resist seismic actions. Table 1 lists the sections of beams and columns.

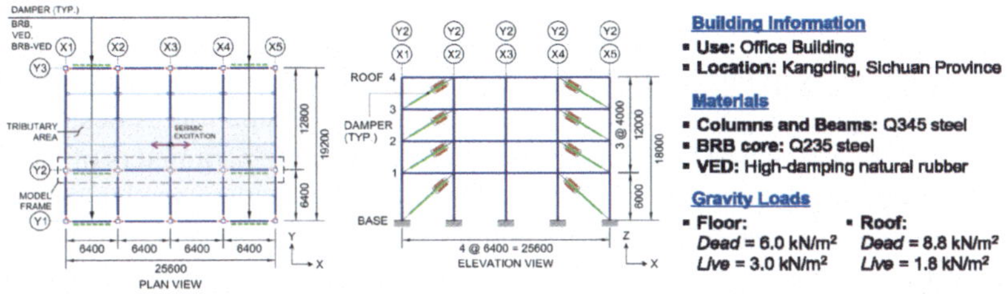

Fig. 8. Prototype structure: 4-story steel frame with BRB-VE hybrid dampers.

Table 1. Steel frame member sizes.

Story/Floor	Box column[a]	H-shaped beam[b]
4	300 × 300 × 20	300 × 250 × 10 × 22
3	350 × 350 × 14	350 × 300 × 12 × 28
2	350 × 350 × 20	400 × 300 × 16 × 28
1	350 × 350 × 25	450 × 300 × 20 × 28

[a] depth × width × wall thickness (mm).
[b] depth × flange width × web thickness × flange thickness (mm).

Seismic Hazard. Minor, moderate, and major earthquakes, termed EQ-L1 (63.2/50), EQ-L2 (10/50), EQ-L3 (2/50), respectively, as defined in GB 50011–2010 [14], were the three seismic hazard levels (SHLs) considered. Three sets of 20 earthquake records were developed for response history analyses, in correspondence with the three SHLs. Figure 9 shows the design response spectrum for EQ-L1, EQ-L2, and EQ-L3, as well as the response spectra of the selected ground motions scaled to EQ-L2.

Design. Due to their multi-phase nature, the design of BRB-VED hybrid dampers can be carried out such that their components are sized and tailored to achieve a specific performance objective under a particular SHL. First, the VED is designed to develop its maximum damping capacity under EQ-L1, so that the control of the response relies primarily on these EDDs. A displacement-based design method [12] was used to determine the sizes of the VEDs in the prototype frames. Second, the BRB is designed to develop its full energy dissipation capacity under EQ-L3. As the contribution in energy

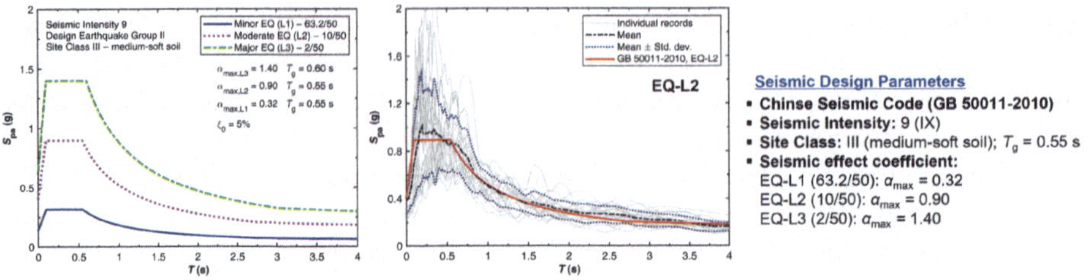

Fig. 9. Design spectrum of three SHLs and acceleration spectra of 20 records scaled to EQ-L2.

dissipation from the VED under EQ-L3 is neglected, the BRBs are primarily responsible for controlling the response through hysteretic energy dissipation. A design method based on performance curves [15] was applied to determine the parameters of the BRBs. Finally, regarding the gap system in Configs. No. 2 and No. 3, the gap size was set as a brace deformation $\Delta_{b,i}$ corresponding to an inter-story drift ratio $\theta_{s,i}$ equal to the code limit under EQ-L1 (i.e., $\theta_{s,i} = \Theta_{L1}$). Table 2 lists the properties of the components of the BRB-VED hybrid dampers added to the prototype frames.

Table 2. Properties of components of BRB-VED hybrid dampers: VED, BRB, Gap System.

		VED component			BRB component			Gap System
Story	No. of dampers	A_v (mm^2)	h_v (mm)	No. of VE layers	A_{sc} (mm^2)	$P_{y,sc}$ (kN)	$K_{e,sc}$ (kN/mm)	Δ_g (mm)
4	2	200 × 450	20	2	150 × 18	793.1	131.0	12.0
3	2	200 × 500	20	2	200 × 20	1175.0	194.1	12.0
2	2	200 × 550	20	2	200 × 22	1292.5	213.5	12.0
1	2	200 × 600	25	4	200 × 25	1468.8	205.9	16.0

VEDs: A_v = VE layer area; h_v = VE layer thickness. Gap System: Δ_g = gap size.
BRBs: A_{sc} = cross-sectional area of steel core; $P_{y,sc}$ = yield strength; $K_{e,sc}$ = elastic stiffness.

4.2 Modeling

Three models were established in correspondence with the design of the prototypes. The modeling approach involved constructing a common numerical model for the steel frame and adding the models representing the BRB-VED hybrid dampers for the three studied configurations, namely BRB-VED, BRB-VEDg, and BRBg-VED.

The steel frame was modeled using a concentrated plasticity approach, that is, *elasticBeamColumn elements* were used to represent columns and beams, with inelastic rotational springs at their ends to account for material nonlinearity due to plastic hinge formation. Rotational springs were modeled using *zeroLength elements* assigned with

Steel01 uniaxialMaterial of OpenSees, defining a bilinear moment-rotation relationship (with post-yielding stiffness ratio of 0.001). Panel zones were modeled as rigid, and the columns were fixed at the base. P-Delta transformations were assigned to the columns to account for geometric nonlinearity. The mass was calculated based on 100% of the dead load plus 25% of the live load, resulting in a seismic weight Ws of 7980 kN. Finally, for nonlinear dynamic analyses, Rayleigh damping was incorporated, assuming a viscous damping ratio of 2% for the 1st and 2nd modes of vibration.

Based on eigenvalue analysis, the fundamental period T_{X1} of the streel frame is 1.433 s. Regarding the prototypes, T_{X1} for the BRB-VED, BRB-VEDg, and BRBg-VED models is 0.573 s, 1.236 s, and 1.224 s, respectively.

4.3 Analysis Results

Nonlinear time-history analyses of the prototype frames were performed in OpenSees (3 models × 20 records × 3 SHLs). Examples of the results at the global (system) and local (BRB-VED hybrid damper) level are provided in Figs. 10 and 11, respectively.

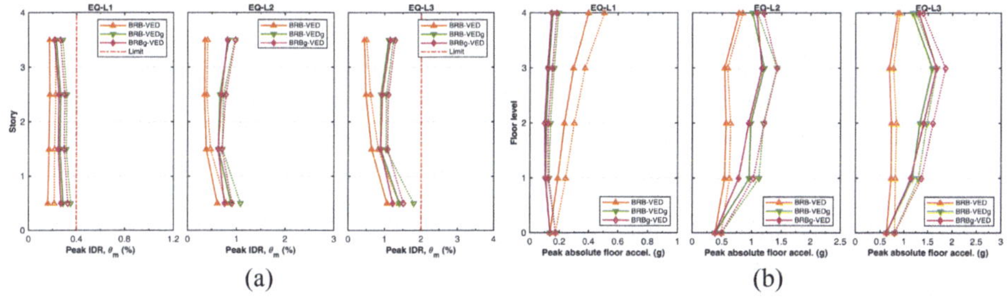

Fig. 10. (a) Peak IDR; (b) Peak absolute floor acceleration. Solid line: average response across the 20 ground motions, $\bar{x}$. Dotted line: average plus one standard deviation $\bar{x}$ + SD.

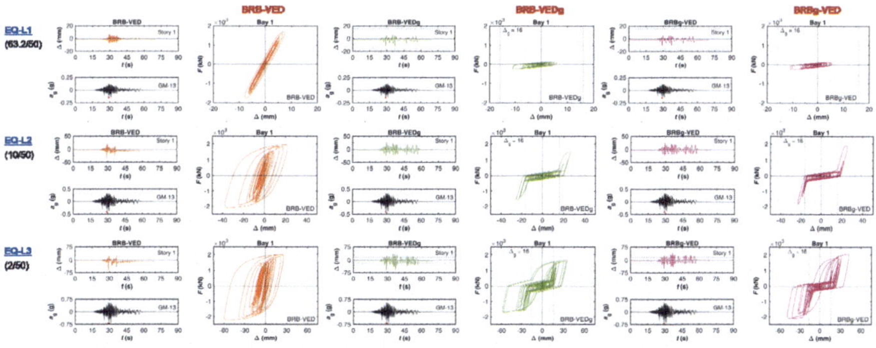

Fig. 11. Hysteretic response of a BRB-VED hybrid damper at Story 1 for the record GM-13.

5 Conclusions

This paper presents the development of modeling schemes for three BRB-VED hybrid damper types in OpenSees: (1) in-parallel; (2) in-series; and (3) a new configuration. The models were tested under dynamic cyclic loading and applied to a 4-story steel frame for response history analyses. Results from the numerical damper test and nonlinear dynamic analyses reflect the multi-stage energy dissipation behavior of BRB-VED hybrid dampers and demonstrate the characteristics of each configuration. In summary, the conceptual working mechanisms of the three configurations for BRB-VED hybrid dampers, including the behavior of their components, were effectively simulated. Therefore, the developed models can be used in future studies to assess the response control effectiveness of different BRB-VED hybrid damper configurations.

Acknowledgments. The authors gratefully acknowledge the financial support from the National Natural Science Foundation of China (Grant No. 52025083, 52208501) and the Shanghai Science and Technology Innovation Action Plan (Grant No. 22dz1201400).

References

1. Arvind, R., Santhi, M.H.: A state of art review on hybrid passive energy dissipating devices. J. Vib. Eng. Technol. **10**(5), 1931–1954 (2022)
2. Maseki, R., Narihara, H., Kimura, Y., Issiki, Y.: Experimental study on hybrid damper composed of visco-elastic element and buckling restrained brace element connected in parallel. J. Struct. Eng. AIJ **57B**, 271–278 (2011). (in Japanese)
3. Kim, D.H., Lee, C.H., Ju, Y.K.: Experimental investigation of hybrid buckling-restrained braces. Int. J. Steel Struct. **17**(1), 245–255 (2017)
4. Zhou, Y., Gong, S.M., Hu, Q., Wu, R.L.: An innovative BRB with viscoelastic layers: performance evaluation and numerical simulation. Struct. Monit. Maint. **5**(2), 205–229 (2018)
5. Zhou, X.Z., Sun, T.S., Sun, B.Y., Ma, N., Ou, J.P.: Vibration-reduction strategy for high-rise braced frame using viscoelastic-yielding compounded BRB. Buildings **12**(8), 1159 (2022)
6. Marshall, J.D., Charney, F.A.: A hybrid passive control device for steel structures: I. Development and analysis. J. Constr. Steel Res. **66**(10), 1278–1286 (2010)
7. Marshall, J.D., Charney, F.A.: A hybrid passive control device for steel structures: II. Physical testing. J. Constr. Steel Res. **66**(10), 1287–1294 (2010)
8. Yamamoto, M., Sone, T.: Damping systems that are effective over a wide range of displacement amplitudes using metallic yielding component and viscoelastic damper in series. Earthq. Eng. Struct. Dyn. **43**(14), 2097–2114 (2014)
9. Mckenna, F., Scott, M.H., Fenves, G.L.: Nonlinear finite-element analysis software architecture using object composition. J. Comput. Civ. Eng. **24**(1), 95–107 (2010)
10. Takeuchi, T., Hajjar, J.F., Matsui, R., Nishimoto, K., Aiken, I.D.: Effect of local buckling core plate restraint in buckling restrained braces. Eng. Struct. **44**, 304–311 (2012)
11. Tong, C., Wu, J., Hua, K., Xie, L.Q.: Low-cycle fatigue estimation curve for buckling-restrained braces based on cumulative plastic deformation. J. Earthq. Eng. **26**(6), 2773–2801 (2022)
12. Zhou, Y., Aguaguiña, M., Beskos, D.E., Gong, S.M.: A displacement-based seismic design method for building structures with nonlinear viscoelastic dampers. Bull. Earthq. Eng. **19**(11), 4535–4585 (2021)

13. Zhou, Y., Chen, P., Gong, S.M.: Mechanical model of a hybrid non-linear viscoelastic material damping device with its verifications. Front. Mater. **6**, 33 (2019)
14. Ministry of Housing and Urban-Rural Development of the People's Republic of China (MHURD): Code for Seismic Design of Buildings (GB 50011-2010). China Architecture & Building Press, Beijing, China (2010)
15. Japan Society of Seismic Isolation (JSSI): Manual for Design and Construction of Passively Controlled Buildings (2nd ed.). China Construction Industry Press, Beijing, China (2008) (in Chinese)

Prediction of Inter-Story Drift Ratio Limits for Reinforced Concrete Frame Structures Based on XGBoost

Y. C. Cao, J. F. Zheng, and C. Xiong(✉)

College of Civil and Transportation Engineering, Shenzhen University, Shenzhen, China
xiongchen@szu.edu.cn

Abstract. The accurate determination of inter-story drift ratio (IDR) limits is crucial for assessing seismic damage in reinforced concrete (RC) frame structures. However, conventional fixed IDR limits may lead to inaccurate damage assessments when applied to structures with varying axial load ratios, steel grades, and beam-column stiffness ratios. To overcome this limitation, this study proposes a novel approach utilizing the XGBoost model to predict IDR limits while considering diverse structural parameters. The results demonstrate a significant improvement over the conventional method, with the prediction accuracies ranging from 90.8% to 97.0%, compared to only 48.3% with fixed IDR limits. These findings promise to enhance the precision of seismic damage assessment methodologies, thereby contributing to improved structural resilience and safety.

Keywords: Inter-story drift ratio limits · damage states · building seismic damage assessment · XGBoost

1 Introduction

Earthquakes are one of the most perilous disasters to humans. Many scholars worldwide have proposed methods for assessing seismic damage to buildings [1–9]. To conduct a rapid damage assessment of structures, clear and accurate damage limit criterion are necessary. The inter-story drift ratio (IDR) is a widely used engineering demand parameter for damage assessment. For example, Vision 2000 specifies the inter-story drift ratio limits for buildings under five damage states [10]. Hazus [11] determines IDR limits for five damage states based on building type (including height) and seismic design level. Other scholars have also provided recommended values for inter-story drift ratios [12, 13].

There is a significant disagreement among different publications regarding the recommended limit value for inter-story drift ratio. Luo et al. [14] found that the maximum IDR for concrete frames, within the elastic deformation range, is highly sensitive to variations in the stiffness ratio between beam and column elements. Qian and Xu [15] found that the yield drift of an RC column is proportional to the tensile steel grade, column section height, and axial load ratio. And the ultimate IDR is related to the axial load

X. Lu et al. (Eds.): EOS 2024, LNCE 638, pp. 28–37, 2025.
https://doi.org/10.1007/978-3-031-90690-9_3

ratio, stirrup reinforcement ratio and shear-to-span ratio. Further research by Qi [12] revealed that when flexural damage occurs in columns, the limit drift of the member is not influenced by the longitudinal reinforcement ratio. Additionally, Sk and Gousia [16] underscored the substantial impact of building height on the IDR limits of structural damage. According to the aforementioned study, the IDR limit of RC frames is influenced by various structural parameters, such as the axial compression ratio, steel grade, beam-to-column stiffness ratio, and others.

The purpose of this study is to use the XGBoost model to rapidly predict the inter-story drift ratio limits of reinforced concrete frames with varying structural parameters. The research methodology involves several key steps. Firstly, a building numerical model is constructed with different design parameters based on relevant codes. Subsequently, nonlinear time-history analyses are conducted on these models using OpenSees to evaluate the nonlinear response of the structure. Based on these analyses, the material-based damage limit is used as a benchmark to determine the structure's damage state, recording the corresponding IDR at each damage limit. A dataset for machine learning analyses is then generated, incorporating the structural design parameters and IDR limits of each damage state. The machine learning model is trained using this dataset. To validate and discuss the proposed method, model testing are conducted using new ground motion series.

2 Dataset Generation

2.1 Building Model Based on OpenSees

Machine learning models require large amounts of data for training, therefore, 162 building models were established. These models are a series of multi-story, multi-span, one-bay frame models. The detailed modeling parameters can be found in Table 1.

Table 1. Building parameters

Modeling parameters	Value
Number of floors	3, 6, 9
Number of spans	2, 4, 6
Story height	3.00 m
Span length	4.50 m, 6.00 m, 7.50 m
Dimension of column	400 mm, 450 mm, 500 mm (3 story) 450 mm, 500 mm, 550 mm (6 story) 500 mm, 550 mm, 600 mm (9 story)
Dimension of beam	200 mm × 500 mm
Steel ratio of column	0.012
Steel ratio of beam	0.009
Dead load + 0.5 × live load	36 kN/m

The OpenSees software is adopted to perform nonlinear time-history analyses of the aforementioned RC frame structures [17]. The building mass was calculated by 1.0 × dead load + 0.5 × live load [18] and applied to beam in the form of line loads. The Rayleigh damping model is used with a damping ratio of 0.05. For the concrete material, the Concrete02 model [17] was used. According to the study of Saatcioglu and Razvi [19], the peak strength of cover concrete is $f_{cp} = 24$ MPa, and the corresponding peak strain is set to $\varepsilon_{cp} = 0.002$; The ultimate strain corresponding to 85% of the peak strength is $\varepsilon_{cu} = 0.0038$ [20]. The Mander model [21] is used for the core concrete of columns. Detailed parameters of concrete material are shown in Table 2. The Steel02 model [17] is adopted for the steel reinforcement, and the steel reinforcement with the yielding strength of 400 MPa and 500 MPa are selected for the simulation. The Young's modulus of the steel is $E_0 = 2.0 \times 10^5$ MPa [22], while its hardening rate is taken as 0.01.

Table 2. Concrete material parameters

	Column cover concrete	Column core concrete	Beam concrete
f_{cp} (MPa)	24.0	29.8	27.6
ε_{cp}	0.002	0.0042	0.0035
f_{cu} (MPa)	20.40	24.89	23.46
ε_{cu}	0.0038	0.0115	0.0087
f_t (MPa)	2.4	2.4	2.4

2.2 Ground Motions

This study adopted the 22 sets (44 records) of ground motions recommended by FEMA P695 [23], as shown in Fig. 1. To improve the generalization performance of the model, each ground motion record was rescaled to five different peak ground accelerations (i.e., PGA = 0.2 g, 0.4 g, 0.6 g, 0.8 g, and 1 g), resulting in 110 ground motion series.

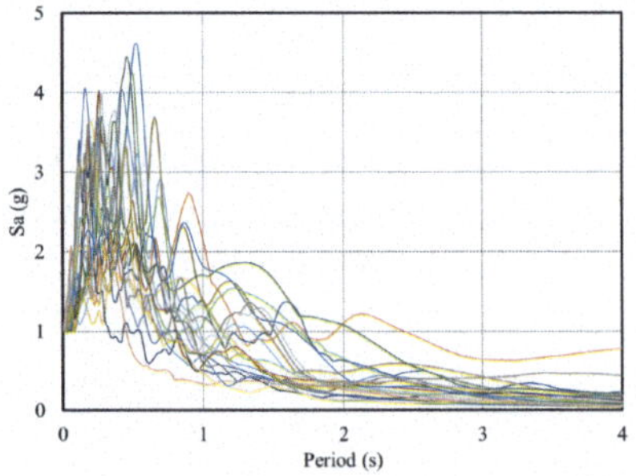

Fig. 1. Response spectra of the 22 ground motions.

2.3 Damage Assessment

Under lateral load of seismic action, RC frame structure normally exhibits shear deformation mode [24] and seismic damage usually initiates at the bottom of the first story. Earthquake reconnaissance studies also show that the first story a RC frame structure usually suffers the most severe damage [25, 26]. Therefore, the following study primarily focuses on investigating the damage of the first story columns. According to FEMA P58 report [2], steel yielding, concrete spalling, and concrete crushing are three typical damage states of flexural members. Therefore, these three damage states are taken as damage states: DS1, DS2, and DS3 (Fig. 2).

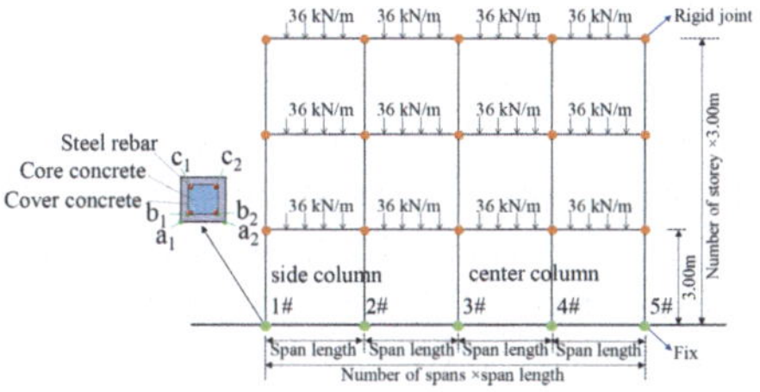

Fig. 2. A representative model for dataset generation.

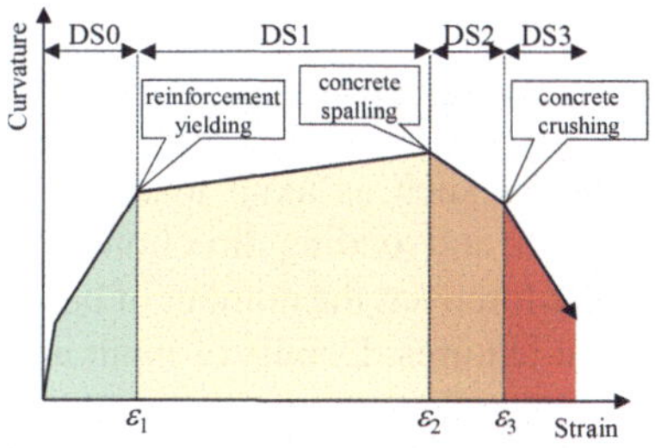

Fig. 3. Damage limit criterion

In numerical simulation, OpenSees was utilized for conducting nonlinear time-history analyses and outputting the fiber strains of edge columns and middle columns of the first story (Fig. 3). Specifically, fiber strain time-history series of steel (c1 and c2), cover concrete (a1 and a2), and core concrete (b1 and b2) were recorded to assess the DS1, DS2, and DS3, respectively. The damage limit of longitudinal steel yielding (DS1) is determined as steel strain reaches $\varepsilon_1 = f_y/E_0$, and the corresponding inter-story drift ratio (IDR) is recorded as θ_1. The damage limit of cover spalling (DS2) is determined as the compressive strain of cover concrete $\varepsilon_2 = 0.0038$, and the corresponding IDR is recorded as θ_2. Similarly, the damage limit of core concrete crushing (DS3) is determined as $\varepsilon_3 = 0.0115$, and the corresponding IDR is recorded as θ_3. In summary, the material level limit criterion of ε_1, ε_2, and ε_3 were adopted to assess the damage states of columns (Fig. 3). At the same time, the story level damage limits IDR (θ_1, θ_2, and θ_3) were recorded for each column.

3 XGBoost Model Training

3.1 Dataset Generation

Based on the analyses in Sect. 2, a total of 17,820 simulation samples need to be computed. Owing to the large number of samples, the Beijing Cloud Supercomputing Center was utilized for batch computation [27]. The number of samples that have reached the three damage states of θ_1, θ_2, and θ_3 were 481, 361, and 317, respectively (two side columns and a middle column for 162 building models). It is worth noting that the number of samples for θ_1, θ_2, and θ_3 is different, this is because some columns do not reach DS2 and DS3. The distribution of IDR is shown in Fig. 4.

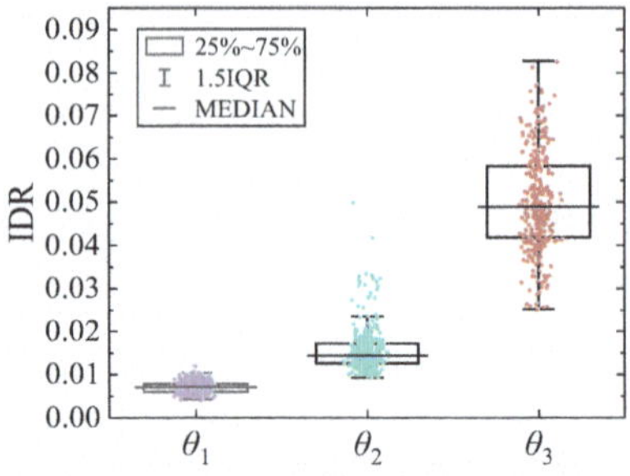

Fig. 4. Distribution of IDR damage limits

3.2 XGBoost Training

Parameters such as axial load ratio, column aspect ratio, span to story height ratio, steel grade and loading directions are adopted as input features for the machine learning model. Moreover, the number of bays and beam to column stiffness ratio are also adopted as input features. Details of input and output features are presented in Table 3.

Table 3. Input and output features of the XGBoost model

Input features	Description
ALR	Axial load ratio
CAR	Column aspect ratio
SHR	Span to story height ratio
SG	Steel grade
LD	Loading direction
Bay	Number of bays
SR	Beam to column stiffness ratio
Output result	Description
θ_1	Damage limit of steel yielding
θ_2	Damage limit of concrete spalling
θ_3	Damage limit of concrete crushing

The dataset obtained in Sect. 3.1 was split into a training set (80%) and a validation set (20%). The optimal parameters for the XGBoost model were determined by performing grid search for parameters of the maximum depth, number of estimators, and learning rate. The performance of each parameter combination was evaluated using k-fold cross-validation (k = 6) [28]. Chicco and Davide [29] argue that R2 is an informative metric in evaluating the performance of regression problems. Therefore, R2 was selected as the model evaluation metric in this paper. Figure 5 depicts the framework of the model training process, while Table 4 shows the search range of hyperparameters and optimal parameter values.

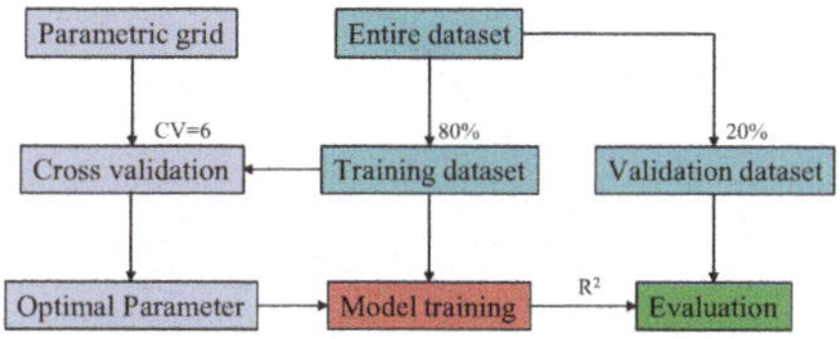

Fig. 5. Model training framework

Table 4. Grid search of optimal hyperparameters for the prediction of θ_1, θ_2, and θ_3

	max_depth range(1,21)	n_estimators range(50,500)	learning_rate range(0.01,0.2)
θ_1	3	350	0.19
θ_2	3	250	0.15
θ_3	4	100	0.17

The final regression results are shown in Fig. 6, with R^2 values of 0.9534, 0.9686, and 0.9583 obtained for predicting θ_1, θ_2, and θ_3, respectively. All R^2 values were greater than 0.95, indicating higher prediction accuracy.

4 Model Validation

4.1 Information of Building Cases

In order to verify the effectiveness of the proposed method, it was compared with the damage assessment method based on fixed IDR limits. The fixed IDR limits are obtained according to the Code for seismic design of buildings (GB 50011-2010) [30] and the work of Cai et al. [31], the IDR limit of θ_1, θ_2, and θ_3 are 1/550, 1/150, and 1/50 respectively. Nine building models selected from Table 2 were adopted in to validate the effectiveness of the method. Detailed information of the models is shown in Table 5, and all other parameters remain the same as presented in Sect. 2.1.

4.2 New Ground Motions

To test the generalization performance of the proposed method, ground motion records were reselected. A dataset of 80 ground motions were selected from the PEER NGA-West2 ground motion database [32], using four different design response spectra as target (corresponding to four different site categories as per the Chinese Code [33]). To ensure a uniform distribution of the damage results, each ground motion records was normalized and rescaled to 10 different peak ground acceleration values (PGA = 0.05–1.4 g), resulting in 800 ground motion series.

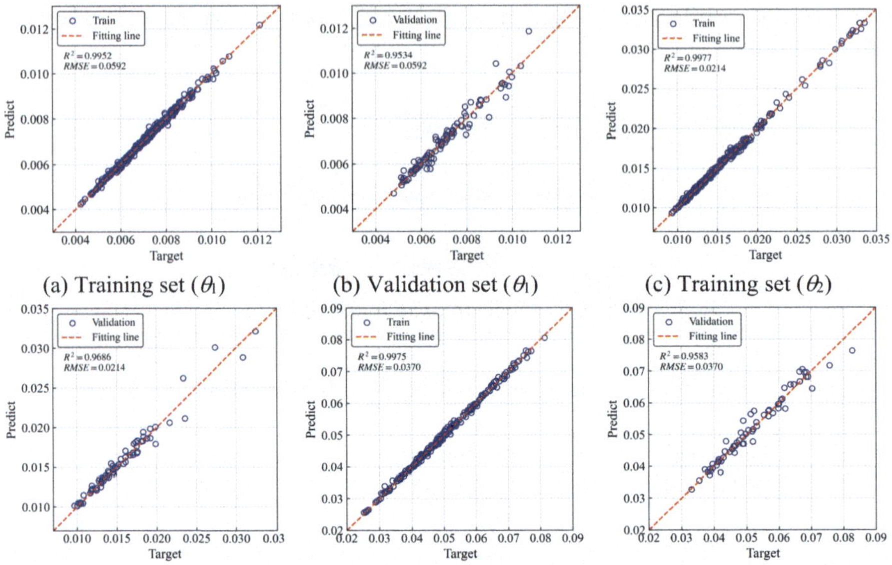

(a) Training set (θ_1) (b) Validation set (θ_1) (c) Training set (θ_2)

(d) Validation set (θ_2) (e) Training set (θ_3) (f) Validation set (θ_3)

Fig. 6. Regression results

Table 5. Selected building models for validation

	Stories	Bays	Column size (mm)	Beam size (mm)	Height (mm)	Span (mm)
Model 3-2	3	2	400 * 400	250 * 500	3000	6000
Model 3-4	3	4	400 * 400	250 * 500	3000	6000
Model 3-6	3	6	400 * 400	250 * 500	3000	6000
Model 6-2	6	2	450 * 450	250 * 500	3000	6000
Model 6-4	6	4	450 * 450	250 * 500	3000	6000
Model 6-6	6	6	450 * 450	250 * 500	3000	6000
Model 9-2	9	2	500 * 500	250 * 500	3000	6000
Model 9-4	9	4	500 * 500	250 * 500	3000	6000
Model 9-6	9	6	500 * 500	250 * 500	3000	6000

4.3 Validation Results

Table 6 show a comparison of the predicted damage states for the nine building models. The true damage states of each building are computed according to the EDP acquired from the nonlinear time-history analysis of each building. The results demonstrated that the proposed method achieved high prediction accuracy for each model, with the highest accuracy of 97.38% for Model 3-6 and the lowest accuracy of 92.25% for Model 9-6.

In contrast, the conventional method based on fixed IDR limits showed less than 50% accuracy, with the highest accuracy of 48.25% for Model 6-6 and the lowest accuracy of 28.88% for Model 3-2.

Table 6. Accuracy of Confusion matrices

Model name	Accuracy of conventional method	Accuracy of XGBoost
Model 3-2	25.88%	95.0%
Model 3-4	28.0%	97.25%
Model 3-6	31.31%	97.38%
Model 6-2	43.38%	94.12%
Model 6-4	38.94%	95.38%
Model 6-6	48.25%	96.19%
Model 9-2	42.19%	92.31%
Model 9-4	46.62%	94.94%
Model 9-6	44.25%	92.25%

5 Conclusions

This paper proposed a prediction model of damage limits criteria for RC frame structures based on XGBoost method. This method can quickly predict the inter-story drift ratio of a building under three damage states: reinforcement yielding (θ_1), cover concrete spalling (θ_2), and core concrete crushing (θ_3). The simulation results of the proposed method exhibit a prediction accuracy of 97.38%–92.25% for different buildings. In the contrast, the conventional method using the fixed IDR limits exhibits a prediction accuracy of 25.88%–48.25%. The results indicates that the proposed method can well consider that influence of different design parameters and yields reliable IDR limits prediction of RC frame structures.

References

1. Kircher, C.A., Whitman, R.V., Holmes, W.T.: HAZUS earthquake loss estimation methods. Nat. Hazard. Rev. **7**(2), 45–59 (2006)
2. FEMA. Seismic Performance Assessment of Buildings Volume 1 –Methodology. Federal Emergency Management Agency, Washington, DC (2012)
3. Almufti, I., Willford, M.: REDi™ rating system: resilience-based earthquake design initiative for the next generation of buildings. Arup Co. (2013)
4. Standard for Seismic Resilience Assessment of Buildings: GB/T 38591-2020. Standards Press of China, Beijing (2020). (in Chinese)
5. Dong, Y., Frangopol, D.M.: Performance-based seismic assessment of conventional and base-isolated steel buildings including environmental impact and resilience. Earthq. Eng. Struct. Dyn. **45**(5), 739–756 (2016)

6. Iryo, T., Iiyama, K., Lalith, M., et al.: On computational costs for Monte Carlo regional-scale simulation for reliable estimate of city and urban area earthquake disaster. J. Earthq. Tsunami **16**(3), 2240003 (2022)
7. Xiong, C., Huang, J., Lu, X.: Framework for city-scale building seismic resilience simulation and repair scheduling with labor constraints driven by time–history analysis. Comput. Aided Civ. Infrastruct. Eng. **35**(4), 322–341 (2020)
8. Cimellaro, G.P., Renschler, C., Reinhorn, A.M., et al.: PEOPLES: a framework for evaluating resilience. J. Struct. Eng. **142**(10), 04016063 (2016)
9. Bozza, A., Asprone, D., Manfredi, G.: Developing an integrated framework to quantify resilience of urban systems against disasters. Nat. Hazards **78**, 1729–1748 (2015)
10. SEAOC. Vision 2000, Performance Based Seismic Engineering of Buildings, vols. I and II: Conceptual Framework. Structural Engineers Association of California, Sacramento (CA) (1995)
11. FEMA. Multi-Hazard Loss Estimation Methodology Earthquake Model. HAZUS-MH 2.1 Technical Manual. Federal Emergency Management Agency (FEMA), Washington, DC, USA (2012)
12. Ghobarah, A.: On drift limits associated with different damage levels. International Workshop on Performance-Based Seismic Design. Department of Civil Engineering, McMaster University, Ontario, Canada, p. 28 (2004)
13. Lv, J., Liu, W., Wang, J.: Quantization research on seismic performance target of reinforced concrete frames. Earthq. Resist. Eng. Retrofit. **33**(5), 80–86 (2011). https://doi.org/10.3969/j.issn.1002-8412.2011.05.014.(inChinese)
14. Luo, W., Qian, J.: Displacement-based seismic design for RC frames. Chin. Civil Eng. J. **5**, 22–29 (2003). https://doi.org/10.3321/j.issn:1000-131X.2003.05.005
15. Qian, J., Xu, F.: Displacement-based deformation capacity design of RC columns. Build. Struct. **288**(12), 30–32 (2007) (in Chinese). https://doi.org/10.19701/j.jzjg.2007.12.009
16. Tehaseen, S.K.G., Kumar, J.D.C.: Effect of change of storey drift and storey height in multi storey building with varying seismic zones. Int. J. Civ. Eng. Technol. **8**(1) (2017)
17. Mazzoni, S., McKenna, F., Scott, M.H., et al.: OpenSees command language manual. Pac. Earthq. Eng. Res. Center **264**(1), 137–158 (2006)
18. Ministry of Housing and Urban-Rural Development of the People's Republic of China, Load Code for the Design of Building Structures: GB 50009-2012. China Architecture & Building Press, Beijing (2012) (in Chinese)
19. Saatcioglu, M., Razvi, S.R.: Strength and ductility of confined concrete. J. Struct. Eng. **118**(6), 1590–1607 (1992)
20. China Academy of Building Research GB50010-2010. Code for Design of Concrete Structures. China Architecture & Building Press, Beijing (2010) (in Chinese)
21. Mander, J.B., Priestley, M.J.N., Park, R.: Theoretical stress–strain model for confined concrete. J. Struct. Eng. **114**(8), 1804–1826 (1988)
22. Karavasilis, T.L., Bazeos, N., Beskos, D.E.: Behavior factor for performance-based seismic design of plane steel moment resisting frames. J. Earthq. Eng. **11**(4), 531–559 (2007)
23. Applied Technology Council. Quantification of Building Seismic Performance Factors (FEMA P695). US Department of Homeland Security, Federal Emergency Management Agency (FEMA) (2009)
24. Xiong, C., Lu, X., Lin, X., et al.: Parameter determination and damage assessment for THA-based regional seismic damage prediction of multi-story buildings. J. Earthq. Eng. **21**(3), 461–485 (2017)
25. Feng, Y., Zhou, J., et al.: Wenchuan earthquake building damage. Sichuan Build. Sci. **35**(6), 139–156 (2009). https://doi.org/10.3969/j.issn.1008-1933.2009.06.036

26. Li, B., Deng, J., Wang, Z.: Implications of the Wenchuan earthquake on the seismic damage of masonry buildings. J. Xi'an Univ. Architect. Technol. **42**(3), 342–347 (2010). https://doi.org/10.3969/j.issn.1006-7930.2010.03.007
27. Beijing Super Cloud Computing Center. http://www.blsc.cn/?bd_vid=8515275390973524037. Accessed 16 April 2024 (in Chinese)
28. Encyclopedia of Database Systems. Springer, New York, NY, USA (2009)
29. Chicco, D., Warrens, M.J., Jurman, G.: The coefficient of determination R-squared is more informative than SMAPE, MAE, MAPE, MSE and RMSE in regression analysis evaluation. PeerJ Comput. Sci. **7**, e623 (2021)
30. Ministry of Housing and Urban-Rural Development of the People's Republic of China, Code for Seismic Design of Buildings: GB 50011-2010. China Architecture & Building Press, Beijing (2010) (in Chinese)
31. Cai, J., Zhou, J., Fang, X.: Comparison of seismic safety degree for RC frame structures between domestic and abroad seismic design codes. Build. Struct. **2**, 1–5 (2006) (in Chinese). https://doi.org/10.19701/j.jzjg.2006.02.001
32. PEER Center. PEER Ground Motion Database (2023). https://ngawest2.berkeley.edu/. Accessed 01 May 2023
33. Deliberating Some Items in Technical Specification for Concrete Structures of Tall Building (JGJ3-2010). China Architecture & Building Press, Beijing (2011) (in Chinese)

Energy-Based Structural Damage Characterization from Shaking-Table Test

Giulio Augusto Tropea[1], Giulia Angelucci[1], Massimo Petracca[3], Giuseppe Quaranta[1](✉), Guido Camata[2], and Fabrizio Mollaioli[1]

[1] Sapienza University of Rome, Rome, Italy
{giuseppe.quaranta,fabrizio.mollaioli}@uniroma1.it
[2] University G. d'Annunzio, Pescara, Italy
g.camata@unich.it
[3] ASDEA Software S.r.l., Pescara, Italy
m.petracca@asdea.net

Abstract. This study evaluates the seismic response of reinforced concrete frame structures with masonry infill walls through advanced analytical and numerical modeling. Utilizing software such as STKO and OpenSees, the study replicates shake table tests to gain a detailed understanding of hysteretic energy dissipation across various structural components. The structure demonstrated significant consistency between experimental and analytical data, particularly in the initial loading phases. The analyses reveal that masonry infills contribute substantially to overall energy dissipation, crucial for reaching key damage states. Detailed examination of concrete and steel behavior indicates that while concrete significantly contributes to energy dissipation, steel remains largely elastic, contributing less to overall damage. A critical finding is the chaotic response behavior of the central column on the first level, attributed to confinement and interactions with surrounding elements. The new methodology focusing on identifying major response variations proves effective, showing coherent results with experimental evidence and enabling multiscale analysis. This approach allows for precise identification of damage states by analyzing energy jumps and intensity variations, crucial for predicting structural behavior under progressive loading and fatigue conditions. The results indicate that the new method provides insights that would be unattainable through traditional energy hysteresis or hysteresis loop backbone analysis alone, validating its application in seismic design and structural assessment.

Keywords: Seismic response analysis · Reinforced concrete structures · Masonry infill walls · Shake table testing · Hysteretic energy dissipation

1 Introduction

Seismic research on reinforced concrete buildings with masonry infills has seen considerable development in recent decades, especially thanks to the use of shaking tables to simulate seismic events. In particular, the importance of analyzing damage states in buildings designed exclusively to respond to gravitational loads emerged, as underlined

X. Lu et al. (Eds.): EOS 2024, LNCE 638, pp. 38–51, 2025.
https://doi.org/10.1007/978-3-031-90690-9_4

by Calvi et al. [2], while Fardis [7, 12] highlighted the importance of evaluating the current state before proceeding with the seismic improvement of reinforced concrete buildings. A relevant study in this field was conducted by Stavridis et al. [13], who performed shaking table tests of a three-story reinforced concrete frame with masonry infills, highlighting how the latter contribute significantly to energy dissipation. Damage States (DS) are a key concept in earthquake engineering, used to describe levels of structural damage following an earthquake. According to the Eurocodes, these states are defined based on the severity of the damage, from total absence of damage (DS0) to structural collapse (DS5) [6]. However, one of the main limitations of current approaches is the difficulty in accurately predicting the behavior of structures during a seismic event, due to a generally statistical approach in characterizing damage states, which, especially in the presence of masonry infills that alter significantly the dynamic response of the system [5] may not take into account the real physical response of the system. In this study, the complexity of the interaction between frame and infills is resolved following the modeling developed in the studies of Decanini et al. [3, 8] and by Liberatore et al. [9], who propose the use of equivalent diagonal props to represent the infill panels inside the frame. This method simplifies the complex interaction between the infill and the surrounding frame, converting the infill into a tensile element carrying only compressive forces. A procedure proposed by Tropea et al. is applied here. [15–17] aimed at identifying the main variations in the seismic response based on the study of the time history of the hysteresis energy. This metric provides an interpretation of the hysteresis work and, according to Mollaioli et al. [4, 11] and Benavent et al. [1], has shown great robustness in characterizing mechanical damage. Energy-Based Seismic Design (EBSD) [18] represents an innovative approach to overcome the limitations of current approaches, relying on the evaluation of the energy dissipated during a seismic event to determine damage states and improve the prediction of structural behavior. Mollaioli et al. [10, 11] demonstrated how EBSD can provide an effective solution to identify critical points in the structural response and more accurately predict damage levels. This approach allows us to analyze in detail how seismic energy is dissipated in the various structural components, at multiple scales and taking into account the mutual interaction between materials and components.

2 Methodology

2.1 Formulation of the Hysteresis Energy for an MDOF System

The calculation of hysteretic energy in a multi-degree-of-freedom (MDOF) system is based on the analysis of the system's response to a cyclic load and the resulting energy dissipation due to the nonlinearity of materials during seismic events. Hysteresis energy can be expressed in terms of forces and displacements for a system with multiple degrees of freedom with degrees of freedom and $\oint$ integration over a complete loading and unloading cycle:

$$E_H = \sum_{i=1n} \oint F_i(u_i)du_i;\ E_{Htot} = E_{Hx} + E_{Hy};\ P_H = \sum_{i=1}^{n} \frac{d}{dt} \oint F_i(u_i)\dot{u}. \quad (1)$$

In a multi-degree-of-freedom (MDOF) system, the hysteretic energy for each degree of freedom is calculated by integrating the product of the hysteretic force (F_i) and the corresponding displacement (u_i) over a complete loading and unloading cycle. Summing the results across all degrees of freedom gives the total energy dissipated by hysteresis in the system. For a reinforced concrete building subjected to an earthquake, the total hysteretic energy (E_{Htot}) is the combined energy dissipated in both the X and Y directions, denoted as E_{Hx} and E_{Hy} respectively. The hysteretic power (P_H) can be derived as the time derivative of the hysteretic energy. In an MDOF system, the total hysteretic power is the sum of the hysteretic power across all degrees of freedom, with each power term being the product of the hysteretic force and the corresponding displacement velocity. Hysteretic energy dissipation can be further broken down for each structural component into the contributions of steel and concrete. By employing beam-column elements with fiber cross-sections and assigning independent uniaxial materials for steel and concrete, the energy densities for each material can be calculated individually. These energy densities are then integrated over each fiber's cross-sectional area and along the element length, using integration weights, allowing a clear division of the hysteretic energy between steel and concrete. The iterative process performed to calculate the energy for each time step of each stage, separating the contribution of steel and concrete, is shown below:

$$\begin{aligned} E_{T,t} &= \sum_{i=1}^{Nelements} \left(\sum_{j=1}^{Ngauss} \left(\sum_{k=1}^{Nfiber} e_d \times A_k \right) \times L_j \right) \\ E_{C,t} &= \sum_{i=1}^{Nelements} \left(\sum_{j=1}^{Ngauss} \left(\sum_{k=1}^{NCfiber} e_{dc} \times A_k \right) \times L_j \right) \\ E_{S,t} &= \sum_{i=1}^{Nelements} \left(\sum_{j=1}^{Ngauss} \left(\sum_{k=1}^{NSfiber} e_{ds} \times A_k \right) \times L_j \right) \end{aligned} \quad (2)$$

The Eq. (2) calculates the total dissipated energy $E_{T,t}$ concrete energy $E_{C,t}$ and steel energy $E_{C,t}$ in a fiber section by summing the contributions of each fiber in the cross-section. Each component depends on the fiber area A_k (of the k-th fiber), the tributary length L_j (of the j-th element) and the energy densities e_d (total), e_{dc} (concrete), and e_{ds} (steel). This methodology enables the identification of variation points in the structural response by analyzing hysteresis energy and power.

2.2 Description of the D I Method

The proposed new methodology is aimed at identifying the main variations in the seismic response of a physical system, focusing on the analysis of hysteretic energy (E_H) and hysteretic power (P_H). The methodology is based on the idea that, by studying the time trend of the hysteresis energy it is possible to identify the main variations in the structural response to a seismic input, which correspond to different damage stages of the system. These variations manifest themselves as jumps in absorbed energy, which are more evident in the event of damage. Energy jumps are analyzed in terms of amplitude and slope, where amplitude represents the amount of energy dissipated in a single hysteretic cycle and slope measures the rate of dissipation. A larger head and steeper slope indicate a rapid and significant increase in energy demand. Two variables D and I are defined which are useful for building a complete indicator of the damage potential (change function, Fig. 1), formulated by relating the energy and power measurements after correct normalization:

$$\overline{E}_H(t) = \frac{E_H(t)}{max(E_H)}; \ \overline{P}_H(t) = \frac{P_H(t)}{max(P_H)}; \ D = \sqrt{\overline{E}_H^2 + \overline{P}_H^2}; \ I = \log\left|\overline{P}_H\right|. \quad (3)$$

The dissipated energy ratio ($\overline{E}_H(t)$) describes the amount of internal energy spent on hysteresis work, while the normalized dissipation rate ($\overline{P}_H(t)$) represents the normalized hysteretic power. The D (dynamic energy indicator) function combines normalized energy and power to provide a cumulative measure of the dynamic state of the system. It represents an indicator of the accumulated damage potential, correlating the dissipative capacity of the structural system with the duration of the seismic excitation. The I (impulsivity index) function is expressed as the logarithm of the normalized power. This logarithmic transformation facilitates the identification of power peaks by emphasizing peak values related to significant changes in energy. D and I are related in a function defined as the change function (Fig. 1) which allows a complete reading of the dynamic response also thanks to the construction of a backbone. The backbone curve is drawn by combining the peak values of the change function, considering exclusively the maximum value reached by I as a function of every i-th value of D. This is an application of the Lagrange polynomial interpolation:

$$P(D_b) = L_1(D_b) \cdot I_1 + L_2(D_b) \cdot I_2 + \cdots + L_n(D_b) \cdot I_n \quad (4)$$

where P is the Lagrange polynomial, D_b are the maximum values of D for each value of I, In is the i-th value of I and L_n are the Lagrange polynomials.

2.3 Critical Points of D and I

Figure 1 shows the fundamental points useful for understanding the proposed methodology, reported below in explanatory form:

1. **Elasticity:** The elastic phase is characterized by $D = 0$ and I which tends to minus infinity when the normalized power $\overline{P}_H$ is close to zero.
2. **Onset of Plasticity:** The onset of the plastic phase is identified by a sudden increase in D and a rapid change in I, indicating a significant change in the structural response.
3. **Key Response Changes:** The backbone derivative of the change function helps identify significant change points. When the first derivative of the backbone function is zero $\frac{\mathrm{d(P(D_b))}}{\mathrm{dt}} = 0$, it indicates a plateau that corresponds to a constant response phase or a new damage state.
4. **Strength deterioration:** Following the maximum peak of I, if the system is subject to degradation of forces and stiffness, there follows a phase in which subsequent cycles do not go back to exceeding the maximum peak reached and in which the curvature of the backbone reverses.
5. **Collapse:** Collapse in analytical models is identified when the final value of the time history of the hysteresis energy is equal to the maximum value of the same

energy expressed during the motion, with a tolerance index that allows considering the possible short reloading half-cycle final that can characterize the response of the model, due to the peculiarities of the constitutive bond chosen to characterize the materials. In essence, it is assumed that if the motion ends with a significant elastic recovery phase, the structure has not fully expressed its potential hysteretic dissipative capacity. The collapse condition therefore corresponds to a final value of D equal to 1, given that at the end of the motion the change in energy is zero and therefore the value of power is equal to zero, it can be deduced that the value of D coincides with the value of $\overline{E}_H$ which is equal to 1 if $E_H(t)_{final} = \max E_H(t)$.

Figure 1 shows three images that allow for a comparison between the hysteresis loop, the change function, and the time history of the backbone derivative, which helps identify the main response variation points. The case displayed is an experimental test conducted on a reinforced concrete column. The elastic section DS1 is marked in blue, an initial damage condition DS2 in green, an additional response variation condition DS3 in yellow, and the two subsequent phases preceding collapse (DS4, DS5) in orange and red. Indeed, the backbone derivative identifies four primary response variation points that follow the backbone plateaus. In the hysteresis loop, the backbone is marked to offer a comparison with the other backbone of the change function on the right. In the bottom left, the history of the derivative of the backbone of the change function displays four peaks where the function equals zero, marking the plateaus and points where a more significant energy change occurred in terms of slope, jump amplitude (and related hysteresis loop), or both. On the right, in the D-I plane graph, the hysteretic energy history is superimposed with inverted axes (time on the y-axis and hysteretic energy on the x-axis), allowing the correlation between the asymptotes at negative infinity of the change function and the energy history plateaus to be visualized. The normalized hysteretic energy values corresponding to the backbone plateaus are marked with red circles. It is necessary to locate the first point on the change function where $P_H = 0$, following a plateau, to determine the energy value where a significant variation in the system's response occurs.

2.4 Geometric Interpretation

In the D-I graph, each hysteretic cycle features four vertical asymptotes at points where $P_H = 0$, marking zero relative velocity or reaction forces. The distance between these asymptotes reflects the amplitude of each hysteretic cycle and the energy dissipated, aligning with the normalized hysteretic energy $\overline{E}_H$. This methodology effectively identifies and quantifies non-linear behaviors and degradation mechanisms in structures, offering critical insights into structural response and durability. It remains versatile across different loads and materials, enabling both global and component-specific analysis by predefining the desired hysteresis energy time history.

3 Experimental Models and Test Procedures

To verify the effectiveness of the new methodology for identifying damage states, a shaking table test was analytically replicated in STKO. The model, based on the 2017 study by A. Stavridis et al. [13], examines a three-story reinforced concrete frame with

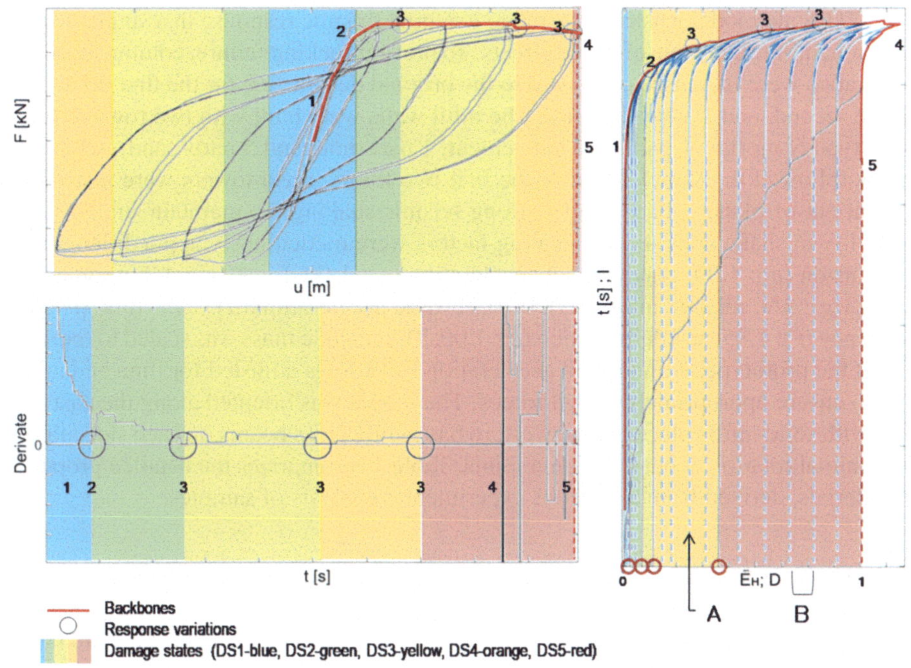

Fig. 1. Comparison between hysteresis loop, change function and time history of the backbone derivative of the change function. The main critical points and damage states (colors) are marked in terms of forces, displacements, normalized energy and time points identified with the derivative.

masonry infills, representing a typical 1920s Californian building. The model accurately reflects the construction features of the time, including a concrete strength of 20.7 MPa, grade 60 steel for reinforcement bars, and design details like longitudinal bar overlaps and lack of transverse reinforcements in beams, replicating common structural limitations of that era (Fig. 2).

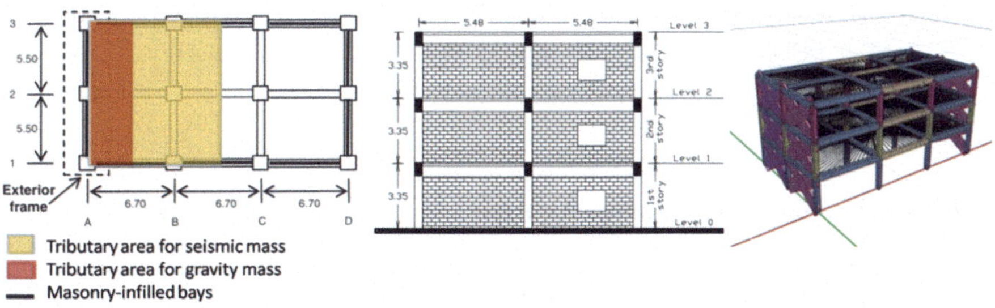

Fig. 2. Prototype structure, plan and lateral view. Mass consideration are showed. Measures in m. Analytical model in STKO.

A 2/3 scale model was constructed to examine seismic response in a single primary direction, minimizing out-of-plane effects. To prevent rocking failure, column reinforcement ratios were increased compared to the original design: 2% for the first floor, 1.5% for the second, and 1% for the third. The infill walls were built with two rows of bricks using type N mortar (1 part Portland cement, 1 part lime, and 5 parts sand), while roof beams followed the same design as the first two floors. Steel towers were installed to prevent out-of-plane collapse under strong seismic shaking. To maintain similarity and reliability in shake table testing, scaling factors were meticulously applied: length (L): 2/3, tension (σ): 1.00, gravitational acceleration (g): 1.00, force (F): 4/9, gravitational mass (Mg): 4/9, seismic mass (Ms): 0.20, seismic acceleration (a): 2.27, time (t): 0.54, frequency (f): 1.85, and deformation (ε): 1.00. The seismic mass was scaled to represent half of the prototype, with ground acceleration recordings adjusted for time and amplitude to ensure appropriately scaled forces. The model was oriented along the east-west axis, with solid infills placed in the western bay, while cross beams on each side allowed for minimal rotation at beam-column joints. Table 1 summarizes the detailed properties of materials, derived from a rigorous experimental analysis of samples.

Table 1. Average properties of concrete and masonry materials across the three floors.

Material	AverageProperty	1st Floor	2nd Floor	3rd Floor
Concrete	ElasticModulus (GPa)	15.14	17.44	16.98
	Compressive Strength (MPa)	37.99	41.95	39.17
	Strain at Peak Stress	0.0032	0.0033	0.0032
	Tensile Strength (MPa)	3.47	3.96	4.04
Masonry	ElasticModulus (GPa)	5.41	6.82	6.52
	Compressive Strength (MPa)	19.80	23.61	22.81
	Strain at Peak Stress	0.0045	0.0041	0.0049
	Brick Unit Compressive Strength (MPa)	48.95	48.95	48.95
	MortarCylinder Compressive Strength (MPa)	4.00	4.10	4.33
	MortarCylinder Tensile Strength (MPa)	1.06	1.24	1.24
	Mortar Cube Compressive Strength (MPa)	10.05	6.41	10.43

3.1 Testing Protocol and Damage Analysis

The dynamic testing spanned 15 days with 44 tests to assess the sample's resilience to various seismic stress levels. This included 15 ambient vibration and 15 white noise tests to analyze the structure's modal properties, followed by 14 seismic events of increasing intensity based on the 1989 Loma Prieta and 1940 Imperial Valley earthquakes. White noise tests before and after each seismic event ensured result accuracy. Only the scaled Gil signal was compared with the analytical model, as subsequent tests caused excessive out-of-plane damage. *Stage 1 (10%, 20%, and 40% of Gilroy 3):* Initial six tests showed no visible damage, with structural stiffness unchanged until test gil67a. *Stage 2 (67% of Gilroy):* Structural stiffness reduced; small cracks appeared on the first floor, particularly around the infill-wall interfaces. *Stage 3 (83% and 91% of Gilroy):* At gil83, base shear reached its peak, showing visible horizontal and diagonal cracks in infill walls and near column tops. Gil91, with higher acceleration, led to larger displacements. *Stage 4 (100% and 120% of Gilroy):* Additional cracking on the first floor; severe diagonal shear cracks observed in the central column. *Stage 5 (250% of El Centro):* Partial collapse of the first-floor infill wall; severe shear cracks in all columns and significant out-of-plane tilt, compromising stability. Each stage progressively demonstrated the sample's structural degradation and behavior under seismic loading (Table 2).

Eurocodes define damage states (DS) based on the severity of structural damage during seismic events, aligning with the EMS-98 scale. The states range from DS0 (No Damage), with no visible damage, to DS5 (Total Damage), indicating near or complete collapse. Intermediate levels include DS1 (Minor Damage), DS2 (Moderate Damage) with major cracks in load-bearing walls, DS3 (Severe Damage) compromising load capacity, and DS4 (Very Severe Damage) threatening stability. In seismic testing, three main damage states were observed: Stage 2 (Gil67) marked the onset of moderate damage (DS1), Gil91 indicated severe damage (DS2), and Gil120 represented near-collapse conditions (DS3), with shear failure in the central column signaling local collapse.

4 Analytical and Numerical Models

4.1 Modeling with STKO

The shake table test was recreated analytically in STKO and evaluated with OpenSees, modified to detail hysteretic energy dissipation across steel and concrete components. The structure was modeled using a distributed plasticity approach with fiber modeling for realistic seismic load behavior. In the 3D model, short-side infills were added for Y-axis response calibration, and central columns were restricted in the X direction, focusing resistance on the infills. The constitutive laws applied included: **Concrete**, Modeled with OpenSees Concrete02, handling monotonic and cyclic behavior with stiffness degradation; **Steel**, Modeled with Steel01 for elastic-plastic behavior and hysteresis; **Infills**, Modeled with HystereticMaterial as two trusses, representing infill walls using equivalent diagonal struts following Decanini and Liberatore's methods. Reduction factors were added for stiffness and strength adjustments based on wall openings. For calibration, a pushover analysis with an inverse triangular load was applied at mass centers, ensuring convergence with the Newmark approach. External multipliers were adjusted for accurate hysteresis, especially for infills, based on concrete and masonry tests from experimental phases (Fig. 3).

To avoid unwanted torsional effects, the GJ value was kept constant in all sections. Furthermore, the mass was distributed uniformly on the beams, with the addition of the linear load carried by the infills on the beams below. The loading phases were carried out progressively: first the distributed gravitational load, then the weight of the infills, and finally the seismic load, following the experimental test procedures.

Table 2. Summary of the seismic test results for a masonry-infilled structure, detailing fundamental periods, acceleration, first-story drift, and observed damage progression

Test	Fundamental Period T1 After Test (s)	Acceleration (g)	First-Story Drift (%)	Observed Damage
gil40b	0.056	0.79	1.27	No visible damage
gil67a	0.060	1.14	1.90	Minor cracks at the first level
gil67b	0.063	1.03	2.24	Minor cracks at the first level and at the interface between the wall and the framing
gil83	0.068	1.16	2.38	First noticeable damage with horizontal and diagonal cracks in the infill walls and minor cracks near the top of the columns
gil91	0.074	1.29	2.36	Greater accelerations, shear forces, and displacements compared to gil83 due to softening of the structure
gil100	0.12	1.59	2.38	Further development of cracks in the first-story infills and crack propagation through the columns
gil120	0.19	1.90	2.43	Significant damage with severe diagonal shear cracks in the middle column at the first level. Compromise of structuralintegrity

(*continued*)

Table 2. (*continued*)

Test	Fundamental Period T1 After Test (s)	Acceleration (g)	First-Story Drift (%)	Observed Damage
ec250	Collapse	1.94	Collapse	Significant collapse of the infill wall at the first level in the east bay and wide shear cracks in all three columns at the first level. Tilting

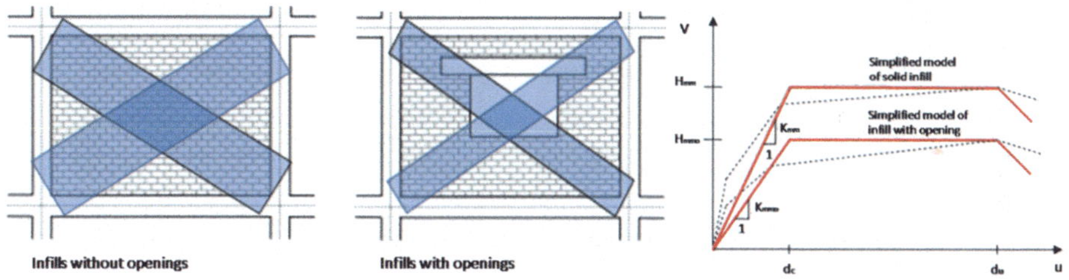

Fig. 3. Shear displacement bilinear constitutive law for infills with reductive factors in case of openings. FEM elements for infills with openings and without openings used in the model.

4.2 Comparison Between Analytical Model and Experimental Data

The seismic response of the analytical model was evaluated globally, at the first level (where soft floor effects and significant damage appeared), and by individual components. The central column between the infills showed the most severe damage in both analytical and experimental tests, with energy dissipation separated into concrete, steel, and infill contributions. Figure 4 shows the time history of global hysteretic energy (E_H) for different seismic signals. The model was progressively loaded with gravitational and infill weights, followed by various seismic signals (Gil67a, Gil67b, Gil83, Gil91, Gil100). Energy dissipation occurred in steps, with marked increases at each peak load, indicating significant energy absorption. The model's hysteresis loop aligns with experimental data and a calibrated backbone, reinforcing its reliability. The red dotted line in the energy function trend illustrates that each new scaled signal affects the structure's response through accumulated damage rather than scaling alone, meaning that energy dissipation increases due to the pre-existing damage. The unscaled Gilroy signal is also displayed, and both the hysteresis loop and time history highlight damage states DS1 (blue), DS2 (green), and DS3 (yellow) identified in previous shake table tests. Axonometric and lateral views show infills contributing significantly to dissipation, with the central column and infills with openings experiencing the most damage, replicating experimental collapse behavior. The backbone of the hysteresis loop (red) demonstrates that main damage states and critical points do not coincide exactly. Figure 5

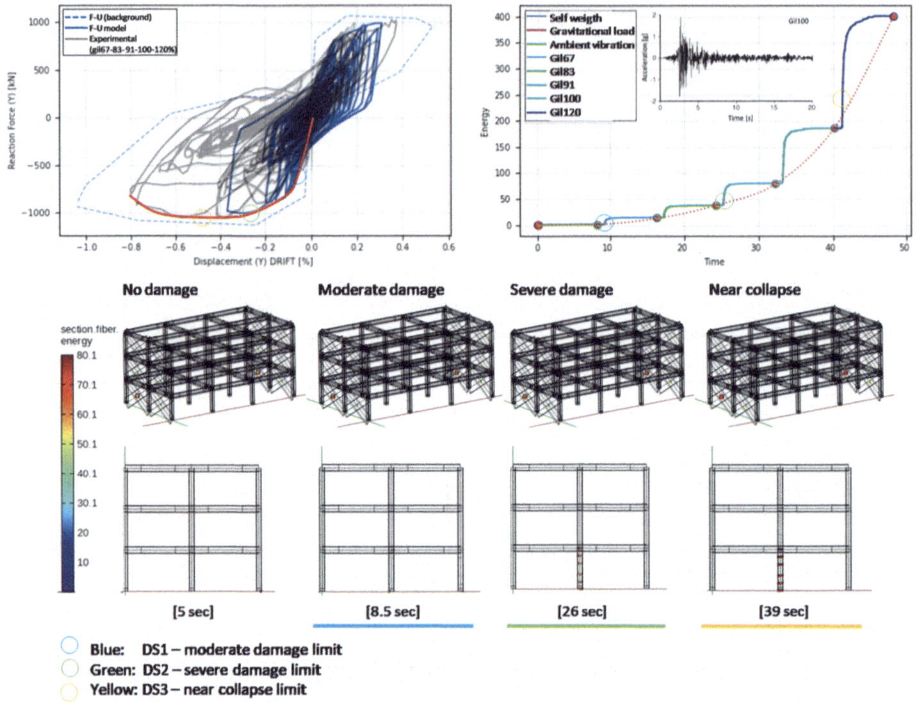

Fig. 4. Dynamic response and energy dissipation of a three-story reinforced concrete frame with masonry infill walls under varying seismic intensities. The left graph shows the hysteresis loops representing the relationship between base shear force and displacement. The right graph plots the cumulative energy dissipation over time for different seismic events. The structural models below illustrate the progression of damage from no damage to near collapse at specified time intervals, highlighting the significant increase in fiber section energy dissipation, particularly in severe and near-collapse conditions.

presents change function graphs for the hysteretic energy (E_H) dissipated by the analytical model at three levels: global, floor, and specific column. The X-axis indicates damage (*D*), while the Y-axis shows response intensity (*I*). Elastic limits (red dots) and primary response changes (black crosses) are noted on asymptotes at negative infinity, aligned with normalized hysteresis energy. No global collapse is observed, as collapse was only reached experimentally with the ElCentro250 signal. The graphs indicate that infills significantly contribute to dissipation, with *D* values for infills much higher than those for steel and concrete, highlighting their role in reaching damage states. Steel remains mostly elastic, while concrete shows substantial damage accumulation. In the global graph, increased distances between asymptotes reflect larger hysteretic cycles and energy jumps, indicating advanced damage. Normalized energies show that while total structure collapse appears with a final *D* value of 1, floor and column analyses suggest that the collapse primarily stems from infill behavior rather than from frame or floor failure. Steel and concrete's dissipative roles are minor in the global analysis but more prominent in specific component analyses. Concrete generally dissipated more energy than steel, which remained below its elastic limit in both the plane and element analyses. Response variations concentrated at the floor level with damage values (D) between 3.5 and 4.5, while variations in the column occurred at D values just below 3, with the column showing more frequent response changes than the floor.

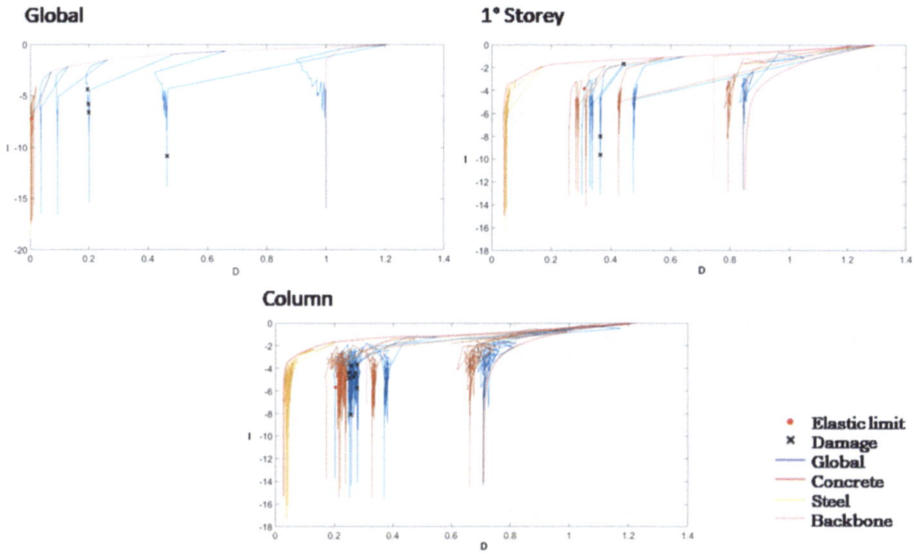

Fig. 5. The figure shows the change functions for the global structure, the first story, and a specific column, highlighting the elastic limit (red dots) and damage states (black crosses). The global, concrete, and steel contributions are depicted in blue, red, and yellow, respectively, with the backbone of the change functions shown in light red. These plots illustrate the response variations and damage accumulation throughout the structural elements.

Figure 6 illustrates a comparison between the load history of scaled signals and normalized hysteresis energy responses, showing steel, concrete, and global responses for the overall structure, the floor level, and the most damaged column. Damage states DS1, DS2, and DS3 are highlighted, with black lines representing experimental data and red lines showing the analytical model, which correspond well overall. Minor discrepancies emerge in the global response at later stages, with analytical data slightly overestimating energy dissipation during DS3, possibly due to fatigue effects not fully accounted for by the experimental inter-story drift method. At the floor level, a chaotic response between DS2 and DS3 indicates advanced damage, with significant, rapid increases in energy up to DS3, after which further variations ceased. For the specific column, numerous chaotic response variations likely stem from its confined state and interactions with surrounding elements, resulting in high structural variability. Some phases of damage in the column align with experimentally identified damage states, suggesting its role in reaching new global damage states. Around DS2, discrepancies in response variations suggest that global damage state DS2 may be driven more by the behavior of infills than the frame itself.

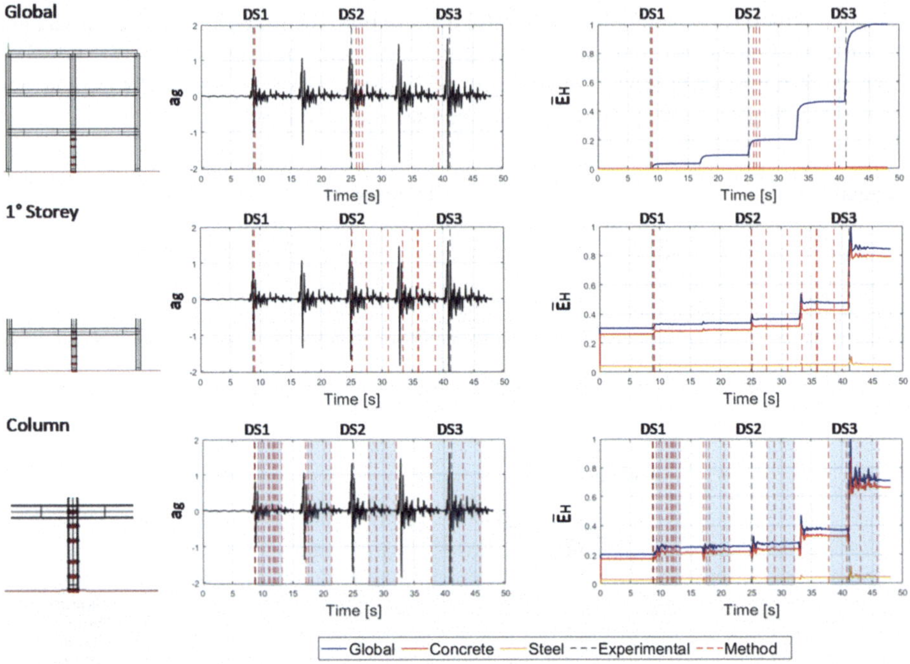

Fig. 6. The figure displays the time histories of seismic signals and the corresponding normalized hysteretic energy responses for the global structure, concrete, and steel and divided by the global analysis, the level analysis and the element (column) analysis. The black solid lines represent the seismic acceleration (***ag***) over time. The red dashed lines mark the damage states (DS1, DS2, DS3) identified through the proposed method, while the black dashed lines indicate experimental observations. The E_H plots illustrate how energy dissipation progresses with time, highlighting the differences in response between the global structure (blue), concrete (red), and steel (yellow).

5 Conclusions

This study demonstrated the effectiveness of analytical and numerical modeling for analyzing seismic responses in reinforced concrete frames with masonry infills. Using STKO and OpenSees, the models accurately replicated shake table tests, providing detailed insights into hysteretic energy dissipation across structural components. The results showed strong alignment between experimental and analytical data, especially during initial loading phases, highlighting that masonry infills significantly contribute to overall energy dissipation and play a crucial role in reaching key damage states. Detailed analysis revealed that concrete dissipates a substantial amount of energy, while steel remains largely elastic with minimal impact on total damage. Steel and concrete behaviors varied based on analysis level (global, floor, or specific component). The central column on the first level exhibited chaotic behavior, likely due to confinement and surrounding element interactions. The new methodology successfully identified primary response variations, yielding results consistent with experimental findings and enabling multiscale analysis. Response variations closely aligned with emerging damage states, a result that could not have been obtained through traditional hysteresis energy or cycle analysis alone. The method allowed for precise identification of damage states by analyzing energy jumps and intensity changes, aiding in predicting structural behavior under progressive loading and fatigue.

References

1. Benavent, A., Zahran, R.: Hysteretic energy dissipation in reinforced concrete structures. J. Earthq. Eng. **24**(9), 1234–1245 (2020)
2. Calvi, G.M., Magenes, G., Pampanin, S.: Experimental test on a three-storey RC frame designed for gravity only. Earthq. Eng. Struct. Dyn. **31**(2), 453–475 (2002)
3. Decanini, L.D., Gavarini, C., Lignola, G.P.: Models of masonry infilled frames for the evaluation of the behaviour factor. Eng. Struct. **23**(3), 280–290 (2017)
4. Decanini, L.D., Mollaioli, F.: An energy-based methodology for the assessment of seismic demand. Soil Dyn. Earthq. Eng. **21**(2), 113–137 (2001). https://doi.org/10.1016/S0267-7261(00)00102-0
5. Dolsek, M., Fajfar, P.: The effect of masonry infill on the seismic response of a four-storey reinforced concrete frame – a probabilistic assessment. Eng. Struct. **30**(11), 3186–3193 (2008)
6. Eurocode 8: Design of Structures for Earthquake Resistance. EN 1998
7. Fardis, M.N.: Seismic Design, Assessment and Retrofitting of Concrete Buildings. Springer Science & Business Media (2009)
8. Liberatore, D., Decanini, L.D.: Modelling of masonry infilled frames. J. Struct. Eng. **127**(2), 105–111 (2021)
9. Liberatore, L., Noto, F., Mollaioli, F., Franchin, P.: In-plane response of masonry infill walls: comprehensive experimentally-based equivalent strut model for deterministic and probabilistic analysis. Department of Structural and Geotechnical Engineering, Sapienza University of Rome, via Gramsci 53, 00197 Rome, Italy
10. Liberatore, L., Mollaioli, F.: Damage potential of the 2009 L'Aquila, Italy, Earthquake. J. Earthq. Tsunami **6**(3), 1250032 (2012). https://doi.org/10.1142/S1793431112500327
11. Mollaioli, F., Bosi, G., Monti, G.: Energy-based seismic assessment of structures. Earthq. Eng. Struct. Dyn. **50**(6), 1573–1592 (2021)
12. Panagiotakos, T.B., Fardis, M.N.: Seismic response of infilled RC frames structures. J. Earthq. Eng. **1**(3), 479–503 (1996)
13. Stavridis, A., Shing, P.B., Willam, K.: Shake-table tests of a three-story reinforced concrete frame with masonry infill walls. J. Struct. Eng. **143**(10), 04017121 (2017)
14. Tropea, G.A., Angelucci, G., Bernardini, D., Quaranta, G., Mollaioli, F.: New Energy Based Metrics to Evaluate Building Seismic Capacity. Department of Structural and Geotechnical Engineering, Sapienza University of Rome, via Gramsci 53, 00197 Rome, Italy
15. Tropea, G.A., Angelucci, G., Mollaioli, F.: Concrete Columns Energy Based Damage Characterization. Department of Structural and Geotechnical Engineering, Sapienza University of Rome, via Gramsci 53, 00197 Rome, Italy
16. Tropea, G.A., Angelucci, G., Quaranta, G., Mollaioli, F.: Structural Damage Characterization via Energy-Based Limit State Definition. Department of Structural and Geotechnical Engineering, Sapienza University of Rome, via Gramsci 53, 00197 Rome, Italy
17. Uang, C.M., Bertero, V.V.: Evaluation of seismic energy in structures. Earthq. Eng. Struct. Dyn. **19**(1), 77–90 (1990)

Inelastic Bar Buckling Models Implemented in OpenSees

Christian Pignanelli(✉), Andrea Lucchini, and Paolo Franchin

Sapienza University, 00197 Rome, RM, Italy
christian.pignanelli@uniroma1.it

Abstract. Modeling of bars' buckling is fundamental in the simulation of the flexural response up to collapse of RC members under seismic loading. When bar buckling occurs, indeed, strength and stiffness degradation starts to significantly increase, limiting the deformation capacity of the member. For typical members' dimensions and layout of the reinforcement, buckling occurs at or after yielding. For these reasons, particular attention must, therefore, be given to the stress-strain model used to represent the behavior of reinforcing steel bars. The present work focuses on the models developed by Gomes-Appleton and Dhakal-Maekawa for inelastic buckling, which are among the most widely used ones and are those implemented in OpenSees. Assumptions adopted by the two models on initiation of buckling, reloading stiffness from post buckling compression state, stiffness of lateral reinforcement and interaction between the bar and concrete are discussed. Modifications introduced in the OpenSees implementations of the models are also reported. To evaluate the models, it is necessary to compare the analysis results with experimental responses. The experimental dataset is expanded with the high-fidelity numerical tests to consider non-standard loading protocols.

Keywords: Flexural response up to collapse · reinforcing bars · low-code members · fiber discretization · uniaxial material · seismic load path · strength and stiffness degradation

1 Introduction

Many members in existing Reinforced Concrete (RC) buildings not designed for seismic actions have sparse stirrups, which, however, combined with low axial loads, are still sufficient to prevent brittle shear failure, as highlighted in the tests by Lucchini et al. (2022) [1]. In these members, flexural collapse, after concrete cover spalling, progresses through bar buckling which causes the lateral strength decay of the member itself.

Phenomenological models can efficiently simulate the flexural response of RC sections but do not account for possible biaxial bending and interaction with axial force. Because of this, fiber models are usually used which require uniaxial material models for concrete and steel. To adequately predict the response up to collapse, the steel model shall not simply describe the constitutive behavior of material but the element response of the bar which can possibly buckle.

X. Lu et al. (Eds.): EOS 2024, LNCE 638, pp. 52–62, 2025.
https://doi.org/10.1007/978-3-031-90690-9_5

The purpose of this article is to evaluate, among the various bar buckling models available in the literature (Table. 1), those by Gomes and Appleton (GA) (1996) [2] and by Dhakal and Maekawa (DM) (2002) [3]. They are among the most widely used ones and, more importantly, the only ones implemented in OpenSees (OS) [4]. It should be noted that the versions implemented in OpenSees by Kunnath et al. (2009) [5] slightly differ from the original ones. The evaluation is conducted by comparing the predictions of these models with the responses obtained from refined finite element (FE) micromodels of bare bars subjected to different cyclic loading conditions.

Table 1. List of bar buckling models proposed in literature

Monotonic		Cyclic	
Authors	**Year**	**Authors**	**Year**
Bae et al.	2005	Monti and Nuti	1992
Berry and Eberhard	2005	Gomes and Appleton	1997
Cosenza and Prota	2006	Dhakal and Maekawa	2002
Gil-Martin et al.	2006	Zong et al.	2014
Urmson and Mander	2012	Yang et al.	2016
Massone and Lopez	2012	Kim and Koutromanos	2016
		Tripathi and Dhakal	2023

2 Factors Influencing Bar Buckling

Bar buckling within a structural member, as opposed to the buckling of a bare bar, occurs under cyclic loading, and it is affected by the interaction with both the core and the cover concrete. Cyclic loading accelerates spalling of the cover, which is a necessary condition for the bar to buckle. Besides, under cyclic bending of the member, bars in tension undergo an outward lateral deformation imposed by the core-bar compatibility, which then contributes to buckling when the bars go in compression. Lateral deformations are caused by shear and axial forces as well (the latter through the expansion of the concrete core) but have negligible effects on buckling as reported in Gil-Martin et al. (2006) [6] and in Korentz (2020) [7]. Previous research, Moyer and Kowalsky (2003) [8], has identified prior plastic deformation in tension as a precondition for the activation of buckling.

Stiffness (diameter D) and spacing, L, of lateral reinforcement determine the extension of the phenomenon, which is usually named local or global (bar) buckling, respectively, if two or more adjacent stirrups are involved. Focusing on the local one, which typically characterizes rebars of lightly reinforced RC members without seismic detailing, the problem of bar buckling can be studied by considering a bare bar only, if the interaction with concrete, for simplicity, is neglected. The inelastic behavior of steel, besides, must be account for because bar buckling usually occurs after yielding, for the usual values of the slenderness ratio $l_{sr} = L/D$. Typically, the geometric slenderness λ of rebars (equal to twice l_{sr} for circular sections) is much lower than the critical value λ_y, which is the slenderness value on Euler's curve (Fig. 1) corresponding to the yielding stress f_y.

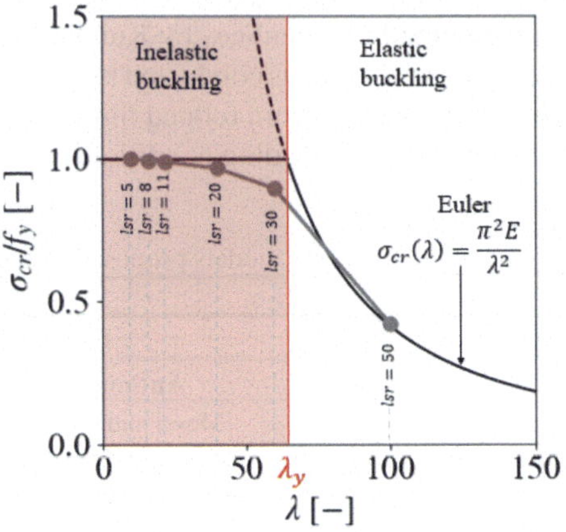

Fig. 1. Engesser-Shanley diagram.

3 Uniaxial Bar Buckling Models in OpenSees

In these models the uniaxial stress-strain response represents, rather than the constitutive behavior of steel, the relationship between the average stress and strain of the entire bar. Gomes and Appleton [1] propose a stress-strain model for the response in compression based on a mechanical approach, while the model by Dhakal and Maekawa [2] is phenomenological. For the cyclic behavior, both models adopt the hysteretic rules of Giuffrè-Menegotto-Pinto (GMP) (1972) [9, 10], with a modification in DM.

3.1 Gomes and Appleton

In GA model the post-buckling behavior is described by a stress-strain curve derived from a limit analysis. According to the kinematic theorem, this limit curve is obtained from the equilibrium of an assumed plastic mechanism for the bent bar, which is a rigid body motion characterized by the formation of three hinges between two consecutive stirrups (Fig. 2).

If the plastic moment, M_p, of the rebar is assumed independent on the axial force, and defining the strain $\varepsilon = \delta/L$, the equation describing the limit curve becomes:

$$\sigma_p = 0.6\frac{f_y}{l_{sr}\sqrt{\varepsilon}} \tag{1}$$

The intersection of this curve with that of the GMP model determines the buckling initiation point (point P in Fig. 3a).

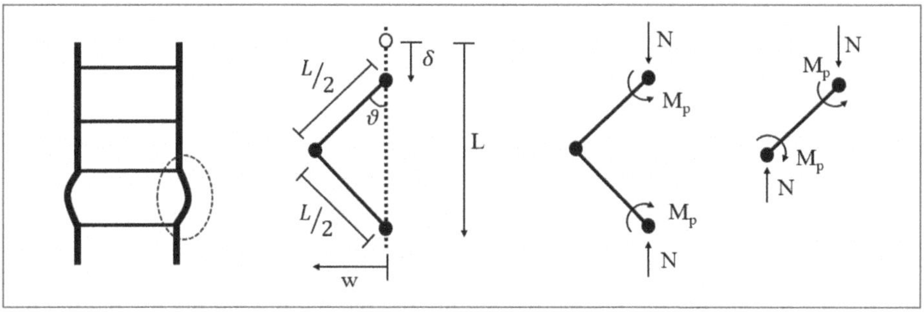

Fig. 2. Equilibrium of a buckled longitudinal steel bar

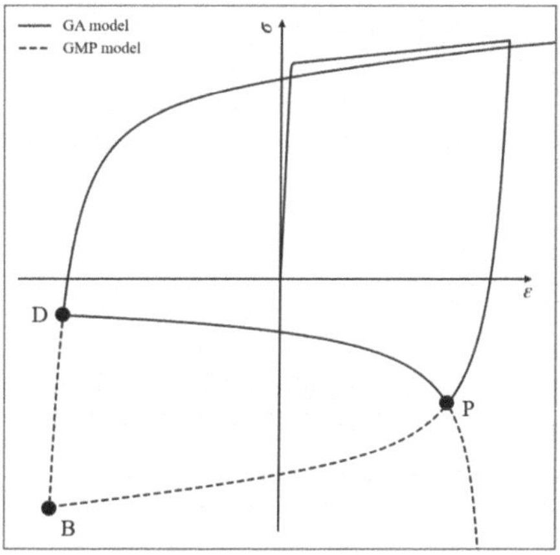

Fig. 3. Gomes and Appleton model

This buckling response is activated in the model after the rupture of the concrete cover, which is detected by checking failure of all concrete cover layers.

After buckling, unloading follows the stress-strain response obtained with the GMP model as described in Fig. 3. This means that at point D after the load reverses, an iterative process to define point B is necessary to obtain the reload path that contains point D.

Stress-strain buckling branch in the OS implementation is described by the following equations:

$$\sigma_b = f_u\gamma - \frac{(\Theta + \gamma)}{(1 + \gamma)}(f_u\gamma - \sigma) \tag{2}$$

where σ is the corresponding stress in absence of buckling. The parameters γ and Θ, respectively, are introduced to avoid kinks and to determine how much the buckling curve should deviate from the initial one. This deviation is controlled by the parameter r which adjust the shape of the curve between the buckled and the unbuckled one, as shown in Eq. (3).

$$\Theta = min(\sigma_p, 1)\beta(1 - r) + r \tag{3}$$

Equation (1) is modified into Eq. (4), introducing an amplification factor β that allows the user to scale the buckling curve. This is useful to adjust the location of the onset point.

$$\sigma_p = 0.6\frac{\beta f_y}{l_{sr}\sqrt{\varepsilon}} \tag{4}$$

Although the parameters β, r, γ are input for the OS version model, it is observed, by inspecting the code, that the parameter β is set equal to one (The recommended values in the OS wiki are $\beta = 1$, $r = 0.4$, $\gamma = 0.5$)

Chang and Mander (CM) (1994) [11] modifications to the GMP hysteresis rules are adopted in this implementation. Unlike the original one, the activation of buckling, for values of r different from one (unbuckled curve), is independent of the surrounding cover concrete fibers.

3.2 Dhakal and Maekawa

In the DM model the monotonic response in compression of the bar that buckles is empirically derived from the results of numerical analyses using a micromodel, in which the effects of the slenderness ratio, l_{sr}, and of the yield strength, f_y, are investigated. The onset of buckling is assumed at yielding point, and the post-buckling response is characterized by a negative slope equal to 2% the elastic stiffness E beyond an intermediate point $(\varepsilon_i, \sigma_i)$, which is a function of the parameter $\lambda_p = l_{sr}\sqrt{f_y/100}$ (f_y in MPa). A residual stress equal to 20% of the yield strength is assumed.

The equations describing the model are:

$$\sigma = \begin{cases} E\varepsilon & if\, \varepsilon < \varepsilon_y \\ \left[1 - \left(1 - \frac{\sigma_i}{f_{it}}\right)\left(\frac{\varepsilon - \varepsilon_y}{\varepsilon_i - \varepsilon_y}\right)\right] & if\, \varepsilon_y < \varepsilon < \varepsilon_i \\ \sigma_i - 0.02E(\varepsilon - \varepsilon_i);\, \sigma_c \geq 0.2f_y & if\, \varepsilon \geq \varepsilon_i \end{cases} \tag{5}$$

where f_{it} is stress in the tension backbone corresponding to the intermediate point.

The coordinates of the intermediate point are given by the following equations:

$$\varepsilon_i = (55 - 2.3\lambda_p)\varepsilon_y;\, \varepsilon_i \geq 7\varepsilon_y \tag{6}$$

$$\sigma_i = \alpha(1.1 - 0.016\lambda_p)f_{it};\, \sigma_i \geq 0.2f_y \tag{7}$$

where parameter α is introduced to normalize the stress of the intermediate point with respect to the nature of strain hardening:

$$\alpha = 0.75 + \frac{\varepsilon_u - \varepsilon_{sh}}{300\varepsilon_y};\, \alpha \leq \frac{f_u}{1.5f_y};\, 0.75 \leq \alpha \leq 1.0 \tag{8}$$

In the OS implementation [4], modifications were made to the definition of the parameter λ_p, to make it dimensionless $\left(\lambda_p = \sqrt{2000 f_y / E}\right)$, thus allowing the model to be used in any reference system (Fig. 4). Also, the CM hysteresis rule is adopted, as opposed to the GMP rule prescribed in the original version, and parameter α is an input value.

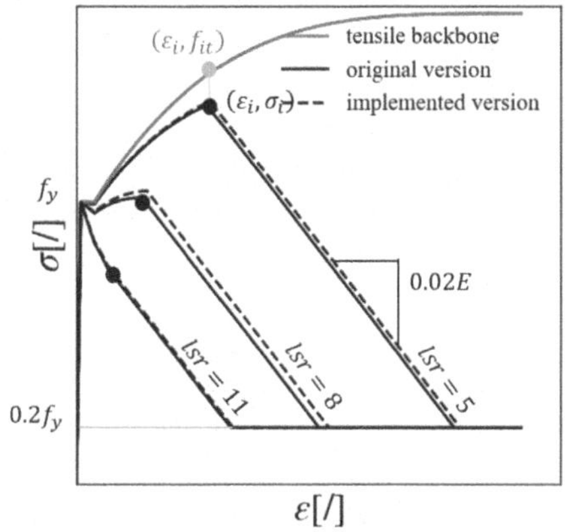

Fig. 4. Monotonic compression envelope of Dhakal and Maekawa model

4 Evaluation of the Buckling Models

This chapter reports the evaluation of the OS models for bar buckling considered in the study. First, the FE micromodel used to numerically simulate reference response of the bare bars is described. Then, the loading protocols adopted for the simulations are introduced. Finally, the results of the comparisons between the responses of the micromodel and the uniaxial buckling models are illustrated.

It is important to note that the micromodel is used instead of experimental tests because most of these latter are monotonic, while bar buckling, as previously commented, is a phenomenon that occurs under cyclic loading conditions. Besides, by using a micromodel, one can consider loading protocols that are more representative of actual load paths than simple protocols characterized by increasing deformation amplitudes.

4.1 Micromodel

Buckling is a highly nonlinear geometric phenomenon, which can be simulated in a FE model through a corotational geometric transformation with the bar discretized into multiple elements (i.e., a micromodel), and assigning a shape with a small eccentricity as initial configuration (cosine for the boundary conditions considered herein). A single beam-column element cannot be used to represent the buckled shape of a bar, since the expansion of the displacement field is incompatible with its deformation. Furthermore, because with the corotational transformation the axial load follows the chord of the deformed bar, to simulate buckling an initial lateral deformation is also needed. Figure 5(a) illustrates the employed model, whose settings, in terms of number of elements, integration points (IP) per element and fiber section discretization and uniaxial steel material model ("ReinforcingSteel" of OpenSees) have been selected based on a mesh sensitivity analysis (not shown).

Figure 5(b) and (c) report the predictions of two available cyclic experimental responses (Monti and Nuti (1992) [12]; Cosenza and Prota (2006) [13]). The match with the test results shows the accuracy of the micromodel, which predicts well key aspects of the buckling response, such as, the peak stress, the post-peak negative slope, and the straightening effect on the reloading stiffness. The latter effect characterizes the post-buckling response in tension when the bar, before returning to a straight configuration, resists the axial force primarily through its flexural stiffness.

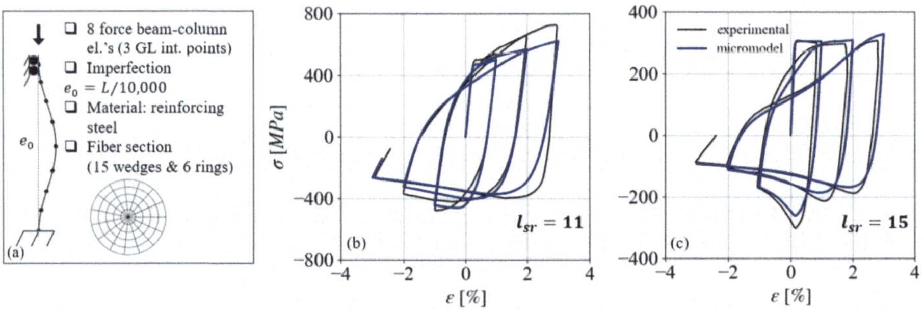

Fig. 5. Micromodeling of the bar: (a) Details, (b) Comparison with Monti and Nuti test, (c) Comparison with Cosenza and Prota test

4.2 Cyclic Loading Protocol

Experimentally recording the response of a rebar that buckles within a structural member is complex because it is impossible to monitor the stress, and strain gauge measurements only capture local deformations rather than global ones. This is the reason why most of the tests in the literature have been carried out on bare bars. Although several characteristics of rebars have been investigated, including strength, slenderness, and ductility, in terms of loading protocols, most test are monotonic (M) or follow the standard symmetric cyclic protocol denoted as C(A) in Fig. 6. Rodriguez et al. (1999) [14] introduced the C(B) and C(C) asymmetric protocols of Fig. 6 to more accurately replicate the deformation of reinforcement bars in columns and beams, respectively. In addition to these protocols, a new one is introduced in this paper, C(D) in Fig. 6, which is characterized by non-increasing deformation cycles. This latter protocol provides a more realistic representation of the deformation path under seismic actions (Table 2).

4.3 Results

The accuracy of the buckling models is evaluated through comparison with results obtained from the micromodel. The goal is to estimate errors in the predictions, and

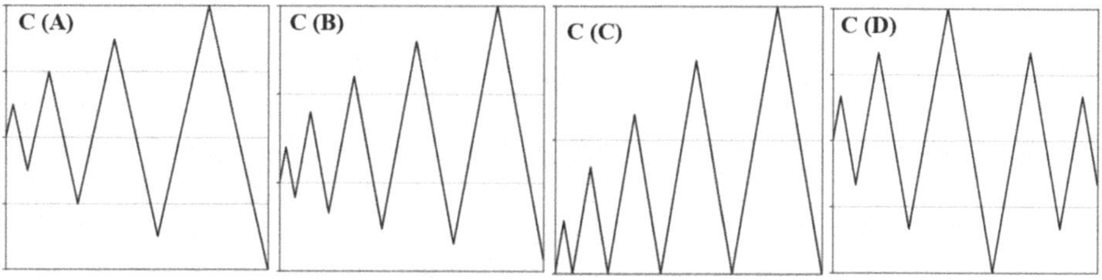

Fig. 6. Load protocols selected

Table 2. Experimental database

Authors	Year	N°	D(mm)	l	f_y (MPa)	Load
Mander et al.	1984	3	16	6, 10, 15	295	M
Monti and Nuti	1990	9	16, 20, 24	5, 8, 11	440	M
Monti and Nuti	1990	9	16, 20, 24	5, 8,,11	440	C(A)
Rodriguez et al.	1999	10	16	2.5, 4, 6, 8	450	M
Rodriguez et al.	1999	8	16	2.5, 4, 6, 8	450	C(A)
Rodriguez et al.	1999	13	16	2.5, 4, 6, 8	450	C(B)
Rodriguez et al.	1999	5	16	2.5, 4, 6, 8	450	C(C)
Bayrak and Sheikh	2001	7	20	4 to 10	520	M
Bae et al.	2005	18	25, 32	4 to 12	440	M
Cosenza and Prota	2006	12	8, 12, 14, 16	5 to 70	350	M
Prota et al.	2009	4	8, 12, 14, 16	5 to 100	350	C(A)

how these errors may vary with the imposed load path. Rebars with slenderness ratio l_{sr} of 5 and 15 are considered, representative of members with close and sparse stirrups, respectively. Figure 7 shows the responses, for all four cyclic protocols, at selected cycles (highlighted in the left panel in blue) for an easier visual comparison. It can be noted that, for both l_{sr} values, the uniaxial models fail to accurately predict the bar's behavior. For l_{sr} = 5, the GA model underestimates the resistance, while the DM model underestimates the post-buckling stiffness. The straightening effect in the reloading phase, evident for the case of $l_{sr} = 15$, is not well predicted by either of the uniaxial models, being not accounted for in their formulations. As a result, the amount of dissipated energy per cycle is grossly overestimated. The level of prediction error changes with the considered loading path and it is maximum in the more realistic protocol C(D) with decreasing cycle amplitudes.

It is interesting to observe that in some cases the response of DM model exhibits a sudden drop of resistance indicative of a possible numerical bug which contributes to the error in the prediction.

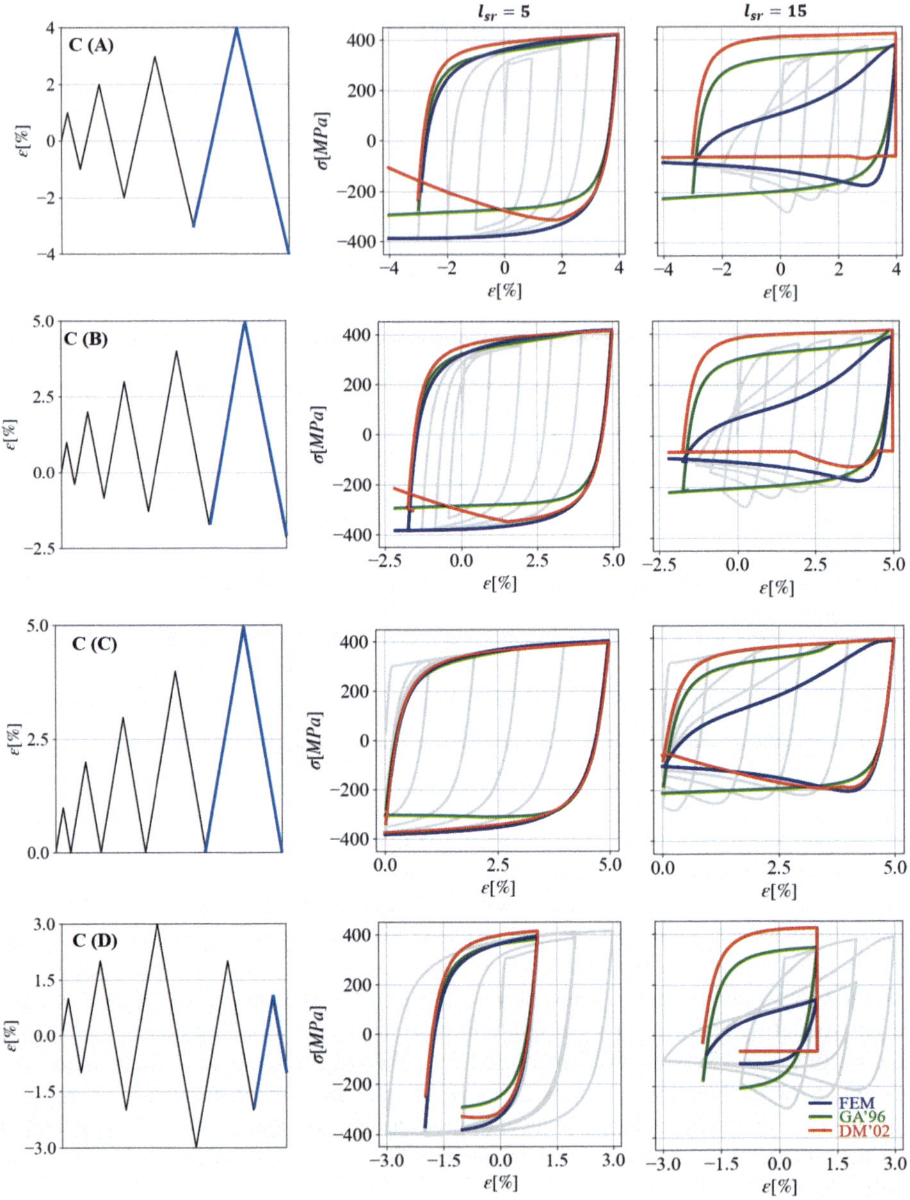

Fig. 7. Comparison between the results of the micromodel and the uniaxial bar models for two slenderness and four load paths

5 Conclusions

Although buckling models for bars are based on the results of tests conducted on bare bars, it is essential to remember that they are not designed to simulate the behavior of these bars, but rather that of the bars within structural members. Furthermore, as previously discussed, from a mechanical perspective, this phenomenon manifests within a structural member only under cyclic loading. Therefore, cyclic models developed for nonlinear dynamic analyses, aimed at replicating the actual behavior of structural members under seismic actions, must exhibit buckling only in relation to cyclic loading. This is not the case for the models implemented in OS and assessed in this study, which predict buckling also in monotonic compression.

Additionally, the Gomes-Appleton model shows problems in describing the post-buckling behavior, with discrepancies in both stiffness and strength. The straightening effect of the bar during the reloading phase is not accurately represented by both models, thus affecting the reloading stiffness and the consequent energy dissipation in the cycles. These issues occur regardless of the loading protocol applied, although the extent of the discrepancies depend on the protocol used. It has been observed that the crucial factor is not the ratio between peak cyclic tensile and compressive strains (i.e., lack of symmetry), but, rather, a decreasing ratio between current and maximum previous strains.

The most recent model proposed by Tripathi and Dhakal (2023) [15] reportedly introduces a significant improvement to correct the stiffness during reloading, providing a promising foundation for future developments. This, indeed, could represent an intermediate improvement step, possibly not a final one because it is still a phenomenological model based only on the standard cyclic load protocol.

References

1. Lucchini, A., Melo, J., Arêde, A., Varum, H., Franchin, P., Rossetto, T.: Load path effect on the response of slender lightly reinforced square RC columns under biaxial bending. J. Struct. Eng. **148**(3), 04021278 (2022)
2. Gomes, A., Appleton, J.: Nonlinear cyclic stress-strain relationship of reinforcing bars including buckling. Eng. Struct. **19**(10), 822–826 (1997)
3. Dhakal, R.P., Maekawa, K.: Path-dependent cyclic stress-strain relationship of reinforcing bar including buckling. Eng. Struct. **24**(11), 1383–1396 (2002)
4. McKenna, F., Fenves, G.L., Scott, M.H., et al.: Open System for Earthquake Engineering Simulation. University of California, Berkeley, CA (2000)
5. Kunnath, S.K., Heo, Y., Mohle, J.F.: Nonlinear uniaxial material model for reinforcing steel bars. J. Struct. Eng. **135**(4), 335–343 (2009)
6. Gil-Martin, L.M., Hernandez-Montes, E., Aschheim, M., Pantazopoulou, S.J.: Slenderness effects on the simulated response of longitudinal reinforcement in monotonic compression. Struct. Eng. Mech. Int. J. **23**(4), 369–386 (2006)
7. Korentz, J.: Influence of geometric imperfections on buckling resistance of reinforcing bars during inelastic deformation. Materials **13**(16), 3473 (2020). https://doi.org/10.3390/ma13163473
8. Moyer, M.J., Kowalsky, M.: Influence of tension strain on buckling of reinforcement in concrete columns. ACI Struct. J. **100**, 75–85 (2003)

9. Giuffrè, A., Pinto, P.E.: Il Comportamento Del Cemento Armato per Sollecitazioni Cicliche Di Forte Intensità. Giornale Del Genio Civile (1970). https://cir.nii.ac.jp/crid/1572261549873498624
10. Menegotto, M., Pinto, P.E.: Method of Analysis for Cyclically Loaded RC Plane Frames Including Changes in Geometry and Non-Elastic Behavior of Elements under Combined Normal ...." Cir.Nii.Ac.Jp. (1973). https://cir.nii.ac.jp/crid/1571980075471506816
11. Chang, G.A., Mander J.B.: Seismic Energy Based Fatigue Damage Analysis of Bridge Columns: Part I. Evaluation of Seismic Capacity. academia.edu. (1994). https://www.academia.edu/download/34954595/chang-fatigue-damage-part1.pdf
12. Monti, G., Nuti, C.: Nonlinear cyclic behavior of reinforcing bars including buckling. J. Struct. Eng. **118**(12), 3268–3284 (1992)
13. Cosenza, E., Prota, A.: Experimental behaviour and numerical modelling of smooth steel bars under compression. J. Earthq. Eng. **10**(03), 313–329 (2006)
14. Rodriguez, M.E., Botero, J.C., Villa, J.: Cyclic stress-strain behavior of reinforcing steel including effect of buckling. J. Struct. Eng. **125**(6), 605–612 (1999)
15. Tripathi, M., Dhakal, R.P.: Constitutive model for reinforcing bars including bar buckling and low-cycle fatigue. Eng. Struct. **295**, 116772 (2023)

Investigation of Rotational Effects of Near Fault Ground Motion Pairs on Structural Response Using OpenSees

Yunlong Chang, Baiping Dong(✉), and Feng Zhou

Tongji University, Shanghai 200092, China
baipingdong@tongji.edu.cn

Abstract. California Building Code (CBC2010) requires that near fault ground motion records should be rotated to the fault-normal and fault-parallel (FN/FP) directions and response history analyses (RHAs) should be performed separately for the ground motions with rotation or not. In the meanwhile, seismic design codes (e.g., ASCE/SEI 7–10) require RHA with ground motion records rotated to the maximum response direction to ensure coverage of the most adverse response. Velocity pulse is the most significant feature that differs near fault ground motions from far-field ground motions, which could impose severe demands on structures in the near-fault region. To investigate the effect of velocity pulse on structural response, this paper rotates near fault ground motion pairs to the direction that has the strongest velocity pulse measured by a maximum pulse indicator (MPI). OpenSees is used for the modeling and dynamic analysis of three dimensional 9-story benchmark steel buildings with ground motions considering the initial records, the FN/FP rotated records, the maximum response rotated records, and the MPI records. The results show that the rotation of records has great effects on structural response and marginal effects on non-structural response.

Keywords: near-fault ground motions · directivity effects · record rotation · pulse direction · 3D buildings

1 Introduction

Ground motions recorded close to the fault have characteristics that are different from those of ground motions recorded further away from the fault. Although the extent of near-fault is still controversial, the near-fault region is generally considered to be confined within a distance of about 20 km from the rupture fault [1, 2]. The features of the ground motions in the near-fault region, which exhibits the dynamic feature of 'forward-directivity' effect, are significantly influenced by the rupture mechanism and slip direction relative to the site.

In the near-fault region, when a fault ruptures towards the site at a velocity close to the shear wave velocity, distinct pulses often appear in the beginnings of ground motion records [3, 4]. Figure 1 shows the forward directivity effect of a unilateral strike-slip fault. Constructive interference of waves concentrates earthquake's energy into one or

X. Lu et al. (Eds.): EOS 2024, LNCE 638, pp. 63–71, 2025.
https://doi.org/10.1007/978-3-031-90690-9_6

two high-intensity cycles within the velocity time history in the direction of fault rupture, which fundamentally differentiates near-fault ground motions from far-field ground motions. As a result, a large number of buildings and bridges in strong earthquakes are directly affected by the velocity pulse [5], and damages to structures in near fault region could be very serious in moderate earthquakes [6].

Pulse-type ground motions have been identified as critical in the design of structures in the near-fault region in some codes. California Building Code (CBC2010) requires that in the region of near faults, usually within 15 km of fault planes, ground motion records should be rotated to the fault-normal and fault-parallel (FN/FP) directions, and then response history analyses (RHAs) should be performed separately for the ground motions with rotation or not. In the meanwhile, seismic design codes (e.g., ASCE/SEI 7–10) require RHA with ground motion records rotated to the maximum response direction to ensure a coverage of the most adverse response. In spite of this, the influence of rotating ground motion on structural response is usually ignored in the near-fault region, and the initial records are widely used in the design and analysis of structures. In this paper, in view of the complexity of nonlinear analysis of 3D model of a building structures, time history analysis was performed using the FN/FP components, MD components, and maximum pulse indicator components of ground motions in addition to the initial records. Case study and overall response show that the structural damage will be underestimated when only considering the original record, and the rotation of records has marginal effects on non-structural damage.

structural damage.

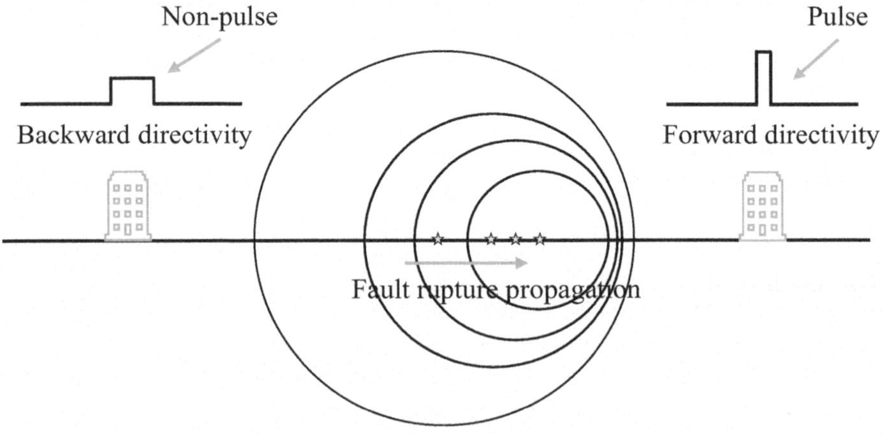

Fig. 1. Schematic drawing of forward directivity effect.

2 Selected Near-Fault Ground Motions

Pulse-type ground motion records can be quantitatively classified by the wavelet-based method proposed by Shahi and Baker [7, 8]. This wavelet-based method uses the wavelet transform to extract pulse-like signals from the velocity time history of ground motion.

After the extraction of the pulses, the ground motion is classified by comparing the original ground motion with the residual ground motion to determine whether the extracted pulses represent a significant part of the original signal. To be specific, the proposed criteria for judgment by Shahi and Baker are used to determine whether there is a pulse in the ground motion and whether the extracted velocity pulse is caused by the forward directivity effect. Shahi and Baker proposed a formula to calculate the pulse indicator (PI) to measure the existence of the pulse. A ground motion is classified as pulse-like if its PI is positive. Besides, they added another criteria to make sure that the pulse is caused by the forward directivity effect. This wavelet-based method is widely used in identifying the pulse of the ground motion.

A set of 12 near-fault ground motions were selected for the subsequent analysis. These ground motions were selected from the PEER (Pacific Earthquake Engineering Research Centre) database, and were taken from 7 events in different countries and regions. They are recorded within 15 km from the rupture fault, and the components obtained by rotation show the velocity pulse. In addition, the average maximum direction (MD) spectrum of the unscaled horizontal ground motions meets the requirement of ASCE7–22. The MD is the direction of the rotating ground motion pair that causes a single lumped mass oscillator, which vibrates freely in both horizontal directions, to have a maximal linear elastic response over exhaustive record rotation. The basic information of the ground motions is summarized in Table 1, and the 5% damped acceleration spectra are shown in Fig. 2.

Table 1. Basic information on selected near-fault ground motions.

No.	event	RSN	Magnitude (M_W)	R_{rup} (km)	Strike (°)
1	1994 Northridge-01	983	6.69	5.43	122
2	1994 Northridge-01	1063	6.69	6.5	122
3	1994 Northridge-01	1084	6.69	5.35	122
4	1994 Northridge-01	1086	6.69	5.3	122
5	1995 Kobe, Japan	1106	6.9	0.96	230
6	1999 Chi-Chi, Taiwan	1492	7.62	0.66	34.6
7	1999 Chi-Chi, Taiwan	1505	7.62	0.32	34.6
8	2003 Bam, Iran	4040	6.6	1.7	354.6
9	2004 Parkfield-02, CA	4107	6	2.5	141
10	2010 Darfield, New Zealand	6906	7	1.2	85.1
11	2010 Darfield, New Zealand	6911	7	7.3	85.1
12	2011 Christchurch, New Zealand	8123	6.1	5.1	59

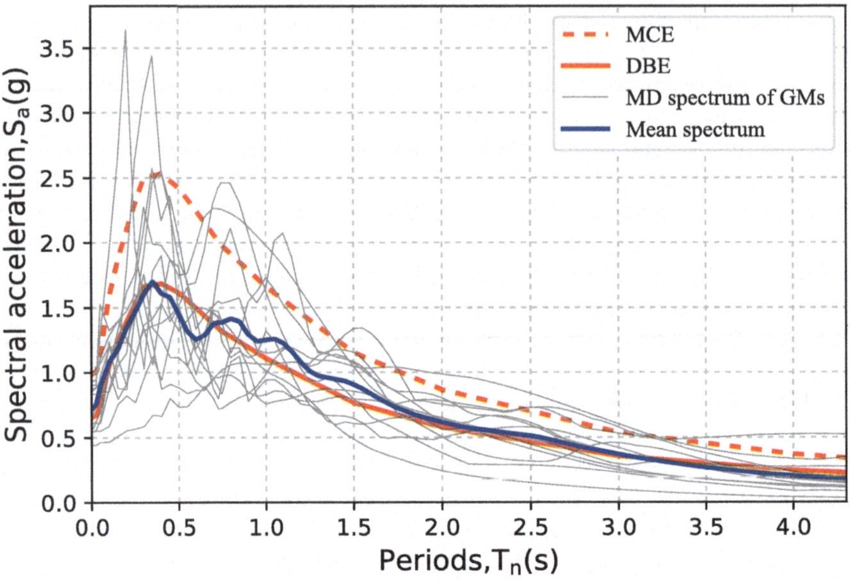

Fig. 2. Elastic response spectra of the selected ground motion records.

If the ground motion in the horizontal plane is understood as a vector, the seismic time-history information measured in different stations are the components of the vector decomposed in different coordinate systems, which means that the time-history information measured by the stations are related to the location of the acceleration measuring instruments in the stations. To get the components of the ground motions in different orientations, the following transformation equations can be used:

$$\ddot{u}_x = \ddot{u}_1 \cos(\alpha_1 - \alpha_{strike} + \theta) + \ddot{u}_2 \sin(\alpha_1 - \alpha_{strike} + \theta) \tag{1}$$

$$\ddot{u}_y = -\ddot{u}_1 \sin(\alpha_1 - \alpha_{strike} + \theta) + \ddot{u}_2 \cos(\alpha_1 - \alpha_{strike} + \theta) \tag{2}$$

where α_{strike} is the strike of the fault, α_1 is the azimuths of the instrument axes, and θ is the angle of rotation as shown in Fig. 3a and 3b. All rotations are positive counterclockwise. Fig. 4 shows the polar plot of identified velocity pulse periods and pulse indicator as a function of the rotation angle, θ, for 12 ground-motion pairs. In these plots 0° is the FP direction and 90° is the FN direction. The red dots indicate (1) pulse periods scaled in polar coordinates and (2) directions in which the velocity pulses are identified. The filled gray area shows the range of θ with velocity pulses. The dashed blue lines show the orientation in which the maximum pulse indicator occur. The solid black line identifies the azimuths of the instrument axes. The numerical values for maximum pulse periods and maximum pulse indicators are presented in the upper-right corner of each subplot. After Examing polar plots of all records, the following observations can be got: Velocity pulses don't always appear in the FN/FP orientation, e.g., GM5 and GM11; pulse periods and pulse indicators change with orientations; and there is little correlation between the maximum of pulse periods and pulse indicators, which appear in different orientations.

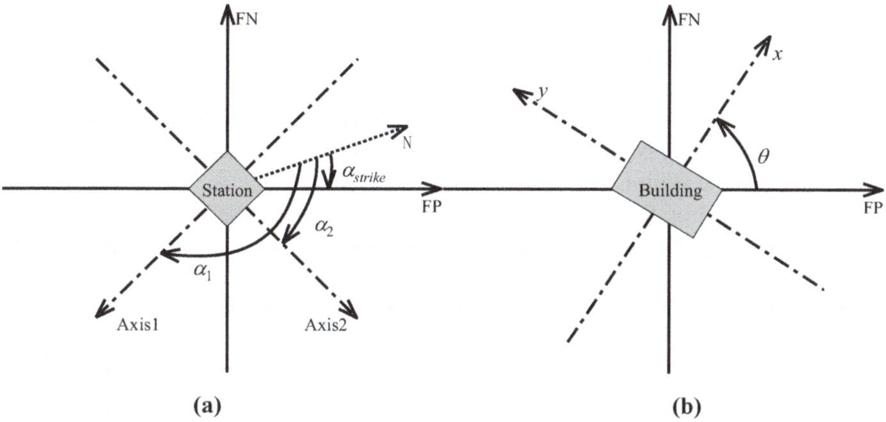

Fig. 3. Reference axes diagram: (a) reference axes for the station, (b) reference axes for the building

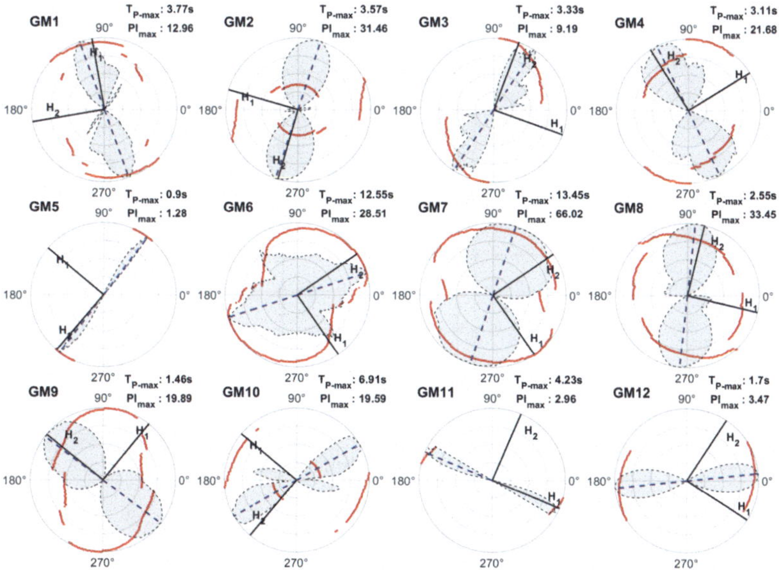

Fig. 4. Polar plots of pulse indicators and pulse periods as the function of the rotation angle, θ

3 Response to Near-Fault Earthquakes in OpenSees

3.1 Nine-Story Benchmark Building

The nine-story steel building designed for the SAC Phase II Steel Project is selected for the subsequent analysis are modelled using the nonlinear dynamic analysis program OpenSees [9]. The structure was designed as office building located on a stiff soil site (Site Class D) in Los Angeles. The lateral load resisting system in both directions consists of two special MRFs on the perimeter of the building. This study focuses on the 3D frame, because actual engineering structures are subjected to multi-directional seismic action and 2D model can only consider horizontal or vertical ground motions at most. Therefore, the model in this paper is subjected to tri-direction ground motions, i.e., the horizontal components under different orientations plus the vertical ground motions recorded in the same station.

As illustrated in Fig. 5, The nine-story building includes a basement level in addition to nine stories above the ground level. The building has five bays of MRF in the NS direction and the EW direction, respectively. In addition, one external connection is designed as pinned connection to avoid bi-axial bending in the corner column. The column lines employ two-tier construction. Column pieces are connected every two levels beginning with the first level, located on the first, third, fifth and seventh levels at 1.83 m above the center-line of the beam to column joint. For this building, the design yield strengths of the beams and columns are 248 MPa and 345 MPa, respectively. More structural design details for the nine-story MRF, e.g., the mass of each floor, can be found in [10, 11].

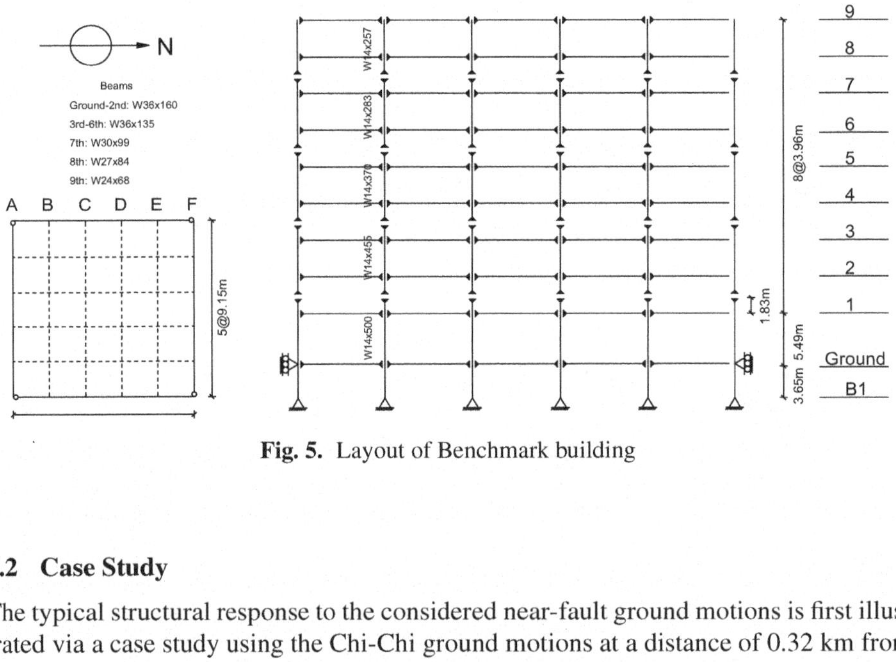

Fig. 5. Layout of Benchmark building

3.2 Case Study

The typical structural response to the considered near-fault ground motions is first illustrated via a case study using the Chi-Chi ground motions at a distance of 0.32 km from the rupture fault (TCU068 station). The roof drift responses of the structure are shown in Fig. 6a. Compared with the roof drift response under the original records of the station, the peak drift and the residual drift of the structure under components of other orientations are obviously larger, with an increase about 50%, and the direction of the peak response appears in the reverse direction. The roof absolute acceleration responses of the structure are shown in Fig. 6b. Compared with the roof absolute acceleration response under the original records, it can be found that the peak acceleration of the structure under components of other three orientations are close.

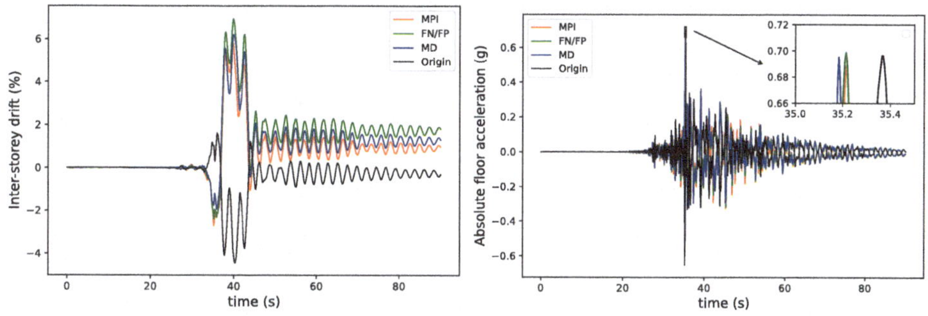

(a) **(b)**

Fig. 6. Case study: (a) roof inter-story drift time histories, (b) roof absolute acceleration time histories

3.3 Overall Response to Near-Fault Earthquakes

Having illustrated the typical response to a specific near-fault earthquake event, an overall view of the behavior of structure is presented herein. Figure 7 shows the maximum inter-story drift and the maximum acceleration response of the nine-story benchmark building subjected to the 12 near-fault ground motions.

For each of the considered orientations of the 12 near-fault ground motions, the location of the maximum inter-story drift is different because of the properties of each ground motion. The values of the maximum inter-story drift ratio of the considered orientations of the 12 near-fault ground motions are different, changing from 4.8% under the original records to 5.8% under the MPI orientation, to 6.2% under the MD orientation, and to 7% under the FN/FP orientation. It can be found that the value of the maximum mean response is 2.74%, 3.17%, 3.1%, and 2.97% under the original, the FN/FP, the MD, and the MPI orientation, respectively. This tells that it's not enough to use the origin records to evaluate the structural response.

The peak absolute floor acceleration is the key factor affecting the non-structural building responses. Unlike the maximum inter-story drift, the mean maximum peak absolute floor acceleration of the 12 near-fault ground motions under the MPI orientation and the origin orientation are larger than that under the original orientation and the MD orientation, reaching a value of 1.1 g in the roof floor. This means the rotation of ground motion records has a marginal effect on non-structural response, though it will make the results more accuracy to consider the different orientations of ground motion records.

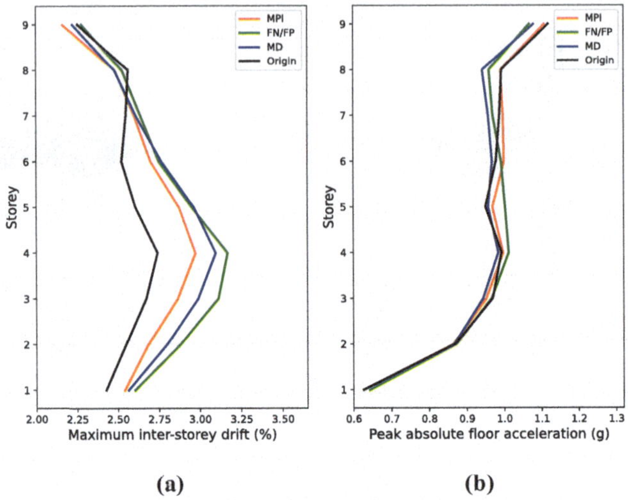

Fig. 7. Mean seismic responses of structures to considered ground motions: (a) maximum inter-story drift, (b) peak absolute acceleration time histories

4 Conclusions

Current seismic design practice in the United States requires as-recorded pairs of ground motions to be rotated to the FN/FP directions before they are used for response history analyses of building structures within 5 km of the active fault. Additionally, ASCE/SEI 7–10 requires RHA with ground motion records rotated to the maximum response direction to ensure coverage of the most adverse response. However, the influence of rotating ground motion on the structural response is usually ignored in the near-field region, and the original records are widely used in the design and analysis of structures. This paper has focused on the overall dynamic behavior of 3D benchmark building under 12 DBE level pulse-like near-fault ground motions in different orientations. The following conclusions are drawn:

(1) Velocity pulses don't always appear in the FN/FP orientation, and there is little correlation between the maximum of pulse periods and pulse indicators, which appear in different orientations.
(2) Rotation has great effects on the inter-story drift response, and the components of the FN/FP orientation will result in the largest structural response. Therefore, it's not enough to use the origin records to evaluate the structural response. Rotation has less effects on the absolute acceleration response, but it will make the results more accuracy to consider the different orientations of components.

References

1. Mavroeidis, G.P., Papageorgiou, A.S.: A mathematical representation of near-fault ground motions. Bulletin of the seismological society of America **93**(3), 1099–1131 (2003)
2. Bray, J.D., Rodriguez-Marek, A.: Characterization of forward-directivity ground motions in the near-fault region. Soil dynamics and earthquake engineering **24**(11), 815–828 (2004)
3. Somerville, P.G., et al.: Modification of empirical strong ground motion attenuation relations to include the amplitude and duration effects of rupture directivity. Seismological research letters **68**(1), 199–222 (1997)
4. Somerville, P.G.: Characterizing near fault ground motion for the design and evaluation of bridges. In: Proceedings of the 3rd national seismic conference and workshop on bridges and highways, pp. 137–149. MCEER, Portland, Oregon (2002)
5. Mollaioli, F., Silvia, B., Decanini, L.D., Panza, G.F.: Characterization of the dynamic response of structures to damaging pulse-type near-fault ground motions. Meccanica **41**(1), 23–46 (2006)
6. Jones, L., Aki, K., Boore, D.: The magnitude 6.7 Northridge, California earthquake of 17 January 1994. Science **266**(5184), 389–397 (1994)
7. Baker, J.W.: Quantitative classification of near-fault ground motions using wavelet analysis. Bull. Seismol. Soc. Am. **97**(5), 1486–1501 (2007)
8. Shahi, S.K., Baker, J.W.: An Efficient Algorithm to Identify Strong-Velocity Pulses in Multicomponent Ground Motions. Bull. Seismol. Soc. Am. **104**(5), 2456–2466 (2014)
9. Mazzoni, S., McKenna, F., Scott, M., Fenves, G.: Open system for earthquake engineeringsimulation (OpenSees). Berkeley, User command language manual, Pacific Earthquake Engineering Research Center, University of California (2006)
10. Gupta, A.: Seismic demands for performance evaluation of steel moment resisting frame structures. Stanford University, California (1999)
11. Ohtori, Y., et al.: Benchmark control problems for seismically excited nonlinear buildings. J. Eng. Mech. **130**(4), 366–385 (2004)

Seismic Performance of Bridges and Similar Systems

Small-Scale PLEXUS 1g Tests on Integral Abutment Bridges: OpenSees Numerical Modelling

Ziyan Huang[1(✉)], Gabriele Fiorentino[1,3], Sha Luo[2], Raffaele De Risi[1], Nicole Metje[2], David Chapman[2], Anastasios Sextos[1], George Mylonakis[1,4], and Flavia De Luca[1]

[1] School of Civil, Aerospace and Mechanical Engineering, University of Bristol, Bristol, UK
ziyan.huang@bristol.ac.uk

[2] School of Engineering, University of Birmingham, Birmingham, UK

[3] Institute of Environmental Geology and Geoengineering, National Research Council, Rome, Italy

[4] Department of Civil and Environmental Engineering, Khalifa University of Science and Technology, Abu Dhabi, UAE

Abstract. Integral abutment bridges (IABs) are widely used worldwide when the crossing span is short-to-moderate. Due to the elimination of bearings and joints, the seasonal expansions and contractions of IABs are absorbed by the backfill and foundation. However, during the cyclic expansion and contraction of the bridge deck caused by temperature variation, the abutment-backfill-embankment system significantly interacts with the superstructure in a way that is difficult to predict analytically. In the context of the EPSRC-funded PLEXUS project, a 1g small-scale experiment was conducted to assess the distribution of the earth pressures behind the abutment. The experiment comprised a soil box filled with sand. A hinged moveable wall was set against the soil to simulate the integral bridge's abutment-backfill interaction, and the thermal-induced deformation was introduced by cyclic movements of the wall. This study aims to numerically model the PLEXUS experiments using OpenSees and its visual toolkit STKO, carrying out a numerical-to-experimental comparison as well as a numerical-to-numerical comparison between different finite element (FE) software packages (i.e. OpenSees and PLAXIS, ABAQUS). In the OpenSees model, the soil-wall interaction interface is represented by "EqualDOF" links, and the backfill is considered a discrete domain with "PressureDependMultiYield02" material. The results indicate good agreement between the numerically predicted and the experimentally measured response when the soil is modelled as nonlinear in OpenSees while highlighting specific differences with respect to the different software packages considered. The strategy using "EqualDOF" for modelling abutment-soil interface and plastic "PressureDependMultiYield02" material showed a relatively satisfactory agreement with experimental results.

Keywords: Integral abutment bridges · Soil-structure interaction · Temperature-induced loading · OpenSees · STKO

X. Lu et al. (Eds.): EOS 2024, LNCE 638, pp. 75–84, 2025.
https://doi.org/10.1007/978-3-031-90690-9_7

1 Introduction

To prevent the development of damage in conventional bridges from superstructure deformation caused by vibrations and temperature variations, expansion devices, such as expansion joints and bearings, are commonly employed to connect various bridge substructures. However, expansion joints and bearings are considered to be weak points that are relatively short-lived due to the penetration and corrosion of liquids and chemicals, leading to extra effort and costs in maintaining, repairing, or replacing [1]. The concept of integral abutment bridge is to construct the bridge deck and the abutment as an integral system, not only fundamentally avoiding the maintenance issues caused by expansion devices but also enhancing IABs with more desirable characteristics compared to conventional bridges: more advanced seismic performance; more effective construction and larger end-span ratios. In the last half-century, IABs have become popular designs widely chosen in construction practice [2, 3].

Due to the elimination of the bearings and joints, the longitudinal deformation of IABs is accommodated by the backfill and the foundation; therefore, the abutment-backfill-embankment system significantly interacts with the superstructure under different loading conditions such as traffic [4], thermal [5] and seismic loading [6]. Especially during the cyclic expansion and contraction of the bridge deck caused by daily or seasonal temperature variation, it has been observed that densification occurs in the backfill, so the earth pressures behind the abutment increase correspondingly over seasons and can exceed the designed bearing capacity of the IAB. This can result in unexpected damage; and the mechanism of this soil-structure-interaction (SSI) effect is difficult to predict analytically.

Efretuei [7] applied PLAXIS 2D to study how the change of backfill soil parameters and the seasonal temperature change can affect the earth pressures developed on IABs. The PLAXIS model simulated a realistic instrumented IAB with available temperature and bridge deformation data. As for the case study, the displacement of abutment can be caused by multiple factors, not just temperature. Thus, the researchers use linear regression for the recorded temperature and displacement data to create an equation for temperature and displacement prediction. The bridge was also simulated to be built in different seasons, and the numerical results indicate that earth pressures mobilised behind the IAB built over the autumn and summer were lower than the pressures developed behind the IAB constructed over the spring and winter.

Caristo et al. [8] tested the performance of a conventional IAB versus a kind of innovative isolation system that contains recycled tyres as a compressible inclusion using PLAXIS 2D. The Mohr–Coulomb model was employed for soil modelling. The temperature change was considered to range from − 10 °C to 50 °C. The whole process was cycled 120 times with displacements of ± 27mm, based on PD 6694–1 [9]. The results indicated that the earth pressures rapidly accumulate behind the abutment wall within the first 30 load cycles, and then there are only slight changes in earth pressure with further cycles. However, the heaving of the backfill soil constantly increased in the whole process. This research verified that this innovative solution using tyres can reduce both the earth pressure and bending moment developed along the height of the abutment.

The EPSRC-funded Priming Laboratory Experiments on Infrastructure and Urban Systems (PLEXUS) project aimed to explore innovative approaches for SSI analysis. A series of 1g small-scale experiments were conducted by Luo et al. in the Heavy and Light Structures Laboratory at the University of Bristol, considering seasonal thermal loading to simulate the abutment-backfill system and assess the development of the earth pressures behind the abutment [10, 11]. Numerical analysis is implemented in accordance with the laboratory tests using FE software OpenSees [12] and its visual toolkit STKO [13], along with comparisons with results from PLAXIS 2D [14] and ABAQUS [15] obtained in a previous study [11]. This enables a desirable numerical-to-experimental comparison as well as a numerical-to-numerical comparison between different software packages in terms of earth pressure development with loading cycles and distribution over the movable wall.

2 PLEXUS Small-Scale Experiments

2.1 Experimental Configuration

The experimental system was composed of four main parts: a soil box, a hydraulic loading system, data acquisition systems, and controlling computers (see Fig. 1). The total size of the testing box was 1525 × 1050 × 1150 mm^3, and the test volume of soil was 1000 × 1000 × 960 mm^3. Since Particle Image Velocimetry (PIV) method was applied in the tests, the walls of the box were fabricated with transparent Perspex, which allows direct observation of the deformation of the backfill.

Fig. 1. PLEXUS 1g small-scale experiments [10, 11].

A 1000 mm high moveable wall was placed in the soil box with a hinge at the base to simulate the integral abutment. Three types of moveable wall configurations (indicated as S1, S2, and S3) were used in the PLEXUS tests (see Fig. 2a). Type S1 was a double-layer wall comprising a 25 mm thick Perspex plate and a 25 mm thick timber plate. Accordingly, the S1 wall can be defined as a flexible integral abutment. Type S2 was a 100 mm single-layer wall made of Polyethylene (PE500), being hinged on a steel pipe at the base. Type S3 was a multi-layer composite wall including a 40 mm thick aluminium frame into the Perspex and timber layers. The experimental campaign considered two groups of loading protocols: group I representing increasing displacements after every two cycles; group C representing constant displacements at each cycle. Referring to the previous research from Luo et al. [11], the numerical results show that the PLAXIS 2D and ABAQUS simulation with a Mohr-Coulomb model can better capture the development of earth pressure on the flexible abutment than those observed in stiffer abutments, and it is unable to capture the densification effect of backfill soil [11]. This paper focuses on the test using the relatively stiff type S2 moveable wall with increasing cyclic loading protocol, and adopts this configuration as a case study in OpenSees platform. The experimental configuration modelled in this study is indicated as S2-I-12 in the PLEXUS tests [11]. The Young's modulus of type S2 Polyethylene moveable wall is 1100MPa [11].

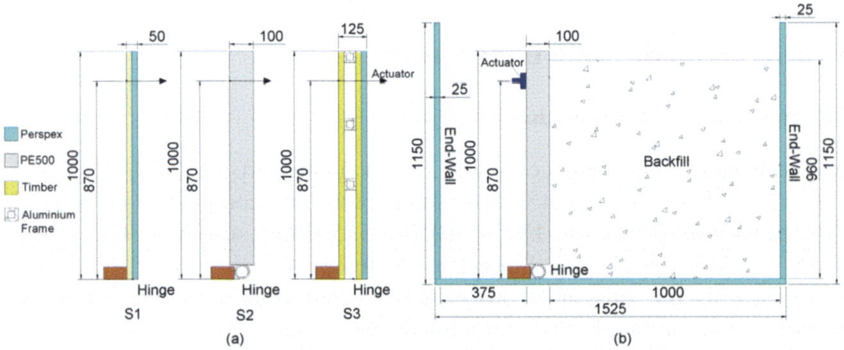

Fig. 2. PLEXUS testing configuration (a) section of moveable wall types S1, S2 and S3, (b) Side view sketch of S2 small-scale test (mm).

The earth pressures were measured by TPC-4000 series Earth Pressure Cells designed to measure total pressure between soil and structures. Three pressure cells were attached to the moveable wall with their centres respectively at elevations of 240 mm, 480 mm and 720 mm from the box base (see Fig. 3).

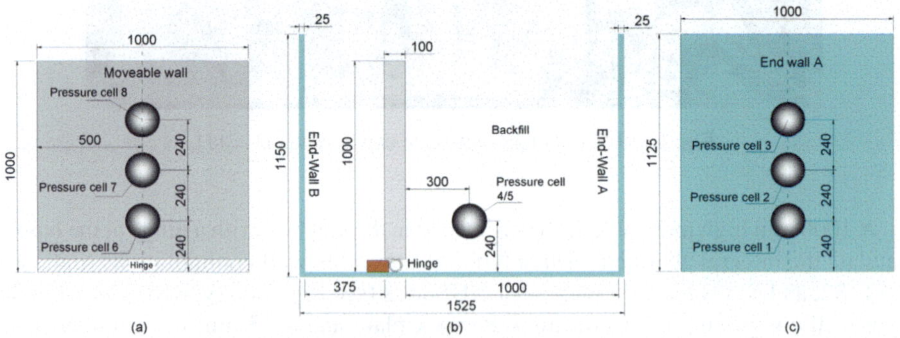

Fig. 3. Setup of pressure cells (mm)

2.2 Backfill Properties

The backfill material employed was dry Leighton Buzzard Sand Fraction B (14/25) (LBSFB), which is a kind of clean, uniform, subrounded, uncemented natural silica sand. The initial dry density of the sand in the test was 1.46 Mg/m^3, which is relatively low because the sand was pluviated with a poorly controlled approach and not compacted afterwards. The specific gravity G_s of the sand is 2.65, and the minimum and maximum void ratios (e_{min} and e_{max}) are 0.635 and 0.83, respectively [16, 17].

2.3 Temperature-Induced Loading

An actuator was set at the height of 870 mm from the bottom of the soil box to control the pull-push pseudo-static movements of the moveable wall, to simulate the thermal-induced bridge deck expansion and contraction. The cyclic displacements imposed by the actuator started at ± 5 mm, increased by ±5 mm every two cycles and finally reached the maximum value ± 30 mm (see Fig. 4a). Figure 4b shows that the densification of backfill leads to increasing load corresponding to the same displacement amplitude.

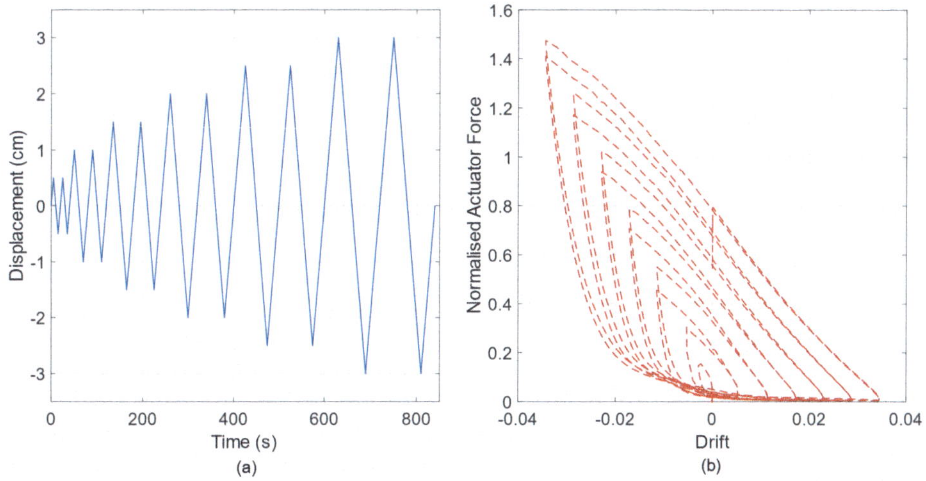

Fig. 4. PLEXUS S2-I-12 test loading input (a) loading input, (b) Normalised actuator force (by $\gamma H^2 L$) vs wall drift (d/H), (adapted from Luo et al. [11]).

3 Previous Numerical Study Using PLAXIS 2D and ABAQUS

The model was put together in plane strain mode in PLAXIS 2D. The movable wall was modelled as a plate with an elastic material, rotating around its fixed hinge, in which displacements of the plate were allowed in both the x-direction and y-direction. To simulate the hinge at the bottom of the moveable wall, a fixed point was added at the bottom of the moveable plate, preventing displacements at the base in both directions while rotation was allowed. The backfill (LBSFB) was modelled using the Mohr-Coulomb model. In the PLAXIS 2D model, the soil-structure interface property was determined by R_{inter} (interface coefficient), which is a reduction factor of the interface strength ranging from 0 to 1, while $R_{inter} = 0$ represents the interface being totally smooth. Based on PLAXIS Manuals [14] and the interface friction tests conducted by Potyondy [18], R_{inter} was defined as 0.7 for the PE500 wall-sand interface.

In the ABAQUS/CAE 2D model, the moveable wall was simulated to be linear isotropic, formed using shell elements. Mohr-Coulomb model and shell elements were adopted for the backfill material. The plane strain boundary condition was considered in the model to minimise container boundary effects. The contact pressure of the wall-backfill interface was simulated by a hard contact model, while the tangential interaction was described by the static-kinetic exponential decay function [11].

The passive earth pressure profiles of the two FE models overlapped around the middle part of the wall. The lower part of the numerical model profiles was less accurate after the second cycle and became significantly larger for both models with respect to the experimental profile, not capturing the "arching" effect observed in the experimental results and predicted by analytical models [19], see Sect. 5. This was due to the Mohr-Coulomb model reaching its limit and not being capable of capturing the densification in the backfill during cyclic loading [11].

4 OpenSees Modelling

4.1 FE Models

In the OpenSees model, the moveable wall was simulated as an "elasticBeamColumn" element. The element property of the backfill soil was represented by 4-node quad elements with plane strain behaviour (see Fig. 5). Since the backfill can be regarded as a homogenous cohesionless domain, the constitutive material model was selected to be "PressureDependMultiYield02" [20] that is based on small-strain shear modulus and built up to investigate cyclic mobility and pressure-dependent strength of pressure sensitive soils, (see Table 1.). According to the empirical equations from Hardin and Black [21] and Kramer [22], the small-strain shear modulus of LBSFB was derived as G_0 = 19800kPa. Considering the typical relationship between soil stiffness and shear strain given by Atkinson and Sallfors [23], the abutment can be regarded as a retaining wall, thus the reference shear modulus (G_r) input was defined as $G_r = 0.4G_0 = 7900$kPa. The command, "UpdateMaterialStage", was used to update "PressureDependMultiYield02" material into the analysis step, and two material stages were respectively applied in the analysis: stage 2 - Elastic with small strain stiffness; and stage 1 - Plastic. The wall-backfill interface is modelled using "EqualDOF" constraints on node-to-node links.

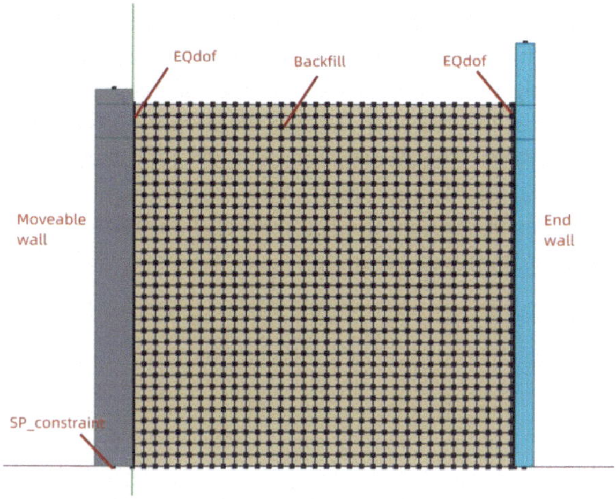

Fig. 5. OpenSees 2D model by STKO.

Table 1. Properties of LBSFB input in OpenSees model for "PressureDependMultiYield02" material [20].

Parameters	LBSFB
Density (ρ)	1.46 Mg/m^3
Reference low-strain shear modulus (G_r)	7900 kPa
Friction angle (φ)	32°
PeakShearStra	0.1
pressDependCoe	0.5
Phase transformation angle	31°
contrac1	0.087
contrac3	0.18
Dilation 1&3	0

5 Results and Comparisons

5.1 Comparison with Experiment

Figure 6a shows the results of the passive earth pressure development on the moveable wall at maximum wall expansion derived from the OpenSees model with elastic "PressureDependMultiYield02" backfill behaviour, compared with experimental results. The figure shows a similar trend of uniform growth in earth pressure during the 12 uniformly increasing cyclic loadings. The maximum value of earth pressure predicted by OpenSees is far larger than the value of the experimental data, especially around the upper part of the wall. This is due to the value of earth pressure being mainly proportional to the wall displacement.

Figure 6b compares passive earth pressures behind the moveable wall derived from the OpenSees analysis with plastic "PressureDependMultiYield02" backfill material and the experiment, indicating that the OpenSees plastic simulation can better capture the cyclic soil densification in the experiment and the arching effect in the earth pressure distribution. The distribution of earth pressure development from OpenSees plastic simulation roughly shows reasonable agreement with experimental results, while the values of the 10th and 12th cycles in the OpenSees model at the location of the middle pressure cell are close to values from the experiment.

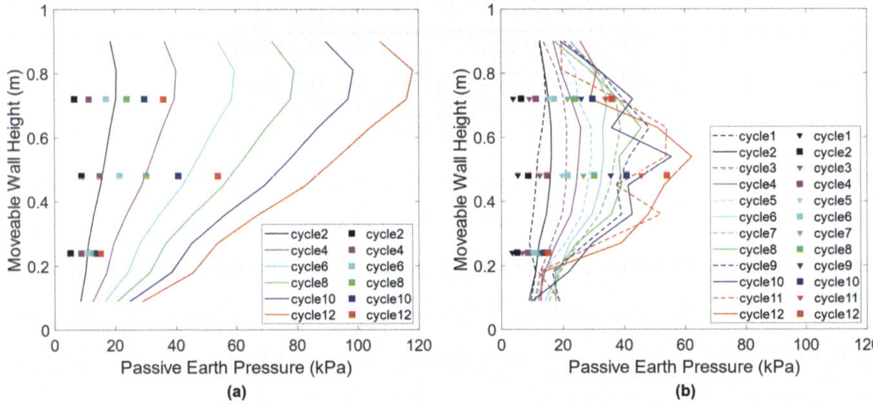

Fig. 6. Passive earth pressure developed behind moveable wall - results from OpenSees model (lines) with (a) elastic and (b) plastic backfill compared with experimental results (dots).

5.2 Comparison with PLAXIS 2D and ABAQUS

Figure 7 shows the comparisons of the passive earth pressures between the three FE software solutions, indicating that the OpenSees model with plastic "PressureDepend-MultiYield02" material has better ability to simulate the soil densification under cyclic loading. The results from PLAXIS 2D and ABAQUS simulations, based on the Mohr-Coulomb model, show significant overestimation of the earth pressures. The OpenSees model still largely overestimates the earth pressures at the location of the bottom pressure cell. After the 4th cycle, however, the upper part of the earth pressure curve from OpenSees shows reasonable agreement with the experimental values and matches better with ABAQUS results rather than the PLAXIS ones.

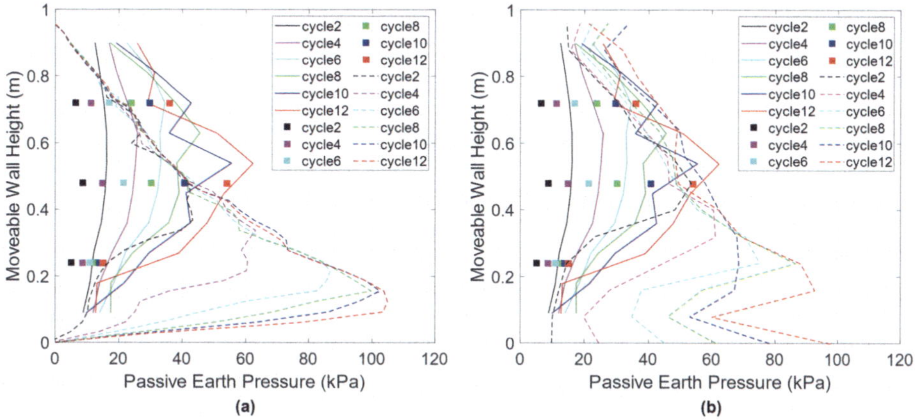

Fig. 7. Passive earth pressure derived from experiment (dots) and OpenSees model with plastic backfill (solid lines) compared with (a) PLAXIS results (dot lines), (b) ABAQUS results (dot lines) from [11].

6 Conclusion

The development and evolution of earth pressures on integral bridge abutments under cyclic thermal loading has presented a significant challenge for the design of IABs. The previous numerical simulations referring to PLEXUS small-scale 1g experiment showed discrepancies in evaluating the earth pressure behind the abutment. In particular, the assumption of Mohr-Coulomb soil model was not capturing in full the experimental behaviour. This paper presented a modelling strategy using OpenSees and its visual toolkit STKO towards simulating one of the setup configurations of the PLEXUS tests. "PressureDependMultiYield02" material was employed to model the soil with both elastic and plastic behaviour with small-strain properties implemented. The results from OpenSees simulation with plastic "PressureDependMultiYield02" model showed that this model is more sophisticated to reproduce the experimentally observed earth pressures, and can capture the soil densification effect. Still, the results from OpenSees plastic model do not match perfectly the experimental results as the numerical values increase faster with the progress of cycles. The contact interface between the soil and abutment was modelled using the "EqualDOF" constraint but an alternative contact modelling strategy could be explored to improve the match with the experiment.

References

1. Wallbank, E.J.: The performance of concrete in bridges. A survey of 200 highway bridges. HMSO/Department of Transport, London, UK (1989)
2. Burke, M.P., Jr.: Integral and semi-integral bridges. John Wiley & Sons, Hoboken, NJ, USA (2009)
3. Hoppe, E.J., Gomez, J.: Field study of an integral backwall bridge. No. VTRC 97-R7. Virginia Transportation Research Council (VTRC). Charlottesville, VA (1996)

4. De Risi, R.: A computational framework for finite element modeling of traveling loads on bridges in dynamic regime. Computer-Aided Civil and Infrastructure Engineering **37**(4), 470–484 (2022)
5. Pétursson, H., Kerokoski, O.: Monitoring and analysis of abutment-soil interaction of two integral bridges. J. Bridg. Eng. **18**(1), 54–64 (2013)
6. Dhar, S., Dasgupta, K.: Seismic soil structure interaction for integral abutment bridges: a review. Transportation Infrastructure Geotechnology **6**(4), 249–267 (2019)
7. Efretuei, E.O.: Thermal impact on soil-structure interaction for integral bridges. University of Leeds, UK (2013)
8. Caristo, A., Barnes, J., Mitoulis, S.A.: Numerical modelling of integral abutment bridges under seasonal thermal cycles. In: Proceedings of the Institution of Civil Engineers-Bridge Engineering, Vol. 171, No. 3, pp. 179–190. Thomas Telford Ltd. (2018
9. Denton, S., Christie, T., Shave, J., Kidd, A.: PD 6694–1: recommendations for the design of structures subject to traffic loading to EN 1997–1. In Bridge Design to Eurocodes: UK Implementation, pp. 434–450. ICE Publishing (2011)
10. Luo, S., et al.: Challenges and perspectives for integral bridges in the UK: PLEXUS small-scale experiments. Proceedings of the institution of civil engineers-smart infrastructure and construction **175**(1), 27–43 (2022)
11. Luo, S., et al.: Physical and numerical investigation of integral bridge abutment stiffness due to seasonal thermal loading. Transportation Geotechnics **42**, 101064 (2023)
12. McKenna, F.: Open System for Earthquake Engineering Simulation. University of California, US (2000)
13. Petracca, M., Candeloro, F., Camata, G.: STKO user manual. ASDEA Soft-ware Technology, Pescara (2017)
14. Brinkgreve, R.B.J., Swolfs, W.M., Engin, E.: PLAXIS 2D User manual. Delft University of Technology & PLAXIS, Netherlands (2019)
15. Khodair, Y.A., Hassiotis, S.: Analysis of soil–pile interaction in integral abutment. Comput. Geotech. **32**(3), 201–209 (2005)
16. Fiorentino, G., et al.: Integral abutment bridges: investigation of seismic soil-structure interaction effects by shaking table testing. Earthquake Eng. Struct. Dynam. **50**(6), 1517–1538 (2021)
17. Fiorentino G., De Risi R., De Luca F., et al.: SSI-induced seismic earth pressures on an integral abutment bridge model: Experimental measurements versus numerical simulations and code provisions. Earthquake Engineering & Structural Dynamics, pp. 1–23 (2024)
18. Potyondy, J.G.: Skin friction between various soils and construction materials. Geotechnique **11**(4), 339–353 (1961)
19. Paik, K.H., Salgado, R.: Estimation of active earth pressure against rigid retaining walls considering arching effects. Geotechnique **53**(7), 643–653 (2003)
20. Yang, Z., Elgamal, A., Parra, E.: Computational model for cyclic mobility and associated shear deformation. Journal of Geotechnical and Geoenvironmental Engineering **129**(12), 1119–1127 (2003)
21. Hardin, B.O., Black, W.L.: Vibration modulus of normally consolidated clay. Journal of the Soil Mechanics and Foundations Division **94**(2), 353–369 (1968)
22. Kramer, S.L.: Geotechnical earthquake engineering Vol 653. Prentice Hall, US (1996)
23. Atkinson, J.H., Sallfors, G.: Experimental determination of soil properties. General Report to Session 1. Proceedings of the 10th ECSMFE, Florence, **3**, 915–956 (1991)

Analysis of Friction-Based Isolated Pier Response Under Vertical Dynamic Load

G. Shid and M. R. Shiravand(✉)

Faculty of Civil, Water and Environmental Engineering, Shahid Beheshti University, Tehran, Iran
m_shiravand@sbu.ac.ir

Abstract. The Friction Pendulum System (FPS) is one of the most widely used sliding-based energy dissipation devices for reducing damage in structures. In this system, the motion of the slider due to external forces, along with its contact with concave or spherical surfaces, generates a friction force. On the other hand, research has shown that variations in normal force and sliding velocity are two critical factors influencing the friction parameter. Notably, changes in the friction parameter can significantly impact FPS performance and, consequently, the response of the structure isolated by this system. Vertical force also affects the coefficient of friction, attributed to variations in the normal force. Therefore, this subtle factor can strongly affect the responses of structures that use friction-dependent devices like FPSs. This study investigates the nonlinear response of a three-span continuous concrete I-girder bridge equipped with FPS. A sinusoidal dynamic load was generated and applied to the structure using Opensees platform. Employing nonlinear time-history analysis, structural responses were obtained and analyzed, including axial force in columns, isolator force, isolator displacement, and column displacement. The results show that variations in vertical force significantly influence the nonlinear dynamic behavior of both the bridge and the isolator, demonstrating that friction and vertical force interactions play a crucial role in the seismic response of FPS-isolated structures.

Keywords: OpenSees · Friction Pendulum System · Seismic Isolation System · Vertical Dynamic Force

1 Introduction

Seismic isolators are available in various types, broadly classified into two main categories: (1) elastomeric isolators and (2) sliding isolators. An isolator acts as a combination of a spring and a damper, characterized by four key properties: initial stiffness, flexible displacement, energy dissipation, and recentering capability. This study specifically focuses on sliding isolators. In sliding isolators, low horizontal stiffness is achieved through the sliding interface, which also serves as the mechanism for energy dissipation. The types of sliding isolators include the Single Friction Pendulum (SFP), Double Friction Pendulum (DFP), Triple Friction Pendulum (TFP), and the XY-FP (a friction bearing designed to resist uplift) [1]. The first iteration of the sliding pendulum isolator was the

X. Lu et al. (Eds.): EOS 2024, LNCE 638, pp. 85–96, 2025.
https://doi.org/10.1007/978-3-031-90690-9_8

Single Friction Pendulum (SFP), also referred to as the friction pendulum system (FPS). Its structure consists of a mechanical slider coated with a Teflon liner, which moves along a concave spherical surface made of stainless steel [1]. This movement generates a frictional force opposing the direction of motion. The coefficient of friction plays a crucial role in determining the energy dissipation behavior of friction-based isolators and is influenced not only by material properties but also by factors such as velocity and normal force [2]. The normal force can vary under vertical loading, leading to fluctuations in frictional force that may significantly impact the performance of friction-based isolators. These variations affect the performance range and overall response of the isolated structure. Vertical loading, therefore, has a substantial influence on structural responses and can induce damage under certain loading conditions [3]. Dynamic vertical loading and its effects on the friction model are critical factors warranting further investigation. Numerous studies have explored friction models and the factors influencing them. Mosqueda et al. [4] investigated the effect of sliding velocity on the friction coefficient through experiments involving four FPS isolators placed beneath a rigid frame subjected to bi-directional loading histories. They observed that the friction coefficient generally depended on velocity, with a slight reduction occurring beyond 25.4 cm/s, consistent with Mokha et al.'s findings. This reduction at higher velocities was attributed to increased temperature and the deposition of composite materials on the stainless steel surface. Experimental results indicated that the steady-state friction coefficient began stabilizing at 2.54 cm/s, earlier than the 12.7 cm/s reported by Mokha et al. The authors concluded that this stabilized coefficient aligns with the constant value assumed in Coulomb's law of friction. Jangid [5] conducted a parametric study on FPS systems to evaluate how the friction coefficient affects the seismic response of buildings and bridges during near-fault earthquakes. His analysis included a multi-story building model and a three-span bridge with a continuous deck subjected to near-fault seismic events. Jangid concluded that an optimal friction coefficient in FPS systems minimizes isolator displacements while reducing superstructure acceleration. He recommended employing friction coefficients between 0.07 and 0.19 for bridge isolators and between 0.05 and 0.15 for building isolators in near-fault regions. Nakajima et al. [6] analyzed a single-degree-of-freedom model with a natural period corresponding to their experimental setup. They highlighted that neglecting the velocity-dependent variation of the friction coefficient in modeling could lead to discrepancies between predictions and experimental observations. In this study, a three-span continuous concrete bridge equipped with an FPS-based isolation system is analyzed. A parameter, denoted as λ, is introduced to represent the ratio of vertical force (VF) in the positive and negative directions. The horizontal dynamic load is modeled as a constant sinusoidal force, while the vertical dynamic load varies according to different λ values applied to the model. Nonlinear time-history analysis (NLTHA) is performed to obtain and compare structural responses, including axial forces in the columns, isolator forces, isolator displacements, and column displacements. The results reveal the impact of VF variations on the nonlinear responses of both the bridge and the isolators.

2 Theoretical Aspects of Friction

Friction is a dissipative force that opposes the relative motion between two bodies in contact. Prevailing theories of friction emphasize the material composition and surface characteristics of solids, acknowledging that no surface is perfectly smooth or flat at the microscopic level. This section examines friction from a macroscopic perspective, an approach particularly relevant to seismic isolation devices and broader engineering applications. The frictional force (F_f) at the sliding interface is expressed as:

$$F_f = \mu.N \quad (1)$$

where μ is the coefficient of friction, and *N* is the normal force at the interface. The frictional force between stationary surfaces is termed static friction, while the frictional force between surfaces in relative motion is referred to as kinetic friction. The classical friction model commonly applied in engineering, known as Coulomb's law, assumes a constant friction level throughout sliding [7]. However, this assumption is not valid for sliding isolators, where the friction coefficient varies significantly with parameters such as sliding velocity and normal force [8]. Constantinou et al. [2] developed a mathematical model to capture these dependencies, demonstrating that the friction coefficient is influenced by material properties, sliding velocity, and the normal force (*N*). The behavior of FPS is particularly sensitive to variations in μ. VF plays a significant role in altering the normal force, thereby affecting μ. Consequently, changes in VF can modify the friction coefficient, with the direction of the vertical force acting on the bearing further influencing the extent of this variation. In this study, a friction model that accounts for changes in both velocity and normal force is employed. This approach provides a more accurate evaluation of the impact of VF on bridge dynamics.

3 Finite Element Model

The selected bridge for the NLTHA is a three-span continuous concrete I-girder deck bridge. The connection between the superstructure and substructure is non-integral and utilizes FPS. A three-dimensional (3D) fiber-based model of the bridge was developed in the Open System for Earthquake Engineering Simulation (OpenSees) [9]. The nodal arrangement and dimensions of the model are shown in Fig. 1. The bridge deck is assumed to remain elastic and undamaged during seismic events. Therefore, it is modeled using equivalent linear-elastic beam-column elements. Nonlinear behavior and plasticity distribution along the elements are captured using the NonlinearBeamColumn element in OpenSees. The pier cross-sections are discretized into fiber sections, representing confined and unconfined concrete as well as reinforcing steel bars. This fiber discretization effectively captures the interaction between axial and bending forces, which is particularly critical under vertical loads. The material properties for concrete and steel are modeled using OpenSees libraries. Concrete behavior is represented by the Concrete07 material model, which incorporates concrete tensile strength. Unconfined concrete is assigned a compressive strength of 33.8 MPa, while the strength of confined concrete is calculated using the Mander et al. model [10]. Reinforcing steel in the bridge piers is modeled with the Steel02 material [11], which has a yield strength of 460 MPa and an elastic modulus of 200 GPa. The connection between the superstructure and substructure is represented by a Single Friction Pendulum Bearing Element. This element employs a friction model that accounts for variations in the coefficient of friction due to changes in both sliding velocity and normal force. These variations are critical to capturing the influence of axial load (N) on the system's seismic response, aligning with the research objectives.

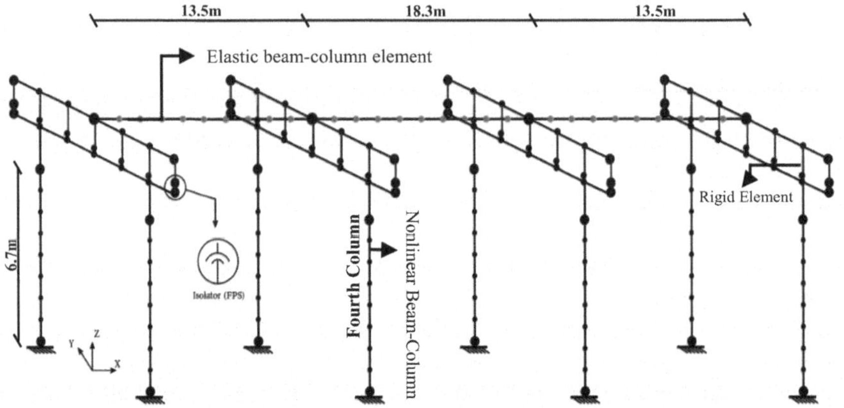

Fig. 1. Nodal scheme and dimensions of the bridge.

4 Generating Dynamic Load

4.1 Definition of Lambda Factor

To account for the impact of VF on the response and performance of FPS and various structural components, as well as to emphasize the significance of inertial force direction in the vertical plane, the following factor, λ, is introduced:

$$\lambda = \left| \frac{VF(+)}{VF(-)} \right| \tag{2}$$

where $VF(+)$ and $VF(-)$ represent the vertical force in the positive and negative directions, respectively. The structural response to dynamic loading is therefore influenced by both the magnitude and direction of the VF. Equation (2) describes three distinct states: (i) $\lambda > 1$: The maximum VF acts in the positive direction. (ii) $\lambda < 1$: The maximum VF is directed negatively. (iii) $\lambda = 1$: The maximum VF is balanced equally in both positive and negative directions.

4.2 Mathematics Model of Generated Dynamic Loads

To accurately study the concept and effects of λ on structural damage, artificial dynamic loads with varying λ ratios were generated. These dynamic loads consist of a sequence of sinusoidal harmonic forced loads. The process begins with the application of a sinusoidal wave with a neutral λ ratio ($\lambda = 1$). Subsequently, a pulse with a selected λ ratio is generated using a combination of three sinusoidal functions. This combined wave repeats periodically with a defined duration and constant time steps. The general form of the initial part of the combined waves composing the overall dynamic load is given by:

$$P_i(t) = A_i \sin(\omega_i t + \theta) \tag{3}$$

In which A represents the amplitude of the i-th wave, θ is the phase of the i-th wave at $t = 0$ (assumed $\theta = 0$), and ω denotes the angular frequency of the i-th wave, derived from $\omega = 2\pi/T_i$. To generate pulses with varying λ ratios, a combination of one complete sinusoidal wave, a half sinusoidal wave, and another half sinusoidal wave (phase-shifted by $T_1/2$) is employed (see Fig. 2). The frequency of the first wave forming the pulse is equal to the frequency of the second wave. This coherent phase relationship leads to an algebraic addition of their amplitudes. The first wave, completing a full cycle during a period of T_1, is defined as:

$$P_1(t) = A_1 \sin(\omega t) \qquad 0 \le t \le T_1 \tag{4}$$

The second wave is described as:

$$P_2(t) = \begin{cases} A_2 \sin(\omega t) & 0 \le t \le \frac{T_1}{2} \\ 0 & \frac{T_1}{2} \le t \le T_1 \end{cases} \tag{5}$$

To achieve the desired λ ratio, the third function is defined with the same frequency but with a phase difference from the first wave. The phase shift enables constructive interference between the second half of the first wave and the third wave, producing an amplitude of $A_2 - A_1$ (see Fig. 2):

$$P_3(t) = \begin{cases} 0 & 0 \le t \le \frac{T_1}{2} \\ A_3 \sin(\omega t + \frac{T_1}{2}) & \frac{T_1}{2} \le t \le T_1 \end{cases} \tag{6}$$

The total dynamic load, combining these functions, is expressed as:

$$P_{Total}(t) = \sum_{i=1}^{3} P_i(t) \tag{7}$$

$$P_{Total}(t) = \begin{cases} A_1 \sin(\omega t) + A_2 \sin(\omega t) & 0 \le t \le \dfrac{T_1}{2} \\ A_1 \sin(\omega t) + A_2 \sin(\omega t + \dfrac{T_1}{2}) & \dfrac{T_1}{2} \le t \le T_1 \end{cases} \quad (8)$$

Based on these equations, the eligible ratio between the maximum positive force and maximum negative force (i.e., the ratio of positive to negative amplitudes in Eq. (8)) can be determined. The sum of the amplitudes of Eqs. (4) and (5) provides the maximum positive force (P^{+}_{max}). The relationship between peak amplitudes in opposite directions is defined by Eq. (2). Thus:

$$\begin{aligned} A_1 + A_2 &= P^{+}_{max} \\ A_1 + A_2 &= \pm\lambda(A_1 - A_2) \end{aligned} \quad (9)$$

The final dynamic load scheme, described by Eq. (8), is illustrated in Fig. 3a. Additionally, a flowchart detailing the dynamic load generation process with the desired λ ratio is shown in Fig. 3b.

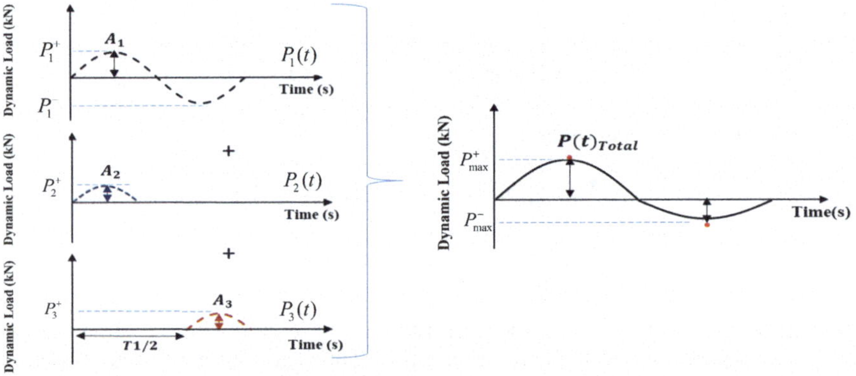

Fig. 2. Schematic diagram of the functions used to generate a pulse with a specific λ ratio.

Thus, the relative ratio of peak to minimum forces inherent in the generated dynamic load can yield different values of the λ parameter. Table 1 provides details of the dynamic loads, illustrating five distinct λ states. Among these, States 1 and 5 represent exaggerated scenarios to facilitate a more comprehensive investigation into the influence of λ on structural damage.

To evaluate the potential impact of λ on seismic responses, an NLTHA was conducted on the bridge model. The results of this analysis were then compared to highlight the effects of varying λ.

Table 1. Details of the specified λ ratios

Dynamic Load Pattern	λ		P^{+}_{max}	P^{-}_{max}
L1	5	a	0.5	0.1
		b	1.5	0.3
L2	1.67	a	0.5	0.3
		b	1.5	0.9
L3	1	a	0.5	0.5
		b	1.5	1.5
L4	0.6	a	0.3	0.5
		b	0.9	1.5
L5	0.2	a	0.1	0.5
		b	0.3	1.5

5 Nonlinear Time History Analysis Results

This section presents the NLTHA results of the bridge model under the dynamic loads detailed in Table 1, with the effects of λ evaluated by recording relevant structural demands during the analysis. Figure 4 illustrates the axial force behavior of the fourth column of the bridge, providing comparisons across different λ states. Figure 4a highlights the comparison between load patterns L1 and L5, which exhibit higher λ factors, while Fig. 4b compares the results for states L2 and L4. In general, L3, represented by blue dashed lines, remains consistent across all states. To ensure accurate comparisons, a specific loading duration is analyzed in each diagram. As shown in Fig. 4, when $\lambda > 1$, the axial force of the column shows minimal variation. In these cases (L1 and L2), the maximum upward force reduces the compressive force acting on the column, introducing the possibility of tensile forces due to the prevailing upward (positive) force. Conversely, when $\lambda > 1$, as seen in L4 and L5, the dominant downward (negative) force combines with the column's existing compressive force, resulting in an overall increase in axial compressive force. This phenomenon is particularly evident at specific peaks occurring between 5.5 and 6.5 s into the loading process for L4a and L4b. When $\lambda = 1$, the applied load at the peaks reduces the column's compressive axial force to its maximum extent, while the contrasting peaks increase it. As shown in Fig. 4a, the axial force varies most notably in L3b.

Figure 5 examines the response history of the isolator displacement located above the fourth column. As illustrated in Fig. 5a, minor differences are observed in the isolator's displacement, whereas Fig. 5b highlights more pronounced variations. Overall, it is evident that when $\lambda > 1$, the isolator displacement increases, accompanied by a reduction in friction. The containment of each load must be considered when comparing demands, leading to comparisons such as L1a versus L5a and L1b versus L5b. For instance, during loading, L2b ($\lambda > 1$) demonstrates greater displacement than L4b, and L2a exhibits a 5% increase in displacement compared to L4a at the start of the displacement

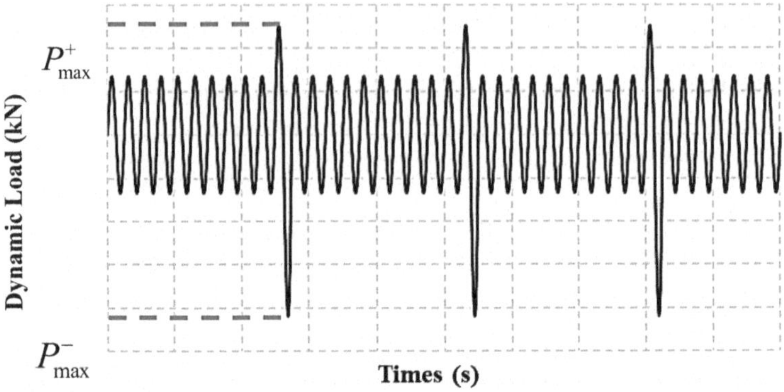

a

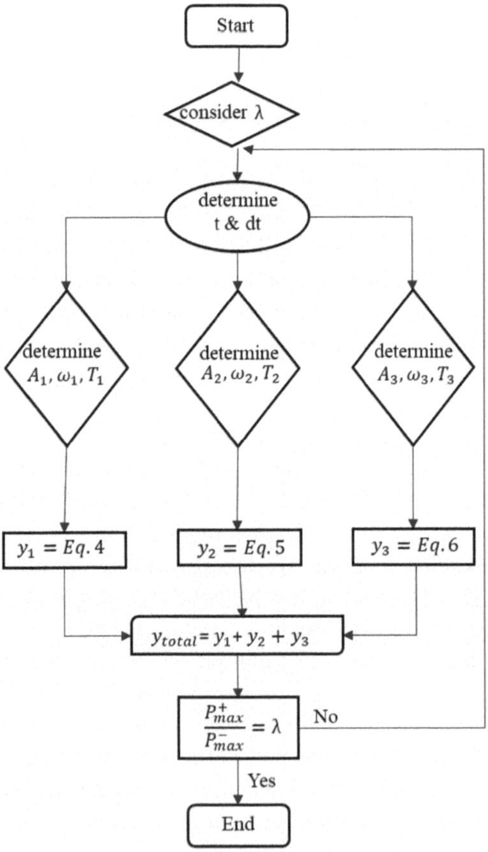

b

Fig. 3. (**a**) Typically dynamic load patterns, (**b**) Flowchart of the dynamic load generation.

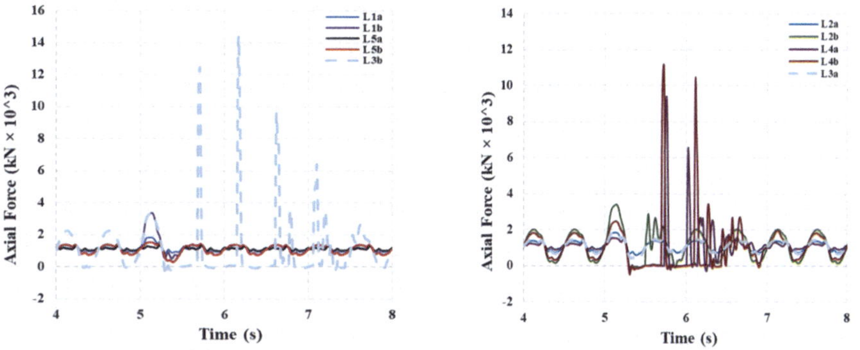

(a) (b)

Fig. 4. Axial Force of the column: (a) L1, L5, and L3b, (b) L2, L4, and L3a.

phase. Additionally, increased displacement is linked to the rise in force, irrespective of the peak force direction. This implies that for $\lambda > 1$, the increase in displacement is achieved through friction reduction, whereas for $\lambda < 1$, it results from increased rigidity. Since the performance of the FPS isolator depends on frictional forces, these results confirm that variations in the normal force significantly influence the isolator's displacement. When $\lambda > 1$, the isolator displacement increases, while $\lambda < 1$ results in lower displacement. Consequently, the isolator's displacement plays a crucial role in affecting the displacement and drift of the columns. Greater isolator capacity to absorb lateral forces and accommodate larger displacements leads to reduced base shear acting on the column, thereby minimizing column drift.

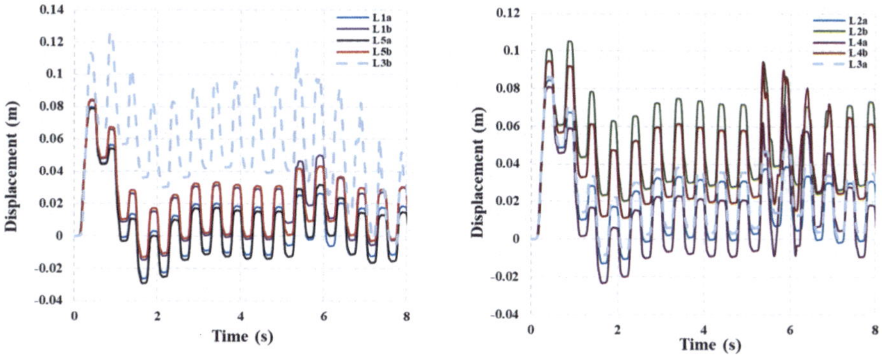

(a) (b)

Fig. 5. Isolator displacement: (a) L1, L5, and L3b, (b) L2, L4, and L3a.

Figures 6a and 6b demonstrate that column displacement is greater when $\lambda > 1$ (cases L1, L2) compared to $\lambda < 1$. As the isolator moves and dissipates energy, column displacement decreases across all cases, with the most notable reduction occurring in the neutral case ($\lambda = 1$), represented by L3b in Figs. 5a and 6a. Similarly, L4b in Fig. 5b, which showed an increase in isolator displacement compared to L4a, exhibits a 31% reduction in the peak displacement of the column. These observations indicate variations in the amount of force absorption and energy dissipation by the isolator, directly influencing column displacement. Furthermore, the response of uplift to vertical forces, particularly regarding the direction of the vertical component, is a critical factor for isolators. To illustrate this, Fig. 7 has been developed. When the isolator has high vertical stiffness and effectively captures uplift, tensile forces within the isolator result in the axial force becoming zero. Consequently, the axial force in the column and the internal force in the isolator exhibit a specific trend. However, when the column experiences tensile forces (represented by negative values in Fig. 4), the internal force in the isolator becomes zero. This can be observed by comparing Figs. 4 and 7. Hence, when the force acts in an upward direction ($\lambda > 1$), there is a potential for generating tensile forces and uplift. In the analyzed example, based on Fig. 4, the ratio between the applied tensile and existing compressive forces was insufficient to induce significant uplift in the isolator. However, in L3b, where the dynamic sinusoidal force reaches substantial magnitudes in both directions (as per Table 1), its combination with sinusoidal horizontal force during specific loading segments generates tensile forces in the column, with corresponding zero-force segments shown in Fig. 7a.

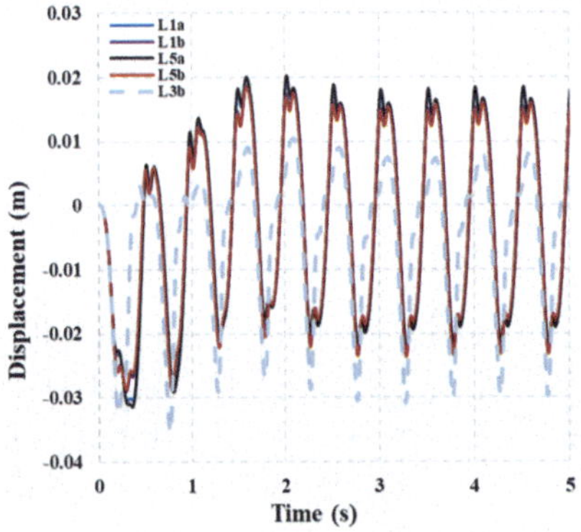

(a)

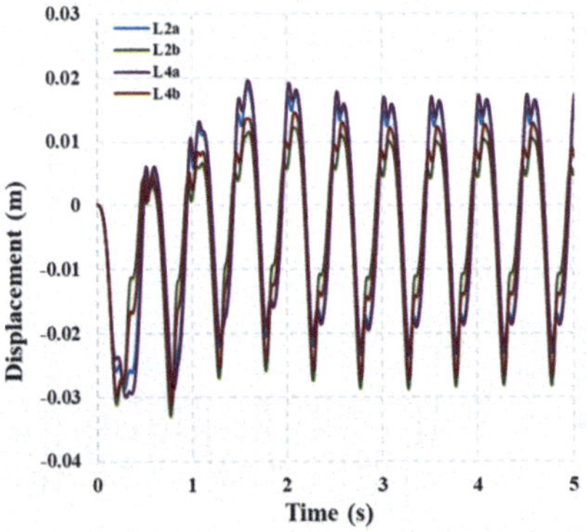

(b)

Fig. 6. Column displacement: (a) L1, L5, and L3b, (b) L2, L4, and L3a.

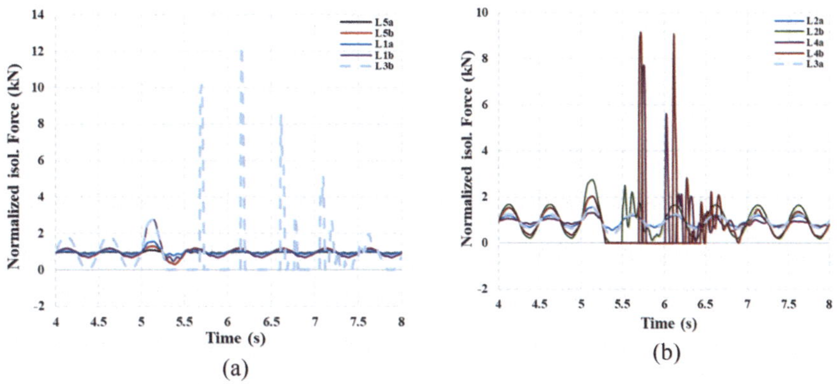

Fig. 7. Isolator force: (a) L1, L5, and L3b, (b) L2, L4, and L3a.

6 Conclusion

In this study, a three-span continuous concrete bridge equipped with an FPS was analyzed to assess the impact of vertical dynamic force variations on the nonlinear response. A parameter, lambda (λ), representing the ratio of vertical force (VF) in positive and negative directions, was introduced to examine the effects of vertical dynamic load on structural behavior. Through nonlinear time-history analysis, axial force in the columns, isolator force, isolator displacement, and column displacement were obtained and compared. The analysis yielded the following general results:

- The nonlinear responses of both the bridge component (column) and the isolator were significantly influenced by changes in the VF, demonstrating the importance of considering different direction of vertical load in the analysis of FPS-isolated structures.
- The dominant type of structural damage varies based on the λ value under conditions of maximum tensile or compressive force.
- Increased compressive forces within the system lead to one category of damage.
- Tension forces result in a different damage profile, impacting the bridge structure in unique ways.
- When λ is equal to one, meaning that the system experiences the same magnitude of maximum force in both positive and negative directions along the vertical axis (as in the L3 state). This results in both compressive and tensile forces causing equal damage to the structure.

This study underscores the need to account for VF changes in the performance analysis of FPS-isolated bridges to better predict potential damage modes.

References

1. Morgan, T.A.: The use of innovative base isolation systems to achieve complex seismic performance objectives. University of California, Berkeley (2007)
2. Constantinou, M., Mokha, A., Reinhorn, A.: Teflon bearings in base isolation II: Modeling. J. Struct. Eng. **116**(2), 455–474 (1990)
3. Chen, Z., Jia, P.: Seismic response of underground stations with friction pendulum bearings under horizontal and vertical ground motions. Soil Dyn. Earthq. Eng. **151**, 106984 (2021)
4. Mosqueda, G., Whittaker, A., Fenves, G.: Characterization and modeling of friction pendulum bearings subjected to multiple components of excitation. J. Struct. Eng. **130**(3), 433–442 (2004)
5. Jangid, R.: Seismic response of isolated bridges. J. Bridg. Eng. **9**, 156–166 (2004)
6. Nakajima, K., Iemura, H., Takahashi, Y., Ogawa, K.: Pseudo dynamic tests and implementation of sliding bridge isolators with vertical motion. In: 12th World Conference on Earthquake Engineering (2000)
7. Naeim, F., Kelly, J.M.: Design of seismic isolated structures: from theory to practice. John Wiley & Sons (1999)
8. Shapiro, D., Rojahn, C., Reaveley, L.D., Smith, J.R., Morelli, U.: NEHRP guidelines and commentary for the seismic rehabilitation of buildings. Earthq. Spectra **16**(1), 227–239 (2000)
9. OpenSees: Open System for Earthquake Engineering Simulation, Pacific Earthquake Engineering Research Center, University of California, Berkeley. (2006). Available at http://opensees.berkeley.edu/
10. Mander, J.B., Priestley, M.J., Park, R.: Theoretical stress-strain model for confined concrete. J. Struct. Eng. **114**(8), 1804–1826 (1988)
11. Menegotto, M., Pinto, P.E.: Method of analysis for cyclically loaded reinforced concrete plane force and bending. In: Proceedings, pp. 15–22. IABSE Symposium on Resistance and Ultimate Deformability of Structures Acted on by Well Defined Repeated Loads, Lisbon (1973)

Advanced Parametric 3D Finite Element Modeling of the Colle Castino Bridge in OpenSeesPy for Structural and Seismic Analysis

Shima Sdeghzadeh(✉), Cristoforo Demartino, and Camillo Nuti

Department of Architecture, Roma Tre University, Largo G. B. Marzi 10, Rome, Italy
{Shima.sadeghzadeh,Camillo.nuti}@uniroma3.it

Abstract. This study introduces the development of a comprehensive parametric 3D model of the Colle Castino Bridge using OpenSeesPy, designed to facilitate extensive statistical analyses via Monte Carlo simulations. Structural engineering challenges, such as uncertainties in material properties, geometric dimensions, and loading conditions, are addressed by parameterizing all aspects of the bridge's geometry and materials. This allows for the efficient simulation of numerous scenarios, enabling robust statistical evaluations. The parametric model systematically varies input parameters, including cross-sectional geometry, reinforcement details, and material properties, to simulate diverse conditions such as corrosion and material degradation. This methodology quantifies uncertainties, assesses their impact on structural performance, and evaluates reliability by calculating failure probabilities under different scenarios. The study provides valuable insights for data-driven maintenance and retrofitting strategies aimed at enhancing bridge safety and longevity. The model's development emphasizes modularity, organizing material properties, geometry, and structural elements into classes and functions, enabling flexible adjustments and updates. This research demonstrates a systematic approach to parametric modeling and statistical analysis of complex structures, contributing to risk-based decision-making in structural engineering. By improving the understanding of uncertainty effects on performance, the study advances the design of resilient, reliable bridge structures and supports informed maintenance strategies.

Keywords: Parametric Modeling · Corrosion and Material Degradation · Colle Castino Bridge · OpenSeesPy · Bridge Reliability

1 Introduction

1.1 Necessity of Statistical Analyses in Structural Engineering

In the field of structural engineering, understanding the variability and uncertainty inherent in material properties, geometric dimensions, and loading conditions is crucial for ensuring the safety and reliability of structures. Statistical analyses enable engineers

X. Lu et al. (Eds.): EOS 2024, LNCE 638, pp. 97–110, 2025.
https://doi.org/10.1007/978-3-031-90690-9_9

to account for these uncertainties by assessing the probabilistic behavior of structural systems under various scenarios (Melchers & Beck, 2018). Specifically, Monte Carlo simulations are a powerful tool for performing such analyses, as they allow for the evaluation of structural responses across a wide range of randomized input parameters, providing insights into the likelihood of different performance outcomes (Rubinstein & Kroese, 2016).

1.2 Importance of Parameterizing of Model

To effectively perform statistical analyses like Monte Carlo simulations, it is essential to have a model where all geometry and material properties are parameterized. Parameterization involves defining the structural model's properties using variables that can be systematically varied within specified ranges. Monte Carlo simulations require the generation of many models to statistically represent the variability in structural properties and loading conditions. Each simulation run involves a different set of parameters sampled from predefined probability distributions, necessitating an efficient method for model generation (Robert & Casella, 2004). Without a parameterized model, creating these numerous instances would be time-consuming and prone to errors. Parameterization automates this process and allows for the automatic generation of multiple model instances with different combinations of parameters, facilitating comprehensive statistical evaluations (Haldar & Mahadevan, 2000; Melchers, 1999). By parameterizing geometry and materials, engineers can efficiently explore the impact of uncertainties on structural behavior, leading to more robust design and assessment practices (Ghanem & Spanos, 1991; Schuëller & Pradlwarter, 2007; Möller & Beer, 2008).

1.3 Development of a Parametric Model for the Colle Castino Bridge

Given the need for statistical analyses and the generation of numerous models for Monte Carlo simulations, developing a parametric model of the Colle Castino Bridge is essential. The bridge, located in a region susceptible to seismic activity and material degradation due to corrosion, presents a complex case study where variability in material properties and geometry can significantly influence structural performance (Biondini & Frangopol, 2008).

Corrosion of steel reinforcement is one of the primary causes of deterioration in reinforced concrete (RC) structures, leading to a reduction in the cross-sectional area of rebars, loss of bond between steel and concrete, and cracking of the concrete cover (Bergami et al., 2019; Bergami et al., 2021a). These degradation processes not only impair the static performance but also adversely affect the dynamic response of structures during seismic events (Nuti et al., 2020a; Bergami et al., 2021b). The collapse of the Polcevera Viaduct in Genova serves as a critical example of the catastrophic consequences that corrosion can have on aging infrastructure (Nuti et al., 2020a; Nuti et al., 2020b).

By creating a parametric model using OpenSeesPy, all aspects of the bridge's geometry and materials can be defined as variables, enabling the simulation of different levels of corrosion and their impact on structural behavior. This approach allows for:

- **Quick adjustments to the model parameters** to simulate varying degrees of corrosion, material degradation, or geometric alterations (Bergami et al., 2022).
- **Ability to perform extensive Monte Carlo simulations** to assess the probabilistic behavior of the bridge under various loading and deterioration scenarios, including the progression of corrosion over time (Bergami et al., 2021a; Bergami et al., 2021b).
- **Improved insights into the sensitivity of the bridge's structural response** to changes in parameters, aiding in risk assessment and decision-making for maintenance or retrofitting strategies (Bergami et al., 2019; Bergami et al., 2022).

To address these challenges, a detailed parametric 3D model of the Colle Castino Bridge using OpenSeesPy was created, enabling the performance of statistical analyses through Monte Carlo simulations.

1.4 Significance of Research

This research contributes to the field of structural engineering by demonstrating a systematic approach to parametric modeling for statistical analyses. The methodology developed can be applied to other complex structures, facilitating risk-informed decision-making in infrastructure management. By emphasizing the need for parameterization in geometry and materials, the study highlights the importance of advanced modeling techniques in addressing uncertainties inherent in structural systems (Der Kiureghian & Ditlevsen, 2009).

2 Location of Case Study: Colle Castino Bridge

The Colle Castino Bridge located in the Teramo province of Italy; a region characterized by high seismic activity due to its proximity to tectonic fault lines within the Apennine Mountain range. Classified as a Zone 1 area in the Italian seismic code. Teramo is subject to stringent design and safety standards, with a Peak Ground Acceleration (PGA) exceeding 0.25 g (Ministero delle Infrastrutture e dei Trasporti, 2019). The bridge, constructed in the mid-20th century, spans approximately 975 m, making it one of the longest bridges in the region and a critical component of the local infrastructure network (Fig. 1).

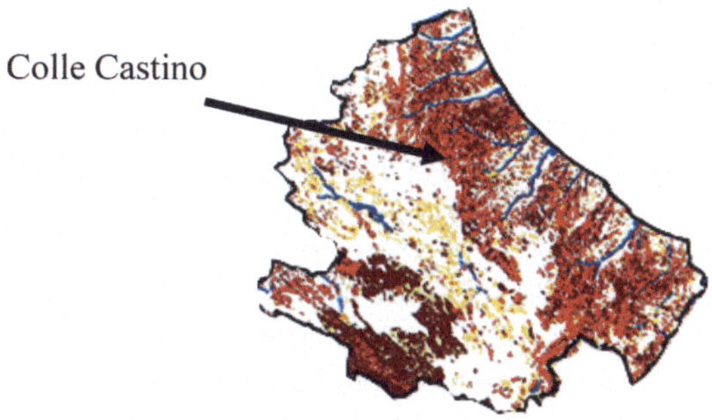

Fig. 1. Hazard map of territory (2020)

Given the importance of the Colle Castino Bridge to the regional transportation network and its exposure to both aging and seismic hazards, it presents a valuable case for assessing the impact of material degradation—especially the loss of concrete cover—on seismic performance (Fig. 2).

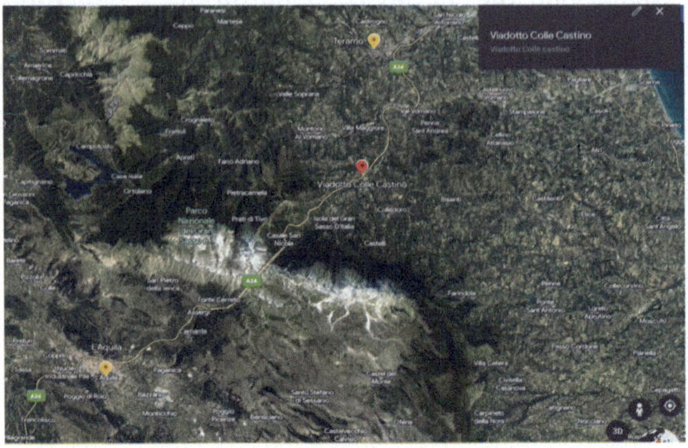

Fig. 2. Location of Colle Castino on map

Fig. 3. Colle Castino Bridge

3 Development of the 3D Parametric Model

To build a parametric model, several core components needed to be defined. First, material properties were established, including the strength and stiffness of concrete and steel. The concrete in the model is represented using the "Concrete01" nonlinear material model, with a mean compressive strength (fcm) of 32.9 MPa, compressive strength of −32.9 MPa, strain at compressive strength of −0.0033, ultimate strength of −32.9 MPa, and ultimate strain of −0.014. The concrete has an elastic modulus (Ec) of 27.656 MPa and a density (ρc) of 2.500 kg/m^3. For the steel reinforcement, the "Steel01" elastic-plastic material model is used to capture both yielding and strain hardening behaviors, with a mean yield strength (fym) 217.39 MPa, yield strength of 217.39 MPa, elastic modulus (Es) of 210 GPa, and strain hardening ratio of 0.01. The steel density(ρs) is

7,850 kg/m^3. These material models provide an accurate representation of the bridge's response to loading conditions, accounting for the inelastic behavior of concrete and the nonlinear response of steel (Fig. 3).

Additionally, the model allows flexibility in modifying the concrete cover, enabling the simulation of corrosion effects in future studies.

By using these material models, the bridge's response to loadings—both inelastic for concrete and nonlinear for steel—is accurately represented, with the added flexibility of modifying the concrete cover to simulate the progression of corrosion in future studies.

The second step in the development of a parametric model is to define the various parameters for the cross-sectional geometry of the bridge's sections, such as width, height, and thickness. These dimensions define the shape and size of the structural elements. In the next step, reinforcement details like rebar sizes, quantities, and positions within the cross-section should be defined to model the internal support provided by the steel reinforcement. Then, the structural elements themselves are defined, comprising the nodes and elements that make up the bridge model. Finally, the analysis procedures planned for use, such as modal or seismic analysis, are determined to guide how the model will be evaluated (Fig. 4).

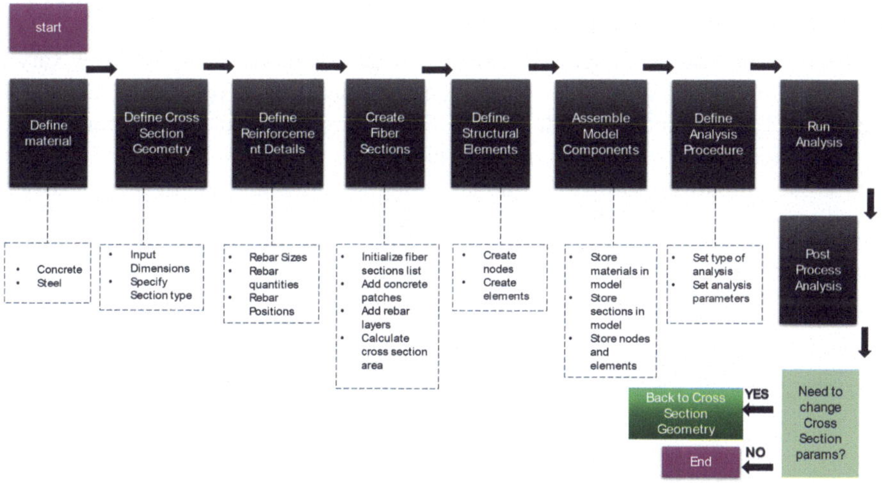

Fig. 4. Procedure of creating a parametric model in Opensees

3.1 Structural Components

The Colle Castino Bridge has 25 simply supported spans, each approximately 39 m in length. These spans rest on bearings placed on the bridge's piers and abutments, allowing for flexibility to accommodate movements resulting from thermal expansion, vehicular loading, and seismic activity. The width of the deck to 11.7 m to support multiple traffic lanes, sidewalks, and safety barriers (Table 1).

Table 1. Overal bridge information

Bridge Details		Bridge Details	
Bridge Length [m]	975	Average Span Length [m]	39
Maximum Deck Width [m]	11.7	Type of Used Sections	A-B-C-D
Number of Columns	24	Longest Column [m]	53
Number of Spans	25	Shortest Column [m]	6

To improve structural efficiency and reduce self-weight, the bridge deck employs hollow box-girder sections, designed in four distinct types labeled A, B, C, and D (Fig. 6) This hollow box-girder configuration represents a strategic balance between material efficiency and structural robustness.

3.2 Columns

Supporting the bridge deck are 25 piers of varying heights, strategically aligned with the simply supported spans. The uneven topography of the bridge site requires piers of different lengths, ranging from 6 m to 53 m, which ensures a consistent deck elevation across the bridge's length. To meet the structural demands imposed by varying pier heights, four types of hollow rectangular cross-sections—types A, B, C, and D—are utilized. Type A sections, with smaller dimensions, are used for the shortest piers, which experience lower load demands. As pier height increases, larger sections (types B and C) are employed to provide the necessary stiffness and load-bearing capacity. The tallest piers, requiring the greatest structural strength, are designed with Type D

Table 2. Summary of bridge piers extracted from original data

Column	Height [m]	Section	Column	Height [m]	Section
25	11.0	A	**13**	12.5	A
24	26.0	A	**12**	10.5	A
23	14.0	A	**11**	13.5	A
22	5.5	A	**10**	17.5	A
21	8.5	A	**9**	22.5	A
20	5.5	A	**8**	19.5	A
19	4.5	A	**7**	26.5	A
18	5.5	A	**6**	39.5	C-B-A
17	6.0	A	**5**	51.5	D-C-B-A
16	6.0	A	**4**	45.5	D-C-B-A
15	6.0	A	**3**	30.5	B-A
14	12.5	A	**2**	15.0	A

cross-sections, which support higher loads and moments due to increased height and deck weight (Table 2).

The piers incorporate specific reinforcement details essential for seismic resilience. Longitudinal reinforcement resists axial loads and bending moments, while transverse reinforcement, in the form of stirrups or hoops, provides shear resistance and confines the concrete core. This transverse reinforcement improves ductility and energy absorption, which is critical for seismic loading scenarios.

4 Development of the 3D Parametric Model

A 3D parametric finite element model (FEM) of the Colle Castino Bridge was developed using OpenSeesPy. The model's geometry was generated from a detailed spreadsheet containing data on column heights, cross-sectional types, material properties, and spatial coordinates. This structured setup allows for easy updates, as any changes in the spreadsheet automatically reflect in the model, supporting efficient testing of scenarios like varying corrosion levels or seismic loads. The model includes the bridge's main components, 25 piers, and 24 spans of 39 m each.

To simulate real-world behavior, bearings and expansion joints were included to represent actual support conditions. Beam-column elements, hinges, and geometric transformations ensured accurate connectivity across structural components, capturing the bridge's response under various loads and boundary conditions (Fig. 5).

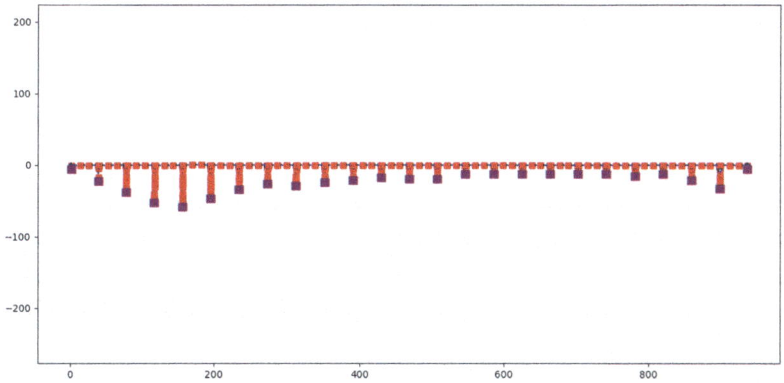

Fig. 5. Generated bridge model based on input spreadsheet data

4.1 Cross-Section Modeling

The cross-section modeling of the Colle Castino Bridge was designed to allow for corrosion effects by enabling selective removal of the concrete cover. As mentioned, four types of hollow rectangular column cross-sections were defined based on the bridge's design.

To develop a parametric model, for each bridge's pier, we defined various parameters, including the width and height of the cross-sections, the number of rebars in each direction, their corresponding sizes, and their coordinates in x and y directions as spreadsheets in excel and we they were called by program. These parameters were used as inputs for generating the cross-sections.

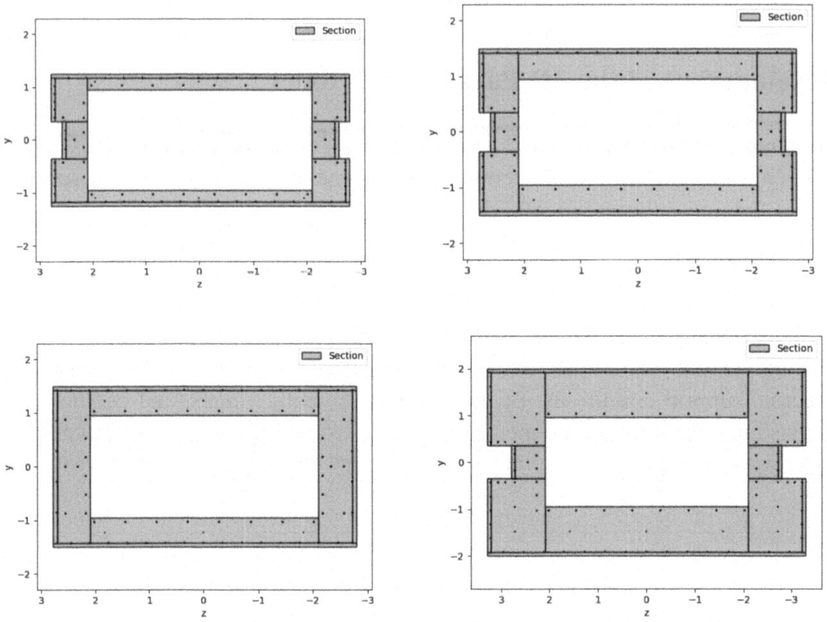

Fig. 6. Column cross sections A, B, C and D

To create these cross sections some libraries were used in openseespy. In summary, the `sectionproperties` and `concreteproperties` libraries are central to the geometry creation process in the code, enabling the definition of complex cross-sectional geometries and the incorporation of material properties. These libraries are essential for the parametric modeling of the bridge's sections, allowing to create detailed and adjustable geometries crucial for simulating different structural scenarios and performing comprehensive analyses.

The `sectionproperties` library provides functionalities like geometry definition through the *`geometry`* class, creation of compound geometries with *`CompoundGeometry`*, customization of parameters such as dimensions and orientation, and preparation for sectional analysis using the *`Section`* class. The *`concreteproperties`* library complements this by managing material properties within the geometric models, offering features like defining realistic material behavior with classes such as *`ConcreteLinear`*, *`RectangularStressBlock`*, and *`SteelElasticPlastic`*, precise placement of reinforcing bars using the *`add_bar`* function, and tools for assessing section strength and ductility through moment-curvature relationships and interaction diagrams. This parametric approach enhances efficiency and reduces potential errors, particularly beneficial for statistical analyses like Monte Carlo simulations where numerous model iterations are required. Practically, these libraries enable the simulation of aging effects due to factors like corrosion, facilitate design optimization by allowing quick exploration of various design options, and provide educational value by visually demonstrating the effects of different parameters on structural performance (Fig. 7).

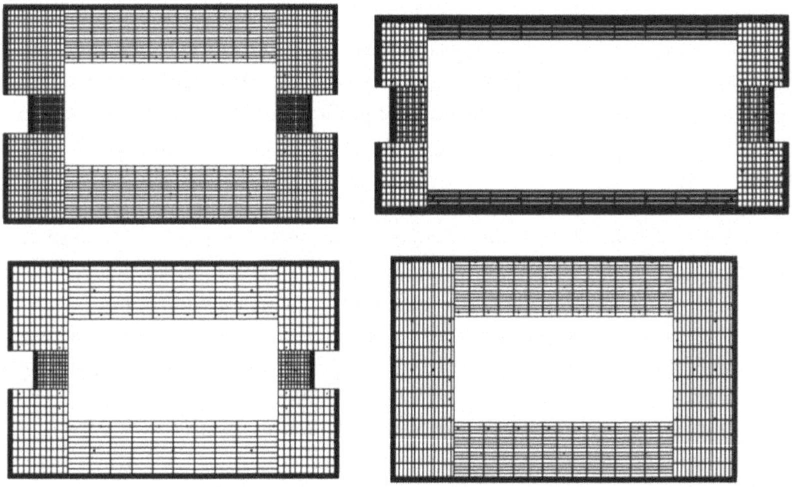

Fig. 7. Colum cross-sections discretization

Automatic generation of Fiber-section discretization is also employed in the modeling process, dividing each cross-section into discrete fibers that represent concrete and steel. This approach enables detailed analysis of sectional responses, such as moment-curvature relationships and stress-strain distributions, providing insights into t_he piers' behavior under various load conditions (Fig. 8).

The deck cross-section was discretized using a triangular mesh to approximate the irregular shape accurately. These cross sections were modelled with two libraries of "Sectionproperties" and "Compoundgeometry" (Fig. 9).

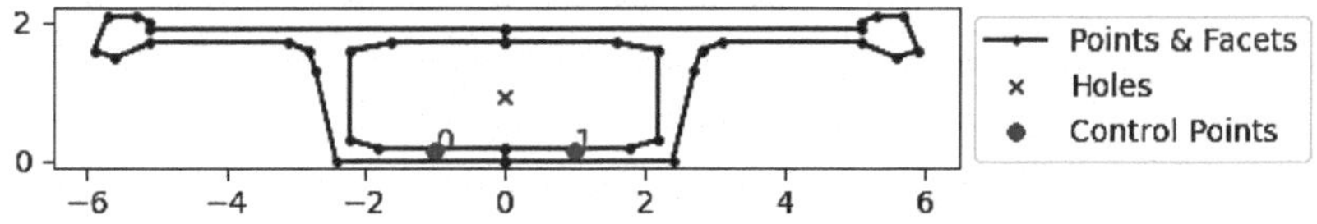

Fig. 8. Deck cross sections

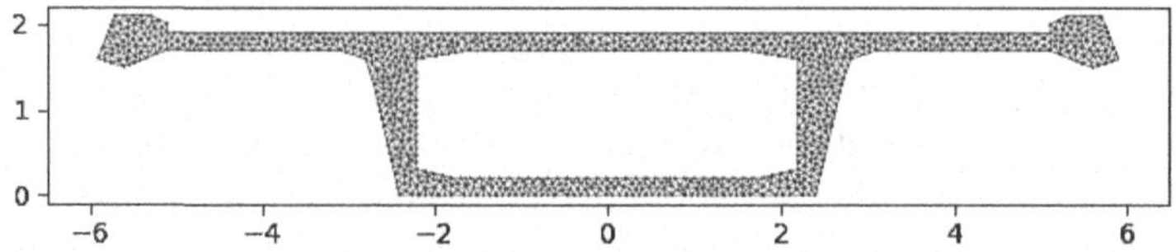

Fig. 9. Deck cross-sections discretization

5 Preliminary Modal Analysis

A preliminary modal analysis was conducted to understand the dynamic properties of the Colle Castino Bridge model developed in OpenSeesPy. The analysis utilized the "eigen command" to compute the natural frequencies and mode shapes of the bridge. Subsequently, the "modalProperties" command in Python was employed to extract and interpret the modal properties (Fig. 10).

The results revealed significant mass participation along the vertical (Z-axis), highlighting the critical role of vertical displacements in the bridge's dynamic response. This finding is particularly important for seismic assessments, as vertical vibrations can substantially influence structural performance during earthquake events.

To facilitate clearer observation and comprehension, the first four piers from the left side of the Colle Castino Bridge were magnified in the analysis. This magnification allowed for a more detailed examination of the modal movements, providing valuable insights into the dynamic characteristics of these specific sections of the bridge.

This foundational phase sets the stage for more advanced modeling, analysis, and simulation in subsequent stages of the research. It provides essential baseline data that will guide the selection of appropriate ground motion records and inform the interpretation of seismic responses.

The modal properties for the first seven mode shapes, along with the corresponding mass participation ratios, are summarized in Table X (replace with actual data). The results highlight major mass participation along the Z-axis, reinforcing the need to account for vertical dynamic effects in the structural analysis.

Table 3 presents the eigenvalues, natural angular frequencies (ω), natural frequencies, and corresponding periods for the first seven modes of the analyzed structure. The eigenvalues (λ) provide an indication of the dynamic characteristics of the structure, including its stiffness and mass distribution. Each mode represents a specific vibration pattern, where the natural angular frequency (ω) corresponds to the square root of the eigenvalue. Additionally, the natural frequency is derived from the angular frequency, while the period represents the reciprocal of the frequency. These results are critical in understanding dynamic behavior and ensuring the structure's safety under dynamic loading.

Table 4 provides the modal participation mass ratios for the first seven modes in terms of translational (MX, MY, MZ) and rotational (RMX, RMY, RMZ) degrees of freedom. These ratios represent the distribution of mass that participates in each mode for a specific direction or rotational axis. Higher values indicate a dominant contribution of that mode to the corresponding degree of freedom. For instance, in Mode 1, the majority of the mass participates in the Z-direction (MZ = 55.02%), with significant

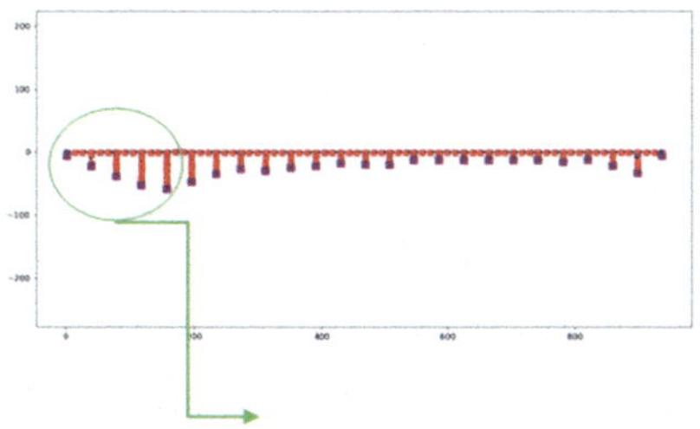

(Displacements are shown with high scale factor)

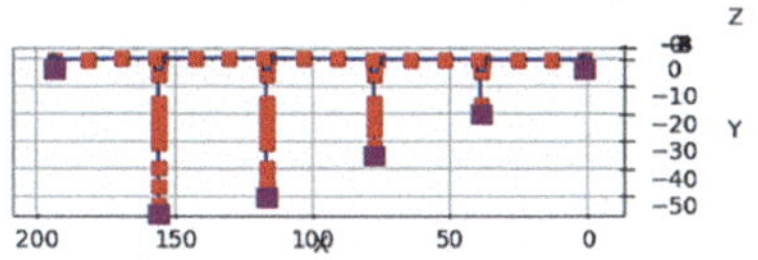

Mode shape1 -xy plane

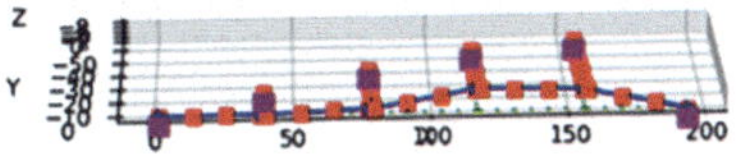

Mode shape1 -xz plane

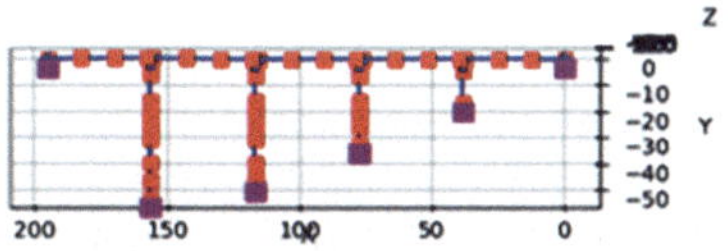

Mode shape2 - xy plane

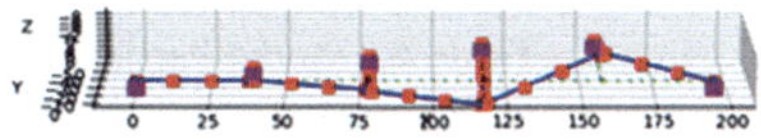

Mode shape2 - xz plane

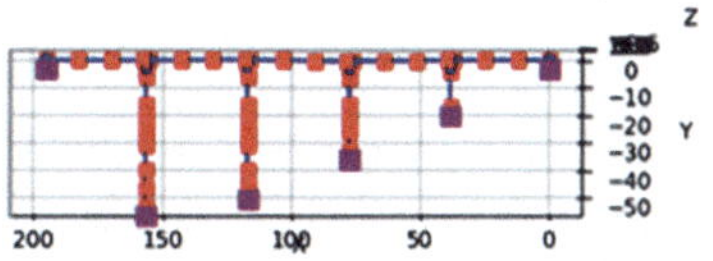

Mode shape3 -xy plane

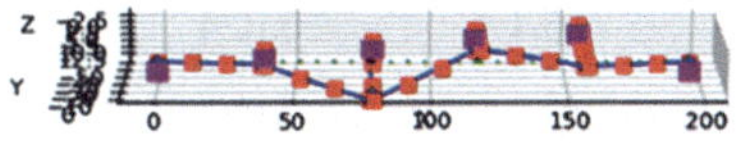

Mode shape3 -xz plane

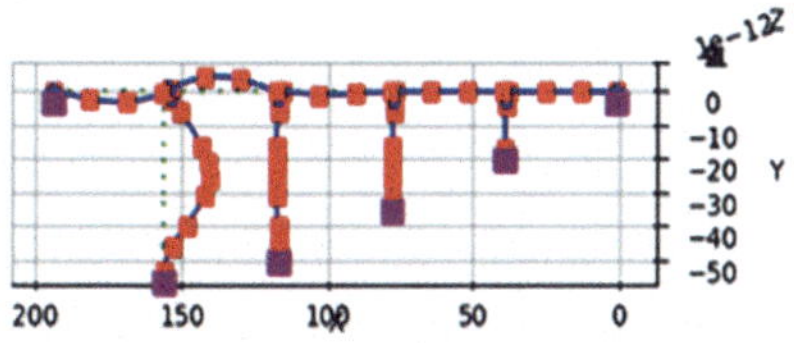

Mode shape4 -xy plane

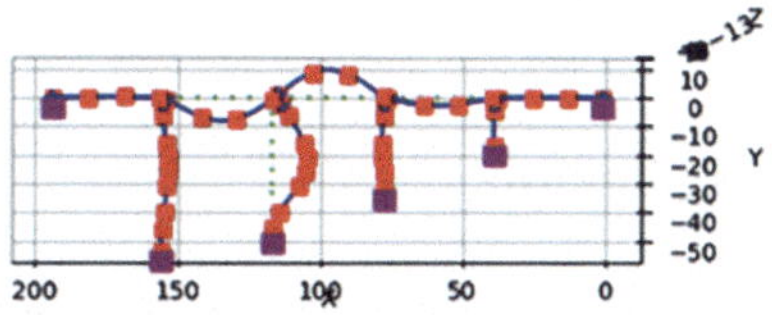

Mode shape5 -xy plane

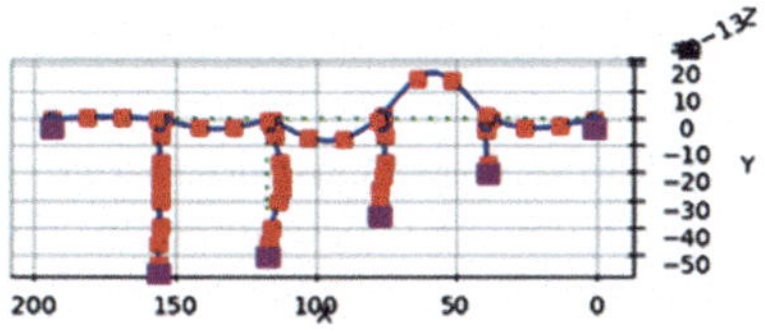

Mode shape6 -xy plane

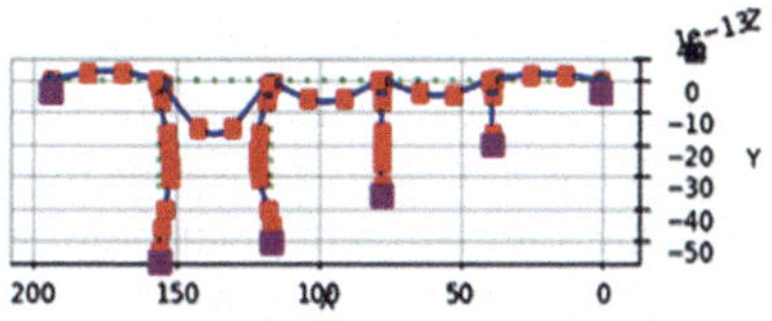

Mode shape1 -xy plane

Fig. 10. Mode Shapes in different planes as xy and xz

Table 3. Eigen values for first seven modes

Mode	Lambda	Omega	Frequency	Period
1.0	27.26	5.22	0.831	1.203
2.0	58.65	7.65	1.218	0.820
3.0	146.54	12.10	1.926	0.519
4.0	244.63	15.64	2.489	0.401
5.0	345.65	18.59	2.958	0.337
6.0	429.68	20.72	3.299	0.303
7.0	485.91	22.04	3.508	0.285

rotational contributions around the X- and Y-axes (RMX = 4.68%, RMY = 12.0%). Such information is essential for understanding the dynamic characteristics of the structure and identifying modes that require attention during design optimization or retrofitting.

Table 4. Modal Participation Mass Ratios (%) for each mode

Mode	MX	MY	MZ	RMX	RMY	RMZ
1.0	0.0	0.0	55.02	4.68	12.0	0.0
2.0	0.0	0.0	0.89	0.49	17.7	0.0
3.0	0.0	0.0	11.22	4.24	7.9	0.0
4.0	15.1	0.024	0.0	0.0	1.0	0.0
5.0	15.12	0.009	0.0	0.0	1.3	0.0
6.0	3.5	0.022	0.0	0.0	2.7	0.0
7.0	0.4	12.79	0.0	0.0	0.26	0.0

6 Conclusion

This study successfully developed a comprehensive parametric 3D model of the Colle Castino Bridge using OpenSeesPy, facilitating flexible analysis and modification to investigate corrosion and other structural effects. By parameterizing all aspects of the bridge's geometry and materials, the model allows for effortless adjustments to cross-sectional dimensions, material properties, and reinforcement layouts. This modular approach is critical for studying aging infrastructure like the Colle Castino Bridge, as it simplifies the simulation of deteriorating structural elements and accommodates a wide range of potential scenarios, as it allows for quick adaptation to changes in geometry, material properties, and reinforcement details by reducing the potential for errors associated with repetitive coding.

Adjusting parameters rather than rewriting code for each case significantly enhances efficiency, enabling the simulation of different scenarios—such as varying levels of corrosion, structural modifications, or material degradation—with ease. This method is particularly useful when assessing the impact of factors like corrosion on bridge structures.

The parametric model's ability to systematically vary input parameters is crucial for performing extensive statistical analyses through Monte Carlo simulations. By doing so, we can quantify uncertainties by assessing the impact of variability on the bridge's structural performance and evaluate structural reliability by determining the probability of failure under different scenarios. This approach informs maintenance and retrofitting strategies with data-driven recommendations, ultimately enhancing the bridge's safety and longevity.

The significance of this research lies in demonstrating a systematic methodology for parametric modeling and statistical analysis of complex structures. By enhancing our understanding of how uncertainties affect structural performance, the developed parametric model leads to more resilient and reliable bridge designs.

Future work could extend this methodology to other types of structures and incorporate more sophisticated models for material degradation. Additionally, integrating real-time monitoring data could further refine the model, improving its accuracy and applicability in structural health monitoring and predictive maintenance strategies.

Acknowledgments. The authors acknowledge the support of ReLUIS (Consorzio della Rete dei Laboratori Universitari di Ingegneria Sismica e Strutturale), Consiglio Superiore dei Lavori Pubblici (Superior Council of Public Works), and the Italian Department of Civil Protection. This research was in the framework of the project 'Agreement ReLUIS - CSLLPP: Experimentation of Guidelines for the Classification and Risk Management, Safety Assessment, and Monitoring of Existing Bridges (Accordo ReLUIS - CSLLPP: Sperimentazione delle linee guida per la classificazione e gestione del rischio, la valutazione della sicurezza ed il monitoraggio dei ponti esistenti).

References

1. Bergami, A.V., Briseghella, B., Nuti, C.: A corrosion model for the interpretation of the cyclic behavior of reinforced concrete sections. Struct. Concr. **22**(6), 3300–3318 (2021). https://doi.org/10.1002/suco.202100257
2. Bergami, A.V., Pelle, A., Briseghella, B., Nuti, C.: Bridge pier corrosion in seismic areas: forecasting future behavior and assessment. Struct. Concr. **22**(3), 1827–1845 (2021). https://doi.org/10.1002/suco.202000736
3. Bergami, A.V., Briseghella, B., Nuti, C.: A corrosion model for the interpretation of the cyclic behaviour of RC sections. Structural Concrete (2019)
4. Nuti, C., Briseghella, B., Chen, A., Lavorato, D., Iori, T.: Relevant outcomes from the history of Polcevera Viaduct in Genova, from design to nowadays failure. J. Civ. Struct. Heal. Monit. **10**(4), 607–635 (2020). https://doi.org/10.1007/s13349-020-00421-1
5. Bergami, A.V., Briseghella, B., Nuti, C.: Simulation of the cyclic behavior of reinforced concrete columns including the effect of the corrosion of rebars: A simple and accurate model. Struct. Concr. (2022). https://doi.org/10.1002/suco.202200555

6. Bergami, A.V., Briseghella, B., Nuti, C.: Understanding the impact of corrosion on the cyclic behavior of reinforced concrete sections. Structural Concrete
7. Nuti, C., Briseghella, B., Chen, A., Lavorato, D., Iori, T., Vanzi, I.: Relevant outcomes from the history of Polcevera Viaduct in Genova, from design to nowadays failure. J. Civ. Struct. Heal. Monit. **10**(3), 337–353 (2020). https://doi.org/10.1007/s13349-020-00404-2
8. Biondini, F., Frangopol, D.M.: Probabilistic life-cycle performance of deteriorating structures: Corrosion effects on concrete decks and piers. J. Struct. Eng. **134**(6), 1003–1016 (2008)
9. Der Kiureghian, A., Ditlevsen, O.: Aleatory or epistemic? Does it matter? Struct. Saf. **31**(2), 105–112 (2009)
10. Haldar, A., Mahadevan, S.: Probability, reliability, and statistical methods in engineering design. John Wiley & Sons, US (2000)
11. Melchers, R.E., Beck, A.T.: Structural reliability analysis and prediction, 3rd edn. John Wiley & Sons, US (2018)
12. Robert, C.P., Casella, G.: Monte Carlo statistical methods, 2nd edn. Springer, Germany (2004)
13. Rubinstein, R.Y., Kroese, D.P.: Simulation and the Monte Carlo method, 3rd edn. John Wiley & Sons, US (2016)
14. Ghanem, R.G., Spanos, P.D.: Stochastic Finite Elements: A Spectral Approach. Springer-Verlag, Berlin (1991)
15. Melchers, R.E.: Structural Reliability Analysis and Prediction. John Wiley & Sons (1999)
16. Möller, B., Beer, M.: Engineering Computation Under Uncertainty—Capabilities of Non-Traditional Models. Comput. Struct. **86**(10), 1024–1041 (2008)
17. Schuëller, G.I., Pradlwarter, H.J.: Computational Stochastic Structural Analysis. Comput. Struct. **85**(5–6), 235–243 (2007)

Numerical Modeling of Dapped-End Beams Subjected to Corrosion and Retrofitting Interventions

Marilisa Di Benedetto(✉), Fabio Di Trapani, Antonio Pio Sberna, Sofia Villar, Mario Ferrara, and Gabriele Bertagnoli

Dipartimento di Ingegneria Strutturale, Edile e Geotecnica, Politecnico di Torino, Torino, Italy
marilisa.dibenedetto@polito.it

Abstract. This study concerns refined numerical modeling of dapped-end beams subjected to corrosion and the assessment of the performance of an external steel retrofitting system. Widely used in infrastructure heritage, dapped-end beams face several challenges due to the complex stress flows in the discontinuous zones caused by the abrupt change in section depth. To gain a better insight into the impact of local stress transfer and overall performance, extensive numerical investigations are essential. In this study refined 3D solid models are built to complement previous experimental tests on a steel retrofitting system for corrosion-damaged half-joints using the STKO software platform for OpenSees. The results from the experimental campaign are replicated numerically following a process of calibration of material parameters. Then, additional numerical simulations are carried out to examine the structural response of the retrofitted half-joint in different configurations not previously assessed in the reference experimental campaign. An improvement of the steel retrofitting system is finally proposed and numerically simulated. The results aim to provide insights into the efficiency and performance of retrofitted damaged half-joints, highlighting the capability of numerical models to complement experimental tests and improve the understanding of structural behavior, and strengthening strategies.

Keywords: Dapped-end beams · Half-joints · Bridges · Concrete · Retrofitting · OpenSees · STKO

1 Introduction

Reinforced concrete infrastructures have shown significant deterioration due to aging, insufficient maintenance, and design flaws [1]. Issues such as carbonation, reinforcement corrosion, and cover spalling severely reduce load-bearing capacity in reinforced concrete bridges, requiring strengthening and retrofitting systems.

Dapped-end beams, also known as half-joint beams, are a notable example of structures often subjected to degradation especially because of corrosion of reinforcement (Fig. 1). These structural elements, widely used from the 1960s to the 1980s for their design simplicity and settlement insensitivity, have shown early degradation

X. Lu et al. (Eds.): EOS 2024, LNCE 638, pp. 111–121, 2025.
https://doi.org/10.1007/978-3-031-90690-9_10

and brittle failures [2, 3]. Several studies have evaluated the performance of dapped-end beams under both undamaged and damaged conditions, employing Strut-and-Tie models (STMs) [4]. Recent research has focused on experimental and numerical investigation of the impacts of reinforcement corrosion and retrofitting solutions for structural deficiencies [5–10].

Fig. 1. Degradation of dapped-end beam bridges caused by steel reinforcement corrosion.

Two common strengthening techniques are Fiber Reinforced Polymers (FRP/CFRP) and external prestressing systems. While FRP/CFRP systems can increase load-bearing capacity by 10–80%, their effectiveness depends on geometric details, reinforcement layout, and fiber orientation [11–14]. On the other hand, external prestressing provides higher performance even if it is more difficult to install [15]. Among recent innovations, precast strain-hardening cementitious composites (P-SHCC) have demonstrated remarkable improvements, achieving up to a 90% performance increment [16].

However, most research on dapped-end beams focused on rectangular cross-sections, with limited attention to T-shaped sections, especially under severe damage conditions. To close this gap, an experimental campaign [17] was conducted to evaluate an innovative steel retrofitting technique consisting of L-shaped plates and transverse prestressing to transfer shear forces through friction, thereby avoiding damaged concrete sections. Tests carried out on scaled T-beams with two reinforcement layouts revealed significant improvements in resistance but also highlighted issues with the flange-web interface [17]. Based on this experimental campaign, the current study incorporates advanced numerical modeling using refined 3D solid models developed with the STKO software platform for OpenSees [18, 19]. Numerical simulations replicate experimental results through material calibration and extend the analysis to explore structural behavior under configurations not previously tested. Based on the numerical results, an improved retrofitting system is proposed and numerically validated.

This study contributes to the understanding of the retrofitting system's efficiency, highlighting the synergy between experimental and numerical approaches, which may be useful for retrofitting corrosion-damaged half-joints and enhancing the durability and safety of aging infrastructures.

2 Reference Experimental Campaign

The experimental campaign [17] involved two T-shaped dapped-end beam specimens scaled to 1:3 of typical bridge girders. Each specimen was 75 cm deep, with a flange width of 80 cm and an overall length of 490 cm, designed to replicate a longitudinal bridge beam connected rigidly to a pier [17]. The beams were cast in two stages: the flanges and webs were cast at different times to simulate the installation of half-joints in real structures. Both specimens had identical geometries but differed in their reinforcement layouts. One beam featured orthogonal reinforcement in the dapped-end region, while the other used inclined reinforcement, as illustrated in Fig. 2.

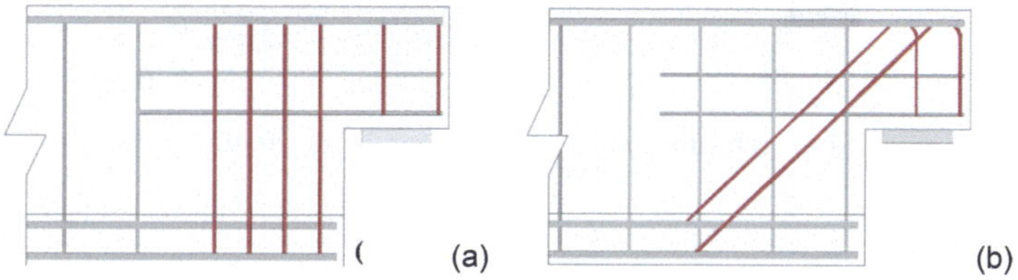

Fig. 2. Reinforcement layout of the dapped-end beam specimens: (a) Beam A; (b) Beam B.

One half-joint of each specimen was designed with reduced reinforcement to simulate the damage due to steel bar corrosion (damaged zone). The other half-joint was designed using the Strut-and-Tie (S &T) method and arranged with full reinforcement as a reference for comparison (undamaged zone). The damaged ends of both specimens (beams A and B) were strengthened using a steel jacketing system consisting of L-shaped steel plates with transverse prestressing bolts, as reported in Fig. 3. The system was designed to restore the load-bearing capacity by transferring shear forces through friction between the plates and the concrete web, bypassing weakened concrete areas. For the sake of brevity, additional details can be found in Bertagnoli et al. [17], which presents the design of the specimens and the retrofitting system.

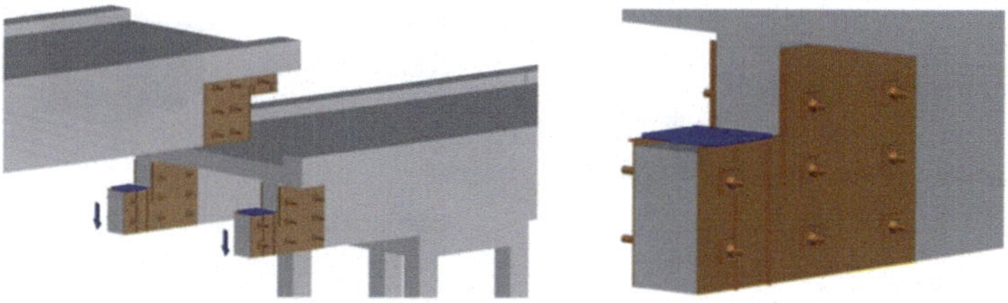

Fig. 3. Steel-jacketing system designed by Bertagnoli et al. [17].

The specimens were tested with vertical loads applied at midspan through a hydraulic jack. Both ends of the beams were supported at the half-joints to replicate the boundary conditions of actual bridge beams. The test setup included two static schemes: in the first configuration, both half-joints were loaded simultaneously, while the second configuration focused on the failure of the undamaged half-joint. Figure 4 illustrates the static schemes adopted for the experimental tests of the two beams: static scheme 1 (SS1) and static scheme 2 (SS2).

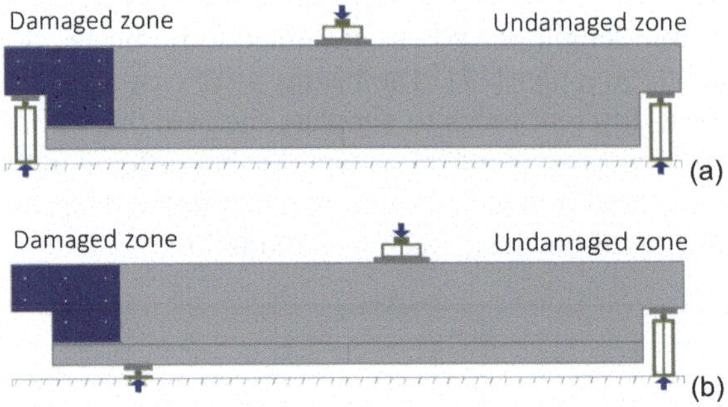

Fig. 4. Static schemes adopted for the experimental program: (a) Static scheme 1 (SS1); (b) Static scheme 2 (SS2).

Preliminary results of the experimental campaign are provided in [17].

3 Refined FE Modeling of the Dapped-End Beams

3.1 Modeling Strategy with OpenSees/STKO

A refined numerical modeling approach was employed to simulate the nonlinear behavior of dapped-end beams by replicating the experimental campaign and extending the analyses to other configurations. The models were developed within the STKO graphical interface for OpenSees [18, 19], by using advanced damage mechanics and 3D finite element modeling to consider the anisotropic response of concrete and its interaction with the retrofitting system. A scheme of the model is illustrated in Fig. 5.

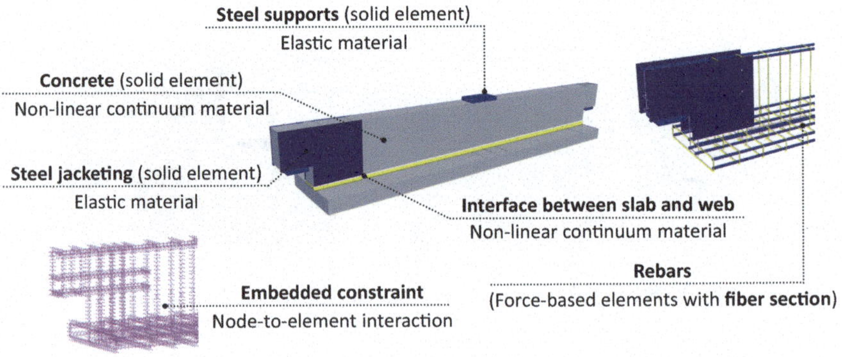

Fig. 5. Refined FE model of the dapped-end beam in OpenSees/STKO.

Concrete behavior was defined using the *ASDConcrete3D* constitutive model [20, 21], which applies distinct damage indices for tension (d^+) and compression (d^-) to compute the update of the effective stress tensor $\boldsymbol{\sigma}$ at each step of the analysis. The latter is evaluated as follows:

$$\boldsymbol{\sigma} = \left(1 - d^{+}\right)\overline{\boldsymbol{\sigma}}^{+} + \left(1 - d^{-}\right)\overline{\boldsymbol{\sigma}}^{-} \tag{1}$$

where $\overline{\sigma}^{+}$ and $\overline{\sigma}^{-}$ are the positive and negative components of the effective stress tensors, respectively.

$V_{d,tot} = V_{d,frame} + V_{d,inf}$ $V_{d,tot} = V_{d,frame} + V_{d,inf}$ Steel reinforcement was modeled as 1D fiber-section *force-based beam-column* elements with a *Hysotropic* material. The rebars are seamlessly connected to the concrete beam using *ASDEmbedded-node* elements to simulate perfect bonding. The steel jacketing retrofitting system, modeled with 3D solid elements, was linked to the beam through *Zero-Length-Implex-Contact* elements simulating frictional and shear interactions using a Coulomb friction law [22, 23].

3.2 Analyses and Model Validation

The boundary conditions of the model replicated the experimental setup. Incremental vertical loads were added to consider gravity, followed by monotonic displacement to simulate failure. A quadratic mesh with an average element size of 30 mm was optimized for computational efficiency and accuracy. The analysis outputs included load-displacement curves (Fig. 6) and crack propagation patterns (Fig. 7), which provided detailed insights into structural behavior.

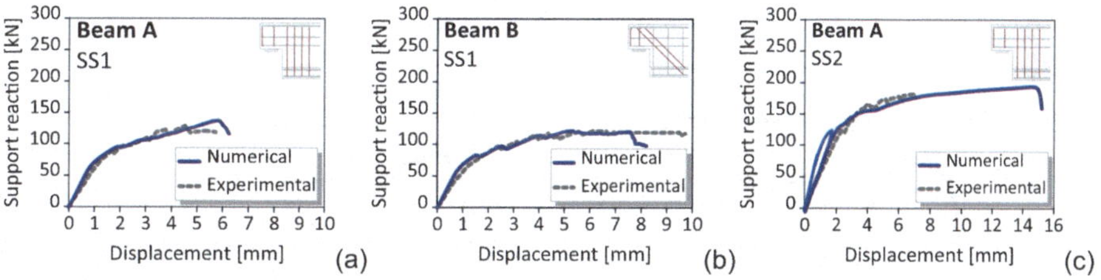

Fig. 6. Comparison between experimental and numerical responses of beam A and beam B in static schemes 1 (SS1) and 2 (SS2).

The validated models demonstrated consistency with experimental results and offered a useful tool to explore retrofitting system performance under different configurations of those tested experimentally.

3.3 Additional Numerical Simulations

Further numerical investigations were performed to assess the actual response of the retrofitted half-joints and make relevant comparisons. Specifically, the structural behavior of the two specimens (Beam A and B), was analyzed under three configurations in static scheme 2:

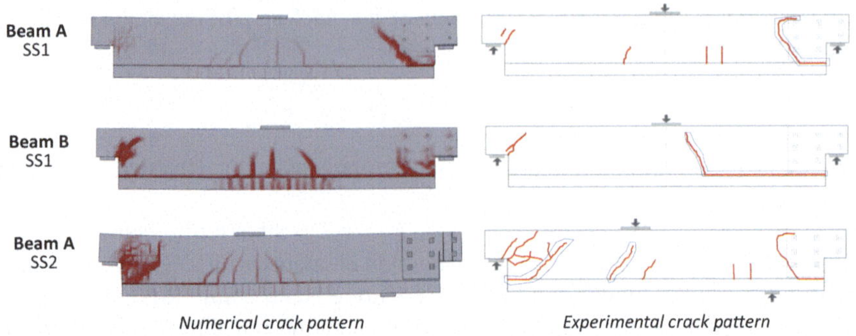

Fig. 7. Comparison between experimental and numerical crack patterns of beams A and B in static schemes 1 (SS1) and 2 (SS2).

- Undamaged half-joint designed using the S&T method;
- Damaged half-joint;
- Retrofitted-damaged half-joint with the steel jacketing system.

Numerical simulations adopted the same mechanical properties and reinforcement diameters used during the experimental campaign to be consistent with the already validated models. Figure 8 shows a comparison of the behavior of beam A and beam B under these configurations.

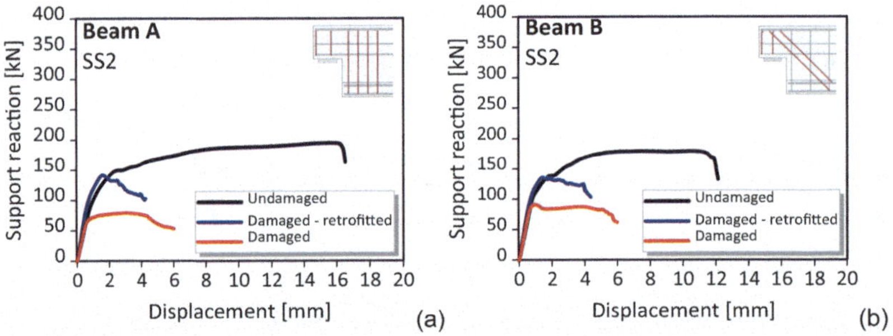

Fig. 8. Numerical responses of dapped-end beams in three different configurations in static scheme 2 (SS2): (a) Beam A; (b) Beam B.

Results indicate that the numerical models predict higher ultimate load-bearing capacities for both undamaged and damaged half-joints than the one predicted with the S&T method. In particular, the numerical predictions of the ultimate load for the undamaged dapped-end beams (A and B) are 30% and 12.5% higher, respectively, compared to the S&T estimations. For the damaged dapped-end beams, the numerical failure loads exceed the S&T predictions by 23% for beam A and 28.5% for beam B. This was because Strut-and-Tie models tend to underestimate the ultimate load, ensuring a conservative and safety-oriented solution.

Despite significantly enhancing the strength of the damaged half-joint, the steel jacketing system does not fully restore the load-bearing capacity to match that of the undamaged half-joint. The Table 1 below summarizes the failure loads across the analyzed configurations.

Table 1. Comparison of the ultimate load determined experimentally, numerically, and using the S&T method across three configurations of the dapped-end beams in SS2.

		Failure load [kN]			
	Half-joint configuration	Experimental	Numerical	S&T	Difference between numerical and S&T [%]
Beam A	Undamaged half-joint	180	195	150	30%
	Damaged half-joint	-	80	65	23%
	Retrofitted damaged half-joint	120	140	-	-
Beam B	Undamaged half-joint	-	180	160	12.5%
	Damaged half-joint	-	90	70	28.5%
	Retrofitted damaged half-joint	120	130	-	-

Failure in the retrofitted saddle occurs due to a block tearing mechanism, while the unreinforced damaged beam fails through localized cracks near the saddle notch without propagating through the entire web, as shown in Fig. 9.

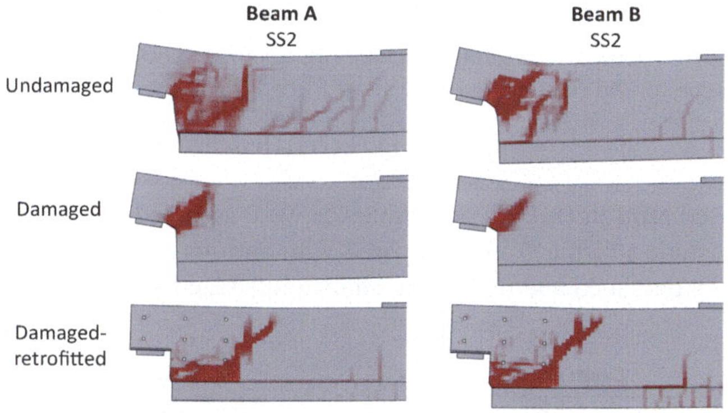

Fig. 9. Numerical crack patterns of the dapped-end beams (beam A and beam B) in three different configurations in static scheme 2 (SS2).

4 Proposal for an Alternative Retrofitting System

The experimental campaign and numerical simulations highlighted the main limit of the original retrofitting system: The stress concentration at the web-slab interface, involving a block tearing failure mechanism. Given this, an alternative version of the retrofitting system was proposed. The latter involves a rigid connection of the steel reinforcement to the slab, using bolts anchored into the slab with a chemical method (Fig. 10a). It seeks to secure a non-invasive installation and improve overall performance with less traffic disruption of roadway operations. The anchors were designed based on the ultimate limit load (V_d) of 200 kN, evaluating the shear force (T_d) acting on the bolts. The latter was computed based on the equilibrium of the damaged retrofitted half-joint, as reported in Fig. 10b, and expressed as follows:

$$V_d \cdot \left(l_R - \frac{l_S}{2}\right) = T_d \cdot (0.8 \cdot h_R) \tag{2}$$

The shear resistance of the bolts was evaluated considering the instructions provided by the anchor manufacturers. Using M20 class 8.8 bolts, the design required four anchors per reinforcement plate, with an anchorage length of 11 cm.

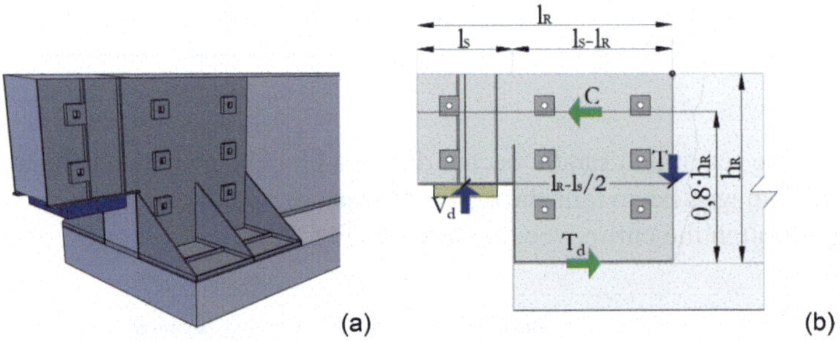

Fig. 10. Alternative retrofitting system: (a) Three-dimensional view of the modified retrofitting system, (b) Scheme for the calculation of the shear force T_d acting on the bolts for a given load V_d.

The proposed reinforcement's performances were assessed through numerical simulations under static scheme 2 and compared with various configurations for beams A and B, including damaged and undamaged half-joints (see Fig. 11).

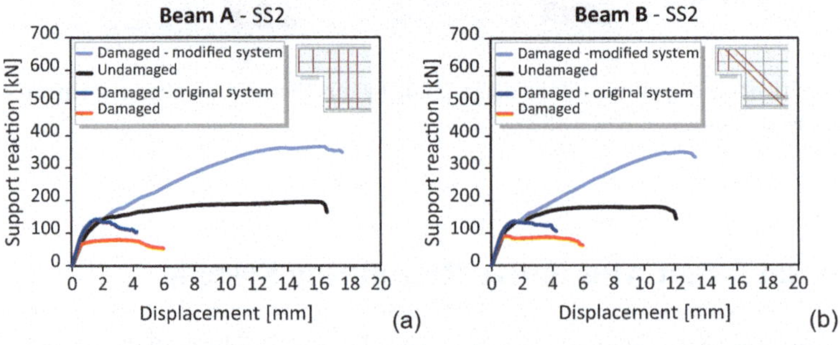

Fig. 11. Comparison of the numerical responses of dapped-end beams under various configurations, including the modified retrofitting system, in static scheme 2 (SS2): (a) Beam A; (b) Beam B.

Load-displacement curves demonstrate that the proposed reinforcement significantly increased the failure load of damaged half-joints, exceeding the resistance of the undamaged ones. The modified retrofitting system, involving chemical anchoring, achieved a load capacity of 350 kN, while the original reinforcement shows a resistance of 120 kN. Figure 12 illustrates the crack pattern propagation in the two retrofitting systems for beam A and beam B.

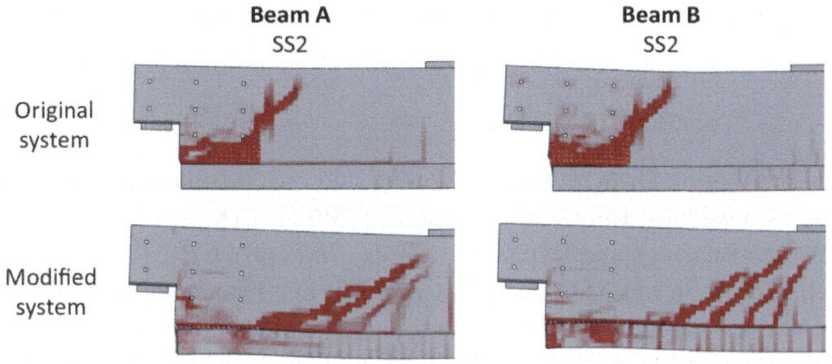

Fig. 12. Comparison of the numerical crack patterns of dapped-end beams (Beam A and B) with the original and modified retrofitting systems under static scheme 2 (SS2).

The alternative retrofitting systems effectively move the failure mechanism from the damaged region to the undamaged one, causing the translation of crack propagation from the web-slab interface to the beam's midspan. Overall, the anchorage systems avoid block tearing failures, ensuring not only better distribution of the applied load but also increased capacity and ductility. These advances showcase what can be achieved using chemical anchoring and fully connected slabs to improve structural response for retrofitted-damaged dapped-end beams.

5 Conclusions

This study provided insights into the structural behavior of corrosion-damaged dapped-end beams, through a combined experimental-numerical approach. The refined numerical models developed on the STKO software platform for OpenSees showed good agreement with the experimental campaign results and allowed the generation of additional tests complementing the experimental program. The models resulted in higher load-bearing capacities for the undamaged and damaged half-joints compared to the value estimated with the S&T method. On the other hand, the steel jacketing system, though increasing the capacity of the damaged half-joint, was not able to fully recover the original load-bearing capacity and ductility, thus requiring further optimization. For this purpose, a modified retrofitting system incorporating anchoring connections to the slab was proposed. Numerical simulation results indicated that the alternative system considerably improved the retrofitted half-joint performance and shifted failure modes from the web-slab interface to the beam's midspan, away from the damaged zone. The proposed modified system shows great potential in mitigating block tearing failure mechanisms, improving load distribution, and enhancing overall structural performance while preserving the original system's primary objective of minimizing disruption to roadway traffic. These results emphasize how experimental data combined with advanced numerical modeling allow obtaining better insight into the behavior of retrofitted-damaged dapped-end beams ensuring the safety and durability of existing infrastructure.

Acknowledgements. This paper is part of the PNRR-NGEU project, which has received funding from the MUR - DM 351/2022.

References

1. Neves, L.C., Frangopol, D.M.: Condition, safety and cost profiles for deteriorating structures with emphasis on bridges. Reliab. Eng. Syst. Saf. **89**(2), 185–198 (2005)
2. Lee, D.J.: Bridge Bearings and Expansion Joints (2nd edn). CRC Press (2015)
3. Di Prisco M., Colombo M., Martinelli P., Coronelli D.: The technical causes of the collapse of Annone overpass on SS.36. Italian Concrete Days, aicap and CTE (2018)
4. Desnerck, P., Lees, J.M., Morley, C.T.: Strut-and-tie models for deteriorated reinforced concrete half-joints. Eng. Struct. **161**, 41–54 (2018)
5. Di Carlo, F., Meda, A., Molaioni, F., Rinaldi, Z.: Experimental evaluation of the corrosion influence on the structural response of Gerber half-joints. Eng. Struct. **285**, 116052 (2023)
6. Rosso, M.M., Asso, R., Aloisio, A., Di Benedetto, M., Cucuzza, R., Greco, R.: Corrosion effects on the capacity and ductility of concrete half-joint bridges. Constr. Build. Mater. **360**, 129555 (2022)
7. Spinella, N., Messina, D.: Load-bearing capacity of Gerber saddles in existing bridge girders by different levels of numerical analysis. Struct. Concr. **24**, 211–226 (2023)
8. Mata-Falcón, J., Pallarés, L., Miguel, P.F.: Proposal and experimental validation of simplified strut-and-tie models on dapped-end beams. Eng. Struct. **183**, 594–609 (2019)
9. Belletti, B., Calcavecchia, B., Ferretti, D., Ravasini, S.: Capacity assessment of uncorroded and corroded dapped-end beams by NLFE and strut-and-tie based methods. Struct. Concr. **25**(2), 1275–1304 (2024)
10. Spinella, N., Messina, D.: Load-bearing capacity of Gerber saddles in existing bridge girders by different levels of numerical analysis. Struct. Concr. **24**(1), 211–226 (2023)
11. Taher, F.: Strengthening of critically designed girders with dapped ends. Struct. Build. **158**, 141–152 (2005)
12. Huang, P.-C., Nanni, A.: Dapped-End Strengthening of Full-Scale Prestressed Double Tee Beam with FRP Composite. Adv. Struct. Eng. **9**, 293–308 (2006)
13. Nagy-György, T., Sas, G., Dăescu, A.C., Barros, J.A.O., Stoian, V.: Experimental and numerical assessment of the effectiveness of FRP-based strengthening configurations for dapped-end RC beams. Eng. Struct. **44**, 291–303 (2012)
14. Quadri, A.I.: Behavior of disturbed region of RC precast beams upgraded with near surface mounted CFRP fiber. Asian J. Civ. Eng. **24**, 1859–1873 (2023)
15. Atta, A., Taman, M.: Innovative method for strengthening dapped-end beams using an external prestressing technique. Mater. Struct. **49**, 3005–3019 (2016)
16. Baraghith, A.T., Khalil, A.H.A., Etman, E.E., Behiry, R.N.: Improving the shear behavior of RC dapped-end beams using precast strain-hardening cementitious composite (P-SHCC) plates. Structures **50**, 978–997 (2023)
17. Bertagnoli, G., Ferrara, M., Giordano, L., Malavisi, M.: Preliminary Investigation on Steel Jacketing Retrofitting of Concrete Bridges Half-Joints. Appl. Sci. **13**(14), 8181 (2023)

18. McKenna, F., Fenves, G.L., Scott, M.H.: Open system for earthquake engineering simulation. University of California, Berkeley, CA (2000)
19. Petracca M., Candeloro F., Camata G.: ASDEA Software STKO user manual (2017)
20. Petracca, M., Pelà, L., Rossi, R., Zaghi, S., Camata, G., Spacone, E.: Microscale continuous and discrete numerical models for nonlinear analysis of masonry shear walls. Constr. Build. Mater. **149**, 296–314 (2017)
21. Di Trapani, F., Villar, S., Di Benedetto, M., Petracca, M., Camata, G.: Seismic response of unreinforced masonry building aggregates: Investigation on the "aggregate-effect" based on an elementary building aggregate. Eng. Struct. **316**, 118301 (2024)
22. Oliver, J., Huespe, A.E., Cante, J.C.: An implicit/explicit integration scheme to increase computability of non-linear material and contact/friction problems. Comput. Methods Appl. Mech. Engrg. **197**, 1865–1889 (2008)
23. Di Trapani, F., Di Benedetto, M., Petracca, M., Camata, G.: Local infill-frame interaction under seismic loads: Investigation through refined micro-modeling. Eng. Struct. **315**, 118088 (2024)

Industrial Equipment and Storage Tanks

Development of Numerical Model for Building Piping Systems in OpenSees

Qingxue Shang[1,2(✉)]

[1] Key Laboratory of Earthquake Engineering and Engineering Vibration, Institute of Engineering Mechanics, China Earthquake Administration; Key Laboratory of Earthquake Disaster Mitigation, Ministry of Emergency Management, Harbin 150080, China
shangqingxue@outlook.com

[2] China Earthquake Disaster Prevention Center, Beijing 100029, China

Abstract. Robust internal building pipeline is one of the most important non-structural components for reducing post-earthquake secondary disasters and maintaining normal use of buildings after an earthquake. The seismic damage of piping systems can result in the loss of firefighting abilities or lead to failure of water supplies and drainage systems in a building which can have serious implications for emergency rescue, post-disaster response and reconstruction work. In this study, moment-rotation hysteresis models of piping joints and force-displacement hysteresis models of piping braces with different configurations are developed using the Pinching4 uniaxial constitutive in the finite element software OpenSees. The calibrated model parameters are obtained using test data from existing literature. The developed hysteresis models can accurately reproduce the hysteresis curves obtained in the experiments and simulate the moment time history during the loading process. And the trend of the accumulated energy dissipation of the numerical results is consistent with the test results as the increase of the number of cycles. On this basis, the generic moment-rotation hysteresis models of piping-joints were further developed. Then numerical model of piping system was developed based on the generic hysteresis model. Comparisons between numerical and experimental responses demonstrate that the developed numerical model can be used to analyze the seismic response of suspended piping systems.

Keywords: hysteresis model · numerical simulation · grooved fit joints · OpenSees · piping system

1 Introduction

Suspended piping system is one of the most critical non-structural systems in a building. Recent earthquakes have demonstrated that non-structural components (NSCs) including piping system of public buildings and critical facilities such as hotels, schools, hospitals, airports, power plants, and industrial units suffered much more damage in comparison with structural components (SCs) [1]. During the 1994 Northridge earthquake, pipelines in different piping systems such as HVAC systems, sprinkler piping systems, and water piping systems experienced widespread failures. Leakage of fire sprinkler piping systems

X. Lu et al. (Eds.): EOS 2024, LNCE 638, pp. 125–132, 2025.
https://doi.org/10.1007/978-3-031-90690-9_11

forced the temporary evacuation of several hospital buildings after the earthquake. After the 2010 Chile Earthquake, four hospitals in the central south region of the country were inoperable, and 12 hospitals lost almost 75% of their functionalities. Most loss was caused by damages to NSCs such as suspended ceilings, light fixtures, and fire sprinkler piping systems. Two largest airports in Chile were closed as well because of non-structural damages and flooding from failed sprinkler piping systems [2].

The seismic damage of piping systems not only resulted in huge economic loss but also mitigated the functionality of the buildings and facilities. Seismic performance of NSCs and seismic design of NSCs has been started since the early 1980s. Component level quasi-static tests were widely used to determine the force-displacement hysteresis response or moment-rotation cyclic response of piping components [3, 4]. Based on the test results, analytical models of piping components can be developed, e.g., numerical models of grooved fit tee joints, threaded tee joints, flexible pipeline connections, cast-iron and copper tee joints, and seismic sway braces for suspended piping systems.

Despite the experimental studies and analytical works have been performed on piping systems, the numerical model of piping system connected by grooved fit joints still needs to be further developed. Considering the fact that there are certain differences in the pipe types, joint types and configuration types adopted in previous studies are selected from different countries, the results are not universal and maybe not suitable for seismic performance analysis of piping systems in China. This paper analyzes the bearing capacity and deformation capacity of the grooved fit piping joints based on the quasi-static test results and the hysteresis model was developed. The pinching4 uniaxial material in OpenSees is combined with the zeroLength element to reproduce the moment-rotation response of the grooved fit piping joint. The finite element model of the whole piping system is developed then. The proposed piping joint moment-rotation hysteresis model is integrated into the finite element model of piping system.

2 Development of a Hysteresis Model for Grooved Fit Joints

2.1 Pinching4 Uniaxial Material

The grooved fit piping joints tested in [4] will be adopted in this study for model calibration. The moment-rotation hysteresis response of grooved fit joints shows high nonlinearity. The Pinching4 uniaxial material can consider the strength degradation, stiffness degradation and pinching effect under cyclic reverse loading, and can accurately simulate various complex hysteresis characteristics. The Pinching4 uniaxial material, along with a zeroLength element in OpenSees platform, was widely used to simulate the force-displacement response of different types of NSCs, e.g., sprinkler piping joints and seismic sway braces for piping systems. The schematic diagram and hysteresis rules of Pinching4 uniaxial material model are shown in Fig. 1. The definition of Pinching4 uniaxial material includes the skeleton curve, unloading-reloading path and three material failure criteria (i.e., the unloading stiffness degradation criteria, reloading stiffness degradation criteria and strength degradation criteria). The solid line in Fig. 1 is the skeleton curve of Pinching4 uniaxial model, which is composed of multiple lines, while the dotted line demonstrates the unloading-reloading path under cyclic loading, which is composed of three-stage lines. The Pinching4 uniaxial material model requires the definition of 39 parameters to completely define the hysteresis behavior, as presented in Fig. 1. Values of (ePdi, ePfi), (eNdi, eNfi) define the shape of the backbone curve, while the rest of the parameters describe the material failure criteria. The pinching parameters (rDispP, rForceP, uForceN, etc.), are based upon the ratio of displacement (Disp) or force (Force) to maximum (P) or minimum (N) historic demands at various points in the unloading (u) or reloading (r) curve. Unloading and reloading stiffness degradation as well as strength degradation can be considered in the model using gKi, gDi, and gFi. In addition, energy degradation parameters (gE) and damage type parameters (dmgType) can be defined.

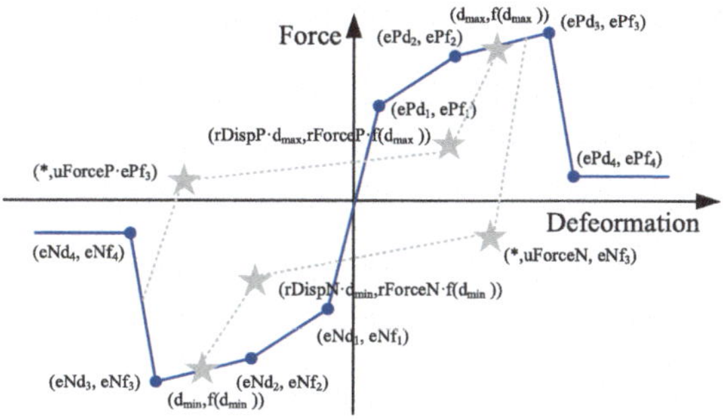

Fig. 1. Pinching4 material properties.

2.2 Validation of the Hysteresis Model with Experimental Data

Based on the hysteresis curve obtained from the quasi-static tests of grooved fit joints, we calibrate the relevant parameters of Pinching4 uniaxial model and assign the calibrated parameters to the zeroLength element to develop the moment-rotation uniaxial hysteresis model of grooved fit joints. Based on previous research experience, the following principles can be considered for the calibration of Pinching4 parameters: (1) The skeleton curve parameters can be determined according to the characteristic points of the skeleton curve obtained by each specimen in the tests. (2) The unloading-reloading parameters can be determined according to the shape characteristics of the hysteresis curves of the specimens. (3) The failure criteria parameters can be determined according to the unloading stiffness degradation, reloading stiffness degradation, strength degradation and energy degradation of the hysteresis curves of the specimens. In the calibration process, the rotation time histories obtained from the tests were used for displacement control loading. Firstly, the reproduction of the hysteresis curve, the moment time history curve, and the cumulative hysteresis energy dissipation error were considered to calibrate the parameters for each specimen. It is considered to meet the calibration requirements when the peak moment and the maximum cumulative hysteresis energy dissipation errors of each cycle of loading are less than 10%. The calibration results show that the values of the 15 parameters defining the material failure criteria and the parameters defining the energy degradation and damage type are found to be similar for all the specimens. The gKi, gDi, and gFi values are set as 0.1, 0.0, and 0.0, respectively. The gKi, gDi, and gFi values are also set as 0.1, 0.0, and 0.0, respectively. In addition, the gE and dmgType parameters are set as 0.0 and cycle.

The hysteresis curves obtained by Pinching4 uniaxial hysteresis model are compared with the test results in Fig. 2. The simulated hysteresis responses of the tested specimens are basically consistent with the test results. The shape of the hysteresis curve obtained by simulation and test is in good agreement. It is indicated that the Pinching4 uniaxial hysteresis model can accurately reproduce the hysteresis response of the grooved fit piping joints. Figure 2 also shows the comparison between the time history of bending moment obtained by simulation results of Pinching4 uniaxial hysteresis model and the test results. It can be found that Pinching4 uniaxial hysteresis model can accurately simulate the time history of moment response change of grooved fit joints during the loading process. There is a certain difference between the test results and the simulation results under relatively small rotation angles. The analytical model can basically reproduce the peak moment value in each cycle of loading accurately and error is relatively small. The comparison between the maximum accumulated energy dissipation obtained by Pinching4 uniaxial hysteresis model and the test results was conducted. The error between the simulation results and the test results is within 8%, which is generally in good agreement. The last row of Fig. 2 shows the accumulated energy dissipation of each loading cycle obtained by analytical results. Compared with the test results, the accumulated energy dissipation increases with the increase of the number of loading cycles, and the results under the same loading cycles are relatively consistent. The comparison indicated that the Pinching4 uniaxial hysteresis model can accurately simulate the accumulated energy dissipation of grooved fit joints and reproduce the change of hysteresis energy dissipation with the increase number of loading cycles. In addition, it should be noted that only one of the three specimens for each type is presented in Fig. 2.

2.3 Development of a Generic Hysteresis Model

In Sect. 2.2, the feasibility and accuracy of the moment-rotation hysteresis model are verified based on the comparison between the uniaxial hysteresis model results of Pinching4 and the test results. However, it can also be found that the hysteresis curve obtained in the tests inevitably has initial defects during the production and installation of the test specimens, which resulted in different test results of the same type of specimens. In ideal condition, the hysteresis curve of the piping joints under positive and negative loading should be symmetrically distributed. However, the test results show that the responses of each specimen under positive and negative loading are different. The test results of the same type of specimens are discrete due to the small size, relatively low bearing capacity and high sensitivity to fabrication and installation accuracy of the test specimens. Therefore, it is necessary to establish a generic hysteresis model to reduce the effects of uncertainties induced from various initial defects in the tests. In this study, there are only three specimens for each type of specimens, the number of specimens is relatively small, and the test results may present inconsistency due to the initial defects. In the calibration process of the generic hysteresis model, a symmetric moment-rotation hysteresis behavior under positive and negative loading of the piping joints were considered. In addition, it should be noted that due to the influence of installation conditions, the relatively small number of test specimens and the assumptions used in this paper, there may be some differences between the calibration results of the parameters and the actual test results. It can be further corrected based on more test results. Figure 3 shows the comparison between the hysteresis curves obtained by the generic hysteresis models and the test results. It can be seen from that the analytical results are in good agreement with the analytical results, indicating that the generic hysteresis model proposed in this study can be used to simulate the hysteresis characteristics of the grooved fit joints.

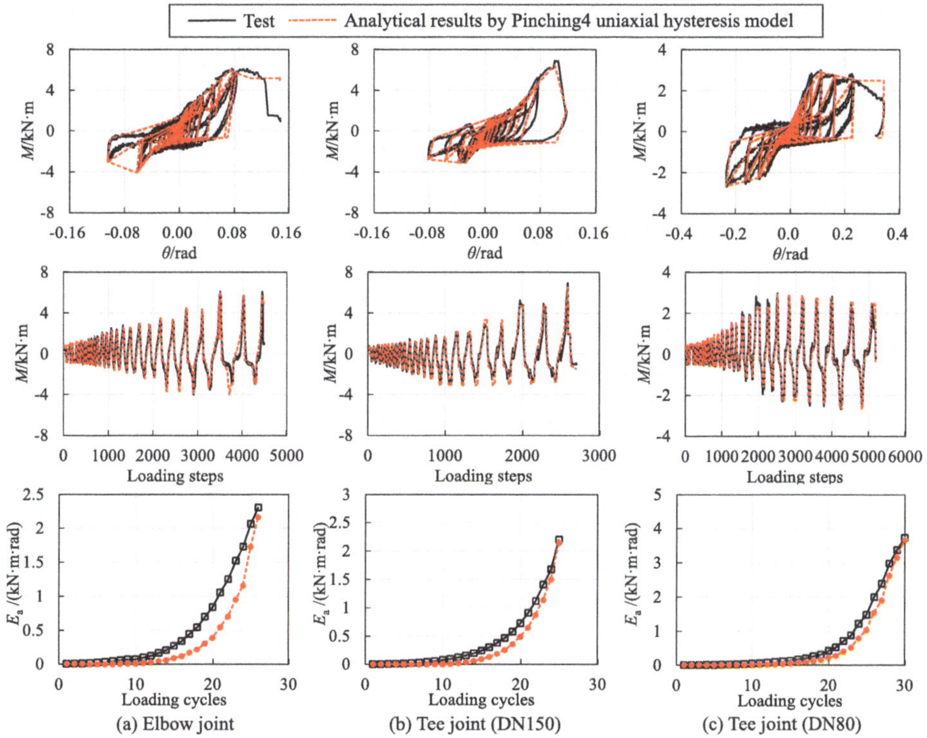

Fig. 2. Comparison of numerical analysis results based on Pinching4 uniaxial hysteresis model and test results.

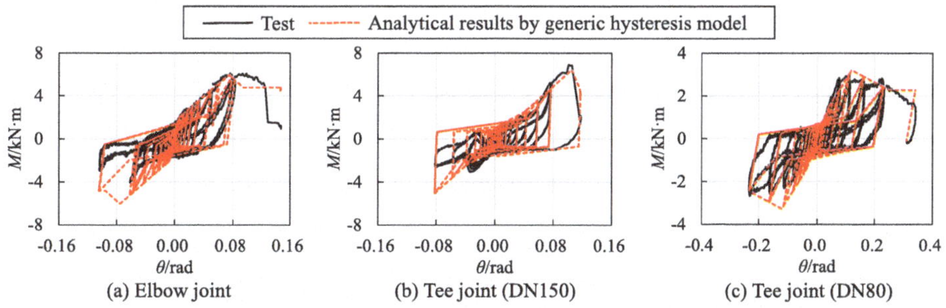

Fig. 3. Comparison of hysteresis curves based on generic hysteresis models and test results.

3 Finite Element Model of Piping-Joint System

3.1 Analytical Models for Primary Components

The generic hysteresis models of piping joints developed and calibrated in Sect. 2.3 didn't consider the combination of piping joints and pipeline segment and cannot realize the numerical simulation of the whole piping system. Therefore, it is necessary to study how to apply the generic hysteresis models in finite element modeling of piping systems. The finite element model of piping systems is further established in OpenSees, and its schematic diagram is shown in Fig. 4. According to the reported research results, the nonlinear behavior of the pipe segments can be ignored since the moment capacity of the pipe section is much larger than that of the grooved fit joint. The nonlinearity of pipe runs is only concentrated at the rotational degrees of freedom of piping joints. The pipe segments are modeled with ForceBased Beam-Column elements using the elastic cross section properties of the pipes and an elastic material with steel material properties. The modulus of elasticity (E) of the pipe material is taken as 200000 N/mm. The inner diameter of DN150 pipe is 155 mm, the outer diameter is 165 mm while the inner diameter of DN80 pipe is 85 mm, and the outer diameter is 90 mm. The Pinching4 uniaxial material along with a zeroLength element were used to simulate the moment-rotation response of a piping joint connecting two piping nodes. The axial and shear directions of the zeroLength element are defined by elastic materials, and its material properties can be determined according to the pipe properties.

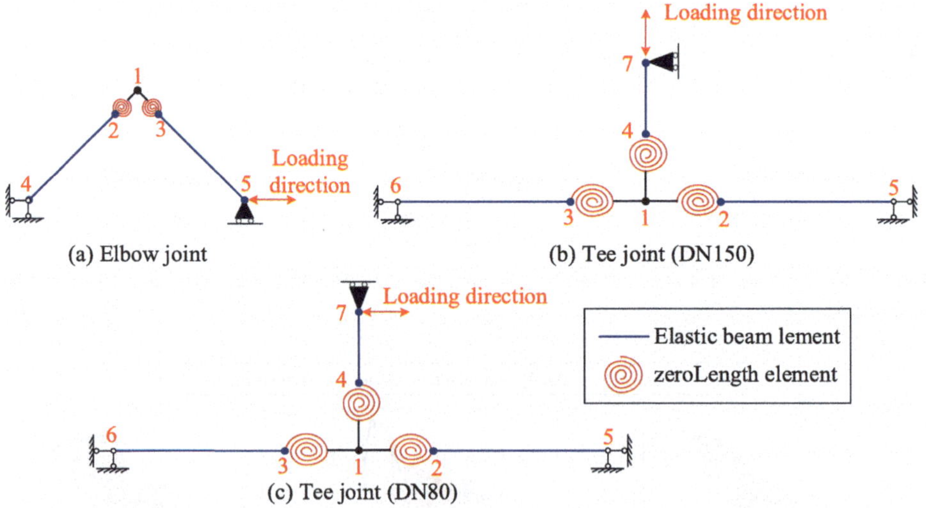

Fig. 4. Finite element models of piping system.

3.2 Validation of the Finite Element Model with Experimental Data

In the finite element model of piping system, the displacement time histories of the actuator in the tests are used as the input to carry out cyclic loading. The comparison between the moment rotation hysteresis curve at the piping joints obtained by the finite element model and the test results is shown in Fig. 5. It can be found that although the control displacement of the loading point is completely consistent with the test, there is a certain gap between the rotation angle at the piping joints obtained by the finite element analysis and the test results. This inconsistency may come from the assumptions used in the modeling process of this study: (1) The tested piping systems are not ideal symmetry system due to the initial defects (e.g., installation error) in the installation process. The failure of the piping joints was all found to be damaged in one side, which resulted in the asymmetric moment-rotation relationship on the left and right sides of the joints. However, the influence of installation error is not considered in the finite element model. It is assumed that the moment-rotation relationship on the left and right sides of the joint is complete symmetrically distributed and unified parameters are used for simulation. (2) In order to improve the applicability of the finite element model of piping system proposed in this study, the generic hysteresis model calibrated in Sect. 3.3 is used to define the moment-rotation relationship at the rotational degrees of freedom of piping joints, which is not completely matched with each tested specimen. However, in general, the finite element model proposed in this study (Fig. 4) can basically simulate the moment-rotation hysteresis response of piping joints.

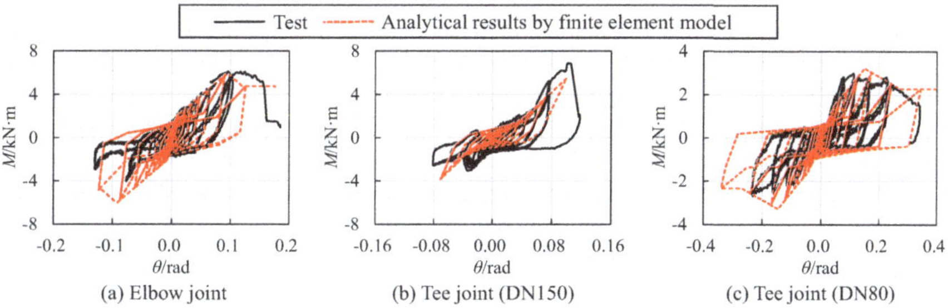

Fig. 5. Comparison of hysteresis curves based on finite element model and test results.

4 Conclusion

The hysteresis model of grooved fit piping joint is developed based on the Pinching4 uniaxial material and zeroLength element in OpenSees. The parameters of the hysteresis model are calibrated based on the test results in a previous component experiment. The results show that the developed hysteresis model can accurately reproduce the nonlinear hysteresis response of piping joint. In order to reduce the uncertainty induced by initial defects in the tests, a generic hysteresis model is further developed based on the calibration results of each specimen. The generic hysteresis model can be used to simulate the hysteresis characteristics of piping joints. The finite element modeling method of piping system is proposed and the finite element model of piping system is generated based on the generic hysteresis model of piping joint. The proposed finite element model of piping system can basically simulate the hysteresis response of grooved fit joint. The developed finite element model can be used for piping system modeling in a real building. The recorded floor acceleration responses in past earthquakes can be used as input to investigate the seismic demand of the piping system under real earthquakes, and the seismic fragility analysis of the piping system can be carried out in future studies. In addition, it is worth noting that the developed generic hysteresis model in this study is only suitable for the specimens investigated in this test, and the parameters are only valid for the specific size in this study. The feasibility for piping joints with different sizes and materials still needs further study.

Acknowledgments. This research was funded by National Key R&D Program of China (2022YFC3803000) and National Natural Science Foundation of China (52208483). Any opinions, findings and conclusions or recommendations expressed in this paper are those of the authors and do not necessarily reflect the views of the sponsors.

References

1. Filiatrault, A., Sullivan, T.: Performance-based seismic design of nonstructural building components: The next frontier of earthquake engineering. Earthq. Eng. Eng. Vib. **13**, 17–46 (2014)
2. Miranda, E., Mosqueda, G., Retamales, R., Pekcan, G.: Performance of nonstructural components during the 27 February 2010 Chile earthquake. Earthquake spectra **28**(1), 453–471 (2012)
3. Tian, Y., Filiatrault, A., Mosqueda, G.: Experimental seismic fragility of pressurized fire suppression sprinkler piping joints. Earthq. Spectra **30**(4), 1733–1748 (2014)
4. Wang, T., Shang, Q., Chen, X., Li, J.: Experiments and fragility analyses of piping systems connected by grooved fit joints with large deformability. Frontiers in Built Environment **5**, 49 (2019)

Derivation of Fragility Functions for Steel Storage Tanks Using Far- and Near-Field Record Sets

Aldo Rapone[1(✉)], Giammaria Gabbianelli[2], Riccardo Milanesi[3], and Roberto Nascimbene[1]

[1] University School for Advanced Studies IUSS Pavia, Pavia, Italy
aldo.rapone@iusspavia.it

[2] Department of Civil Engineering and Architecture, University of Pavia, Pavia, Italy

[3] European Centre for Training and Research in Earthquake Engineering, EUCENTRE, Pavia, Italy

Abstract. Liquid-storage tanks are among industrial plants' most important and critical elements, particularly during a seismic event. The leakage of the content could lead to substantial economic losses, disruption of the plants and, no less importantly, environmental pollution. Therefore, an accurate design and assessment of their performance is crucial in reducing the risk of catastrophes. Yet, past earthquake events have shown the need for a more reliable comprehension of tanks' seismic behaviour. To fulfil this scope, it is necessary achieving a clear understanding of the seismic input selection, and the optimal ground motion sets to excite the storage tanks properly. This paper aims to study the seismic behaviour of a wide taxonomy of storage tanks with different sets of records, considering both far-field and near-field earthquakes. The tanks were developed considering several geometric dimensions, including the diameter-height ratio, the wall thickness, the liquid height and tank radius. The tanks were assumed to be located in three Italian sites, representative of low, medium and high seismic zones. Nonlinear time-history analyses were performed to derive fragility curves and assess the tanks' seismic performance, taking advantage of suitable limit states. The results of this study demonstrate the significant impact of choice between far- and near-field record sets on the seismic performance of the tanks, particularly when considering the fragility curves dispersion. Furthermore, the fragility curves derived from the analyses can be a valuable tool for conducting risk scenario assessments for government bodies and stakeholders, thereby enhancing the safety and resilience of industrial plants.

Keywords: steel storage tanks · time-history analyses · fragility curves · far- and near-field ground motions · parametric analyses

X. Lu et al. (Eds.): EOS 2024, LNCE 638, pp. 133–142, 2025.
https://doi.org/10.1007/978-3-031-90690-9_12

1 Introduction

The recent studies and aftermath inspections after strong earthquakes demonstrated the efficiency of the new standard codes in reducing human losses; once life safety was ensured, the research community focused on another aspect, i.e. the influence of economic losses due to the downtime and nonstructural elements damages, especially in industrial structures [1–3]. Steel storage tanks are one of the most common structures in industrial plants for liquid storing. They play an important role in the functionality and security of these facilities. A large number of studies have developed after past earthquakes to describe and monitor the behaviour of steel tanks and the liquid stored in them [4, 5]. The post-earthquake surveys have shown the recurrence of some types of damage, such as liquid leakage or instability of the steel tank walls known as Elephant Foot Buckling (EFB). The failure or damage of a tank, either sloshing-related or buckling-driven, may cause a contaminating substance to leak or spill over, which causes downtime of facilities and, in some cases, a catastrophic event in terms of environmental pollution and human life losses. To study the behaviour of storage tanks under seismic events, Malhotra [6, 7] has developed a simplified model able to predict the seismic behaviour of cylindrical tanks through mechanics-based mass-spring-dashpot analogies that decouple the impulsive and convective components of motions. The goal of this article is to describe the seismic vulnerability of Italian industrial plants using a large set of steel tanks modelled by Malhotra's simplified model in anchored and unanchored configurations. Both types of ground-structure connections were studied for three Italian seismic zones (low, medium and high seismicity).

2 Numerical Modelling of Tanks

2.1 Anchored Liquid-Storage Steel Tanks

A set of 60 cylindrical steel storage tanks anchored to the ground was first selected for this study. This set consists of tanks with a diameter/height (D/H) ratio in the range from 0.8 to 3.5 and a wall thickness that varies along their height. The developed tank models take inspiration from the Malhotra [6, 7] simplified model for cylindrical tanks; the numerical models were created in OpenSeesPy [8] in order to perform dynamic time-history analyses. The Malhotra model is also adopted by European standards in the EC8 Part 4 – Annex A [9]. It should be noted that the Malhotra [6, 7] system is equipped with few uncoupled degrees of freedom. Each degree of freedom represents a different vibration mode of the fluid-structure system; the response of each is calculated independently and is then combined using a chosen modal combination rule. The single-degree-of-freedom (SDoF) model considers the impulsive mode given by the impulsive mass, the convective masses required to reach at least 98% of the participating mass of liquid inside the tank, and the mass of the steel tank structure. Each of these masses has been placed at a given height as described in the simplified model and connected to the tank walls via springs and dashpots. The walls and the base plate of tanks were instead modelled by means of rigid links. A distinction between the Malhotra model [6, 7] and the one adopted for this study is that the adopted model has a higher number of convective modes (three) and, therefore, is more accurate. All tanks of the set are composed of S275 class of steel, and the liquid stored in them was assumed to be water with a density (r) of 1.0 [t/m^3]. This set of tanks has been considered anchored on the ground, so the nodes at the base have been fully fixed to prevent uplift.

2.2 Unanchored Liquid-Storage Steel Tank

The second configuration of the ground-structure connection is modelled as described by Bakalis et al. [10, 11]. For this case, a set of 60 tanks was considered, without any consideration to the D/H ratio, since the unanchored configuration, due to overturning self-stability, generally requires squat tanks. This model uses a multilinear relation for the spring elements at the base of the tank to describe the uplift of the base plate. For the calibration of the multilinear springs, an accurate nonlinear analysis was conducted on each tank's base plate of the set. The first step consists of a static pushover analysis, where a strip of the base plate was modelled with fibre elements, as illustrated in Fig. 1a, to describe its uplifting behaviour (Fig. 1b). Note that to simulate the tank wall-base plate interaction, axial and rotational springs are used at the end of the base plate element.

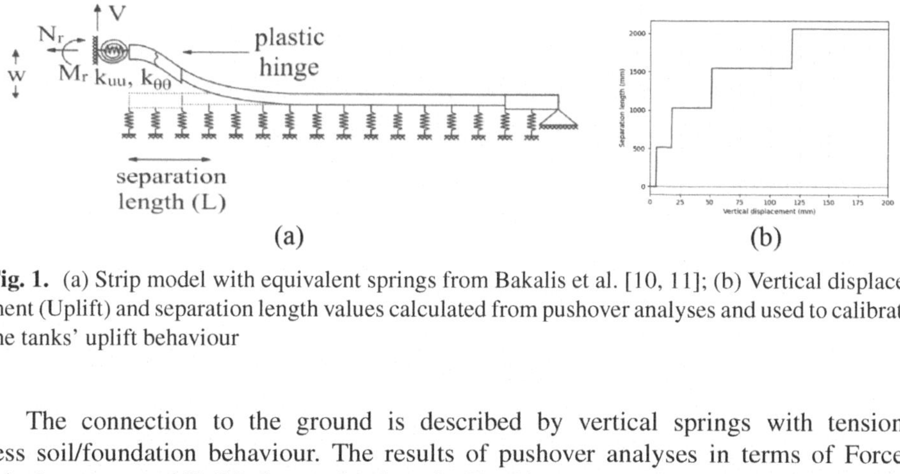

Fig. 1. (a) Strip model with equivalent springs from Bakalis et al. [10, 11]; (b) Vertical displacement (Uplift) and separation length values calculated from pushover analyses and used to calibrate the tanks' uplift behaviour

The connection to the ground is described by vertical springs with tensionless soil/foundation behaviour. The results of pushover analyses in terms of Force-Displacement and Uplift-Separation length (Fig. 1b) were used to calibrate the springs at the base of "Joystick" 3D model of tanks (Fig. 2), where impulsive and convective masses were described in the same way as described for anchored tanks. The correct discretization of the plate strip for the joystick model allows a better understanding of the ground-tank interaction behaviour, in particular in the case of three-dimensional analyses, in which the ground motion excitation occurs in both horizontal directions simultaneously.

2.3 Definition of the Damage States for the Tanks

Past studies and post-earthquake surveys have shown the typical damage in steel storage tanks. The convective waves (also referred to as sloshing waves) can cause the leakage of the fluid stored, whereas the increase of pressure on the tank walls can cause instability

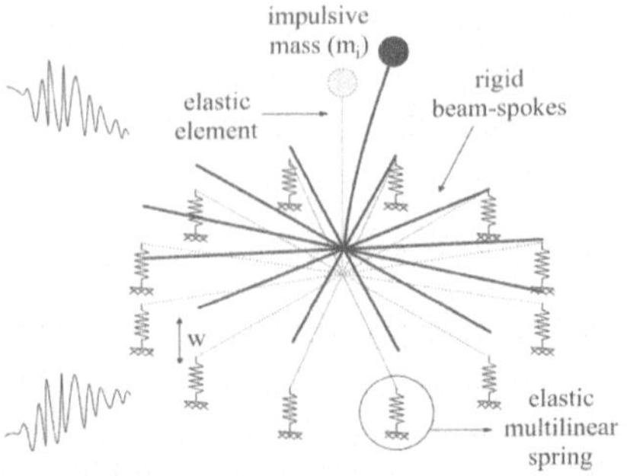

Fig. 2. Three-dimensional joystick model to simulate unanchored steel storage tanks [10, 11].

phenomena such as elephant foot buckling or diamond buckling. It should be noted that the sloshing wave is related to the excitation of the convective modes; on the contrary, albeit the stress in the tank walls is induced by both modes (impulsive and convective), it can be stated that the elephant foot buckling is well correlated to the impulsive mode excitation [12]. Therefore, the damage states considered in this research are summarised in the following:

- convective wave exceeding the freeboard;
- exceeding the 80% of the Elephant Foot Buckling (EFB1);
- exceeding the 100% of the Elephant Foot Buckling (EFB2);
- exceeding the 140% of the Elephant Foot Buckling (EFB3);
- uplift of the tank (for the unanchored configuration);
- overturning of the tank (for the unanchored configuration).

It should be noted that the last damage state, i.e. the overturning of the tank, has never been observed in the considered PGA range.

3 Seismic Hazard and Record Selection

As mentioned earlier, the two sets of tanks (anchored and unanchored) were studied for three Italian seismic zones, representing low, medium and high seismic areas, with a reference value of peak ground acceleration (PGA) at a return period of 475 years of 0.06g, 0.125 g and 0.267 g, respectively. The PGAs have been retrieved by the Italian National Standard [13]. The FEMA P695 [14] Far-field and Near-field sets of ground motions were used to maximize, in turn, the impulsive and convective mass excitation and, thus, their associated damage states. The FEMA's sets are composed of 22 and 28 records for the Far-field and the Near-field, respectively. Each of these accelerograms was scaled so that their PGA is identical to the PGA according to the NTC18 [13]. For this study, the PGA was selected as the intensity measure (IM). However, it is well known that adopting a different IM could lead to diverse results. In lieu of such considerations, future work will focus on selecting other types of IMs, such as the Average Spectra Acceleration (AvgSA). In order to perform the vulnerability study, different return periods (ten) were selected, i.e. 30, 50, 72, 98, 224, 475, 975, 2475, 4975 and 9975 years. Thus, nonlinear time history analyses were conducted with the 50 pairs of records for ten return periods for the three seismic locations, and a total of 90000 time-history analyses were performed.

4 Results of the Analyses

Based on the outcomes of the time-history analyses, fragility curves were derived for both tank configurations. The intensity measure (IM) used was PGA, in g, and the damage states previously described (Sect. 2.3) were used as limit states (LSs). The derived fragility curves are a lognormal function that allows us to identify the probability of exceedance of the correspondent LS; two parameters are provided, i.e. the median, μ, for which IM there is a 50% probability that the LS occurs, and σ, the dispersion of the curve. The results shown are disaggregated by the diameter-to-height (D/H) ratio and then incorporated, including all the tanks. Due to space limitations, the results for all the seismic zones are reported aggregating all the tank D/H configurations, whereas for the high seismic zone, also the results with D/H higher than 2 are shown. It should be noted that, for both types of tanks, the Freeboard LS should be considered as a light damage level, since it implies the impact of the liquid stored in the tank with its roof; this impact, albeit could lead to damages to the roof structure, does not indicate a complete failure of the tank. As such, the following results focus mainly on the elephant foot buckling limit states.

4.1 Results for Anchored Tanks

As a general result, valid for all the analyses conducted, the fragility curves associated with the different levels of elephant foot buckling (EFB1, EFB2 and EFB3) show, as expected, an increasing value of IM as the severity of the LS increases. For the low seismicity zone (Fig. 3), the Freeboard LS is characterised by μ values of 0.09 g and 0.11 g and σ values of 1.1 and 0.83 for Near-field and Far-field, respectively. The EFB1 has μ values of 0.06 g and 0.05 g and σ values of 0.43 and 0.35 for Near-field and Far-field, respectively. Very similar values, but slightly higher, are obtained for EFB2 and EFB3. For both sets of records, it is evident that the higher vulnerability is represented by the EFB, as often observed in similar studies [12].

In the medium seismicity zone (Fig. 4), similar considerations can be made. The EFB1 occurs for μ equals to 0.14 g and 0.11 g and σ to 0.42 and 0.37 for Near-field and Far-field, respectively. The EFB2 and EFB3 curves reached μ values of 0.16 g and 0.19 g and σ value of 0.4 in both cases for the Near-filed set, whereas for Far-field set μ is equal to 0.13 g and 0.15 g, while σ is 0.36. The Freeboard LS occurs for a median (μ) of 0.21 g and 0.24 g and a dispersion (σ) of 1.12 and 0.84 for Near-field and Far-field, respectively. It is interesting to note how the dispersion of all the results decreases for the results obtained with the Far-field accelerogram set. This result indicates that the Far-field set excites more accurately the different damage states of the tanks.

In the high seismic zones (Fig. 5 and Fig. 6), the trend of exceedance of the different LSs is always the same. It can be clearly noted that, as previously described, the most vulnerable limit state is reaching 80% of the elephant foot buckling (EFB1), for which the tanks have a 50% probability of achieving this LS (Fig. 5) for a PGA of 0.21 g with

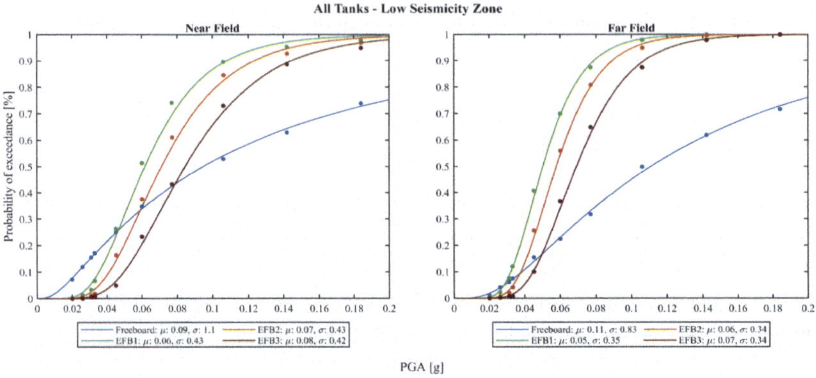

Fig. 3. Near-field and Far-field fragility functions for all tanks in the low seismic zone and anchored configuration.

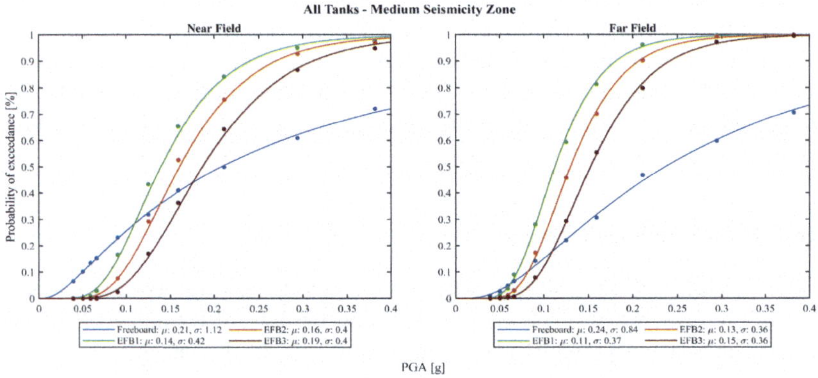

Fig. 4. Near-field and Far-field fragility functions for all tanks in the medium seismic zone and anchored configuration.

the Far-field set and for a PGA of 0.26 g with the Near-field set. Although the difference in terms of PGA between the two sets is not remarkable, a lower dispersion of results is observed for the Far-field set, particularly for tanks with a D/H ratio > 2 (Fig. 6). In this case, the Freeboard occurs for very similar values of μ (0.69 g and 0.7 g), but σ decreases from 1.13 of the Near-field set to 0.8 of the Far-field set.

In general, it is possible to highlight the fact that the choice of a set of accelerograms like the FEMA Far-field, which is characterised by a higher frequency content, excites more the impulsive mass of the liquid. This causes the (μ) values to be reached for a lower PGA. On the contrary, the near-field set allows for slightly more excitation on the convective masses, although with a greater dispersion of the results.

4.2 Results for Unanchored Tanks

For the low seismic intensity zone (Fig. 7), the conservative design, combined with the low seismic action, led to very few exceedances of the Uplift and Freeboard LSs.

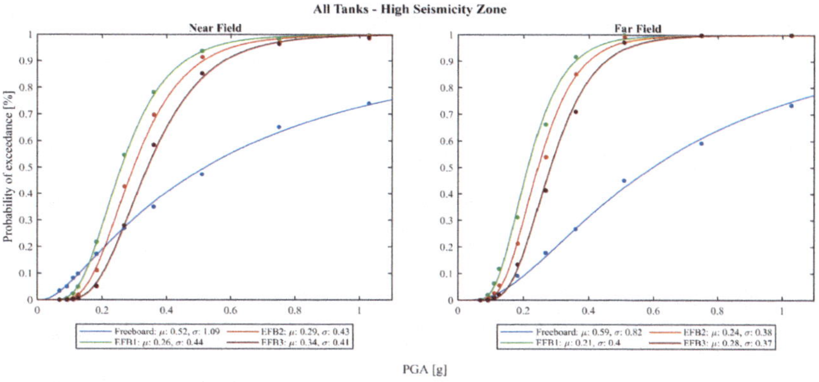

Fig. 5. Near-field and Far-field fragility functions for all tanks in the high seismic zone and anchored configuration.

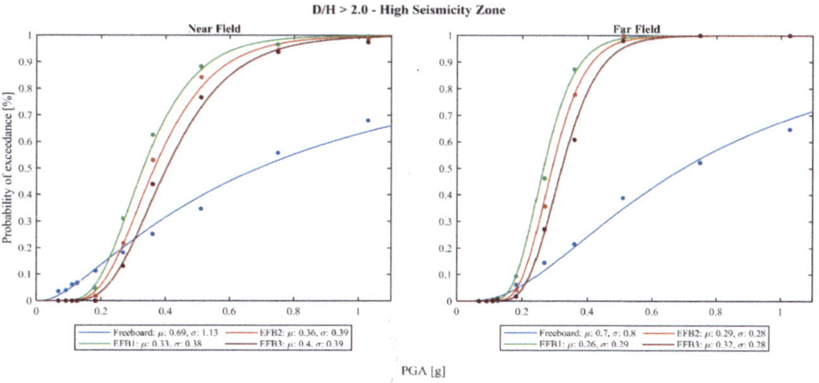

Fig. 6. Near-field and Far-field fragility functions for tanks with a D/H ratio higher than 2.0 in the high seismic zone and anchored configuration.

As a consequence, considering that the maximum PGA for this site is lower than 0.19 g, and noting that the median values (μ) of the above-mentioned LSs are higher than 0.2 g (0.79 g for the Uplift in the Near field case), it can be concluded that the probability of obtaining these LSs is very limited. On the other hand, the EFB LSs have medians around 1 g, showing that the unanchored tanks are prone to such kinds of failures.

For the tanks in the medium seismic zone, it can be observed that for both sets of records, the fragility curves show a steep trend for all limit states except for Uplift. In Fig. 8, it can be noted that the probability of Uplift occurs for μ values of 0.17 g and 0.19 g, and it is significantly lower, in terms of probability, compared to the other LSs, which range between 0.04 g and 0.07 g if the median (μ) is considered.

Regarding the tanks in the high seismic zone (Fig. 9), it is observed that the Freeboard limit state has a 50% probability of exceeding for median (μ) values of 0.49 g and 0.56 g, which is significantly higher than the Uplift LS, which occurs at μ equals to 0.35 g and 0.32 g, for the Near-field and Far-field set, respectively.

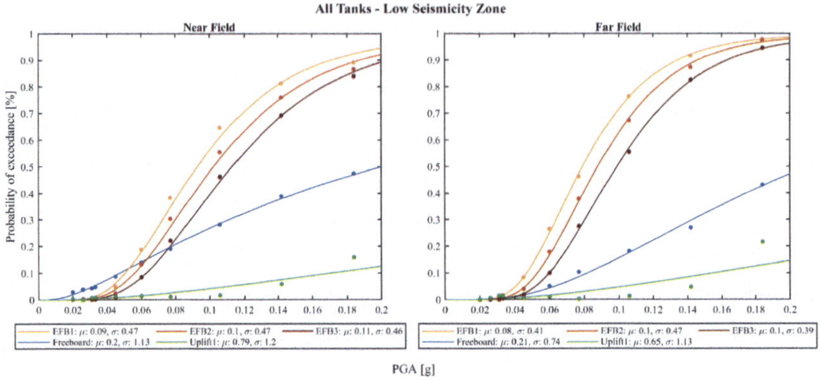

Fig. 7. Near-field and Far-field fragility functions for all tanks in the low seismic zone and unanchored configuration.

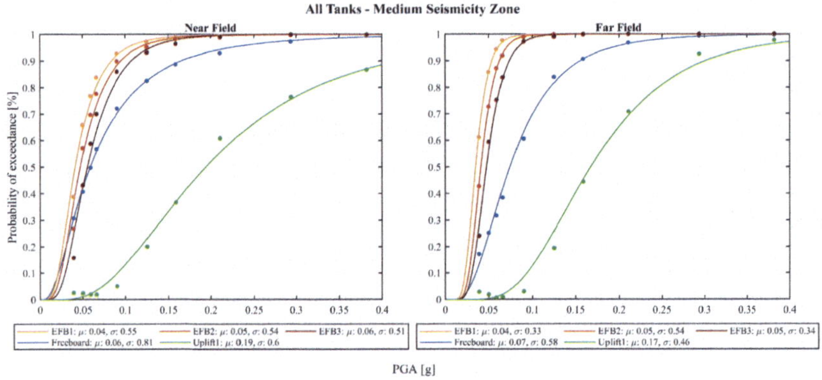

Fig. 8. Near-field and Far-field fragility functions for all tanks in the medium seismic zone and unanchored configuration.

This change in the sequence of limit state exceedances, i.e. the Uplift becoming more vulnerable compared to the Freeboard if a juxtapose is made with respect to the medium seismicity zone, can be attributed to the conservative design approach for the freeboard height of the tanks in case of high seismic actions. As a matter of fact, the choice of the Freeboard height directly influences the Freeboard LS; the approach adopted to define such height is oversimplified by the standard approaches, and if time history analyses are conducted instead, different seismic responses can be obtained. In general, trends similar to those obtained for anchored tanks can be found for unanchored tanks, i.e. the tanks are more prone and vulnerable to Elephant foot buckling rather than Freeboard or Uplift LSs; a higher dispersion (σ) of the results can be found for the Near-field set, and slight changes in terms of medians (μ) can be retrieved.

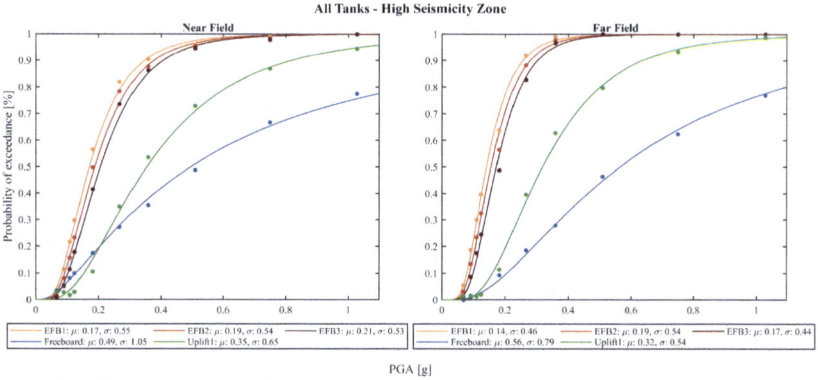

Fig. 9. Near-field and Far-field fragility functions for all tanks in the high seismic zone and unanchored configuration.

5 Conclusions

The present research aimed to assess the differences in seismic vulnerability assessment of steel storage tanks adopting two different record sets, namely, Far-field and Near-field, provided by FEMA P695 [14]. Different damage states were identified; it should be noted that such states are strictly correlated to the various modes of vibration of the tank, namely impulsive (short periods) and convective modes (larger periods). The adoption of record sets with different characteristics will lead to a distinct seismic assessment due to the inherent frequency content of accelerograms obtained from Near- and Far-field sets.

For this purpose, two typologies of tanks were selected and designed, with a distinction in terms of connection to the ground. The first set of tanks is anchored to the ground, whereas the second set is unanchored. For the latter, the overturning stability is provided by the self-weight of the tank.

The preliminary results obtained in this research highlighted that the seismic vulnerability, in terms of medians of fragility curves, is not subjected to significant changes. On the other hand, the dispersion of the results is notably reduced in the case of Far-field records.

Further analyses are necessary to corroborate the results herein obtained, in addition to adopting additional intensity measures, such as the Average Spectral Acceleration (AvgSA) in combination with Far- and Near-field record sets.

Acknowledgements. The authors gratefully acknowledge the Italian Department of Civil Protection (DPC) for their financial contributions to this study through the 2024–2026 Project (WP7).

References

1. Van Loenhout, J., Below, R.: Human cost of disasters: An overview of the last 20 years 2000–2019, Centre of Research on the Epidemiology of Disasters – United Nations Office for Disaster Risk Reduction. Belgium, Brussels (2021)
2. Miglietta, C., Perrone, D., Gabbianelli, G., Micelli, F., Ciucci, M.: Release of Hazardous Substance Due to Seismic Loads: Influence of Pipe to Tank Connection in Fragility Estimation. Chem. Eng. Trans. **111**, 127–132 (2024). https://doi.org/10.3303/CET24111022
3. Grajales-Ortiz, C., Melissianos, V.E., Bakalis, K., Kohrangi, M., Vamvatsikos, D., Bazzurro, P.: Relationships between earthquake-induced damage and material release for liquid storage tanks, In: Proceedings of the 18th World Conference on Earthquake Engineering. Presented at the 18th World Conference on Earthquake Engineering, Milan, Italy (2024)
4. Bevere, L., Ewald, M., Wunderlich, S.: A decade of major earthquakes: lessons for business, Swiss Re Institute, Zurich, Switzerland (2019)
5. Campedel, M.: Analysis of major industrial accidents triggered by natural events reported in the principal available chemical accident databases, JRC Scientific and Technical Reports, JRC42281 (2008)
6. Malhotra, P.K.: Seismic Response of Soil-Supported Unanchored Liquid-Storage Tanks. J. Struct. Eng. (1997). https://doi.org/10.1061/(asce)0733-9445(1997)123:4(440)
7. Malhotra, P.K.: Practical Nonlinear Seismic Analysis. Earthq. Spectra (2000). https://doi.org/10.1193/1.1586122
8. Zhu, M., McKenna, F., Scott, M.H.: OpenSeesPy: Python library for the OpenSees finite element framework. SoftwareX (2018). https://doi.org/10.1016/j.softx.2017.10.009
9. EN 1998–4: Eurocode 8: Design of structures for earthquake resistance - Part 4: Silos, tanks and pipelines. European Committee for Standardisation, Brussels, Belgium (2006)
10. Bakalis, K., Vamvatsikos, D., Fragiadakis, M.: Seismic risk assessment of liquid storage tanks via a nonlinear surrogate model. Wiley (1997). https://doi.org/10.1002/eqe.2939
11. Bakalis, K., Fragiadakis, M., Vamvatsikos, D.: Surrogate modelling of liquid storage tanks for seismic performance design and assessment. In: Papadrakakis M., Papadopoulos V., Plevris V. (eds.) Proceedings of the 5th International Conference on Computational Methods in Structural Dynamics and Earthquake Engineering (COMPDYN 2015), pp. 2066–2080. National Technical University of Athens, Greece (2015). https://doi.org/10.7712/120115.3523.1375
12. Gabbianelli, G., Milanesi, R.R., Gandelli, E., Dubini, P., Nascimbene, R.: Seismic vulnerability assessment of steel storage tanks protected through sliding isolators. Earth. Eng. Struct. Dyn. **52**(9), (2023). https://doi.org/10.1002/eqe.3885
13. Norme Tecniche per Le Costruzioni: DM 17/1/2018 (NTC18). Gazzetta Ufficiale della Repubblica Italiana. Italian Ministry of Infrastructure and Transport (2018)
14. Federal Emergency Management Agency (FEMA): Quantification of Building Seismic Performance Factors (FEMA P695). Federal Emergency Management Agency, Washington, DC (2009)

Seismic Assessment of Different Pressurized Tank–Pipeline Configurations on Industrial Plants

Bryan Chalarca[1(✉)], Giammaria Gabbianelli[2], Daniele Perrone[3], and Mariano Ciucci[4]

[1] Department of Civil Engineering and Architecture, University of Pavia, Pavia, Italy
bryan.chalarcaecheverri@unipv.it

[2] Department of Civil Engineering and Architecture, University of Pavia, Pavia, Italy

[3] Department of Engineering for Innovation, University of Salento, Lecce, Italy

[4] Department of Technological Innovations and Safety of Plants, INAIL, Rome, Italy

Abstract. Natural Hazard Triggering Technological Disasters (Na-Tech) are becoming a key aspect of the improvement of the resilience capacity of society. In this regard, the reduction of the seismic vulnerability of industrial plants helps reduce detrimental effects such as life losses, economic losses, and environmental damage. However, the large variability of equipment, connections, materials, size, etc., makes the definition of the seismic capacity of industrial plants a difficult task. This preliminary study provides a seismic assessment of the interaction between pressurized tanks and connecting pipelines subjected to far-field ground motions. The seismic assessment was carried out on three configurations of pressurized tanks and pipelines subjected to nonlinear time-history analysis using the FEMA P695 far-field ground motion set scaled to three increasing intensity levels. The results highlight the importance of considering the dynamic interaction between the tanks and the pipelines, usually neglected in design requirements.

Keywords: Na-Tech · industrial plants · seismic assessment · tanks · pipelines · nonlinear analysis · seismic vulnerability

1 Introduction

Recent middle and strong seismic events worldwide have demonstrated that although current building regulations can significantly reduce human losses, earthquakes can still cause important economic losses, affecting the resilience capacity of communities [1]. A significant portion of these direct and indirect economic losses are generated by affectation to business and industrial facilities [2]. The latter are of particular interest due to their high vulnerability to diverse natural hazards such as earthquakes, floods, storms, severe temperatures, etc., leading to the so-called Na-Tech events (Natural Hazard Triggering Technological Disasters) [3]. Due to the complexity and vast possibilities of components and configurations, industrial plants exhibit a high seismic vulnerability that can lead to structural collapse, explosions, fire, toxic emissions, etc. [4]. Among the different sources of industrial plant vulnerabilities, the connection and interaction between pipelines and

X. Lu et al. (Eds.): EOS 2024, LNCE 638, pp. 143–151, 2025.
https://doi.org/10.1007/978-3-031-90690-9_13

tanks represent an important aspect of seismic evaluation due to the high probability of being affected during seismic events [3]. Failure of tank-pipeline connections can cause loss of containment, leading to indirect damage and the affectation of other components due to cascade effects [5].

Although the importance of an appropriate seismic assessment of tank-pipelines connection is well known, the complexity of the different components, connections, geometry, and differences of the order of magnitude of the structural properties of thanks and pipelines make it challenging to achieve accurate, complete, and manageable numerical models. This paper explores a simplified modelling approach for the seismic response assessment of tank-pipelines interaction. To achieve this, two liquid storage tanks, as well as three different geometries of pipelines connecting both tanks were modelled and subjected to nonlinear time-history analysis using the FEMA P695 far-field ground motion set [6] scaled to three increasing intensities. The response was measured in terms of median peak axial forces and median peak moment at the beginning, middle span, and end of each pipe.

2 Numerical Model of the Tank-Pipelines and Seismic Input

2.1 Numerical Model

The case study setup is composed of two water storage steel tanks (tank 1 and tank 2) with a diameter of 40 m and 20 m, total height of 28.75 m and 20.1 m, and wall thickness of 0.029 m and 0.015 m, respectively. On the other hand, the pipelines are composed of 8-in. (219 mm) typical industrial steel pipes made of API 5L Gr. X52 steel (nominal yield and ultimate strength: 418 MPa and 554 MPa, respectively) and were assumed to be filled with water. Three configurations of the tank pipelines were assessed and are shown in Fig. 1. The tanks and pipelines were modelled considering the following assumptions:

- Only the impulsive mass of the liquid storage tanks contributes to the total displacement of the tanks.
- The pipelines are fixed-connected to the tanks.
- The damping is only given by the impulsive damping component of the tanks.
- Both tanks were assumed fixed-base.

The numerical models were developed using OpenSeesPy version 3.5.1. The impulsive behaviour of the tanks was modelled using a *zeroLength* element characterized by an elastic spring and a dashpot in parallel that represent the impulsive stiffness and impulsive damping coefficient, respectively [7]. Table 1 contains the dynamic properties of the case study tanks. The pipelines were modelled using *forceBeamColumn* elements with fibre sections characterized by a *Steel01* material. In order to account for the modified properties of the pipe elbows, the modification proposed by Bursi *et al.* [8] was implemented. Flexible elbows represent critical elements due to their geometrical irregularity. To take into account this phenomenon, an equivalent straight elbow element was adjusted to that of the relevant curved elbow according to EN 13480–3 [9]. The used elbows were characterized by a radius, R, equal to 1.5 times the outer diameter of the connecting pipe. In addition, it was assumed that the effect of flexibility of an elbow

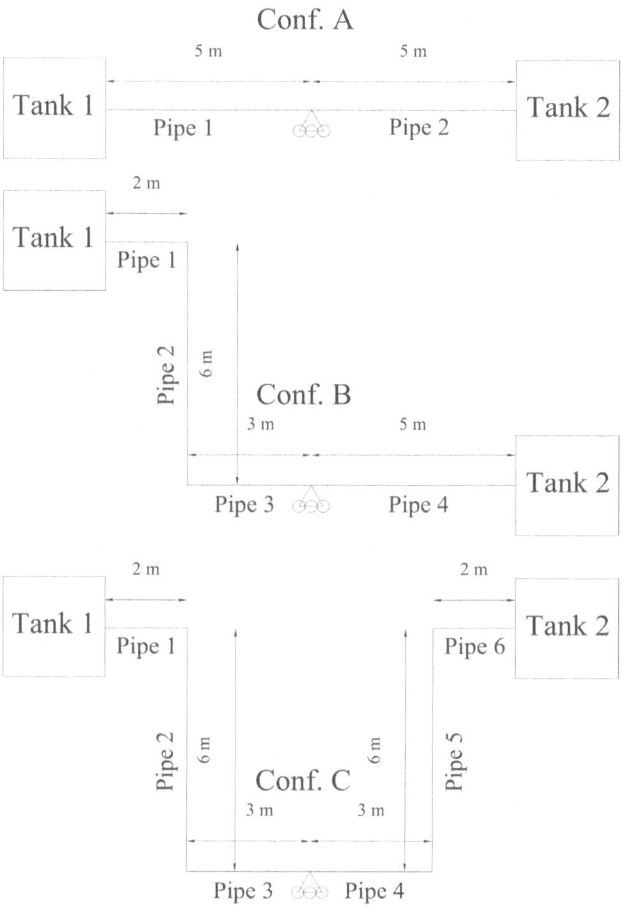

Fig. 1. Case study tank-pipeline configurations.

spreads across a distance, L, equal to two times the mean diameter of the pipe as shown in Fig. 2. More information about this modelling approach can be found at [8]. Table 2 reports the properties of the pipes and the equivalent elbows.

Table 1. Dynamic properties of the case study tanks.

Properties	Tank 1	Tank 2
Impulsive mass (kN/g)	24160.8	5050.3
Impulsive damping constant (kNs/m)	674558.0	178547.8
Impulsive stiffness (kN/m)	11075847136	3945270717
Impulsive period (s)	0.00957	0.00711

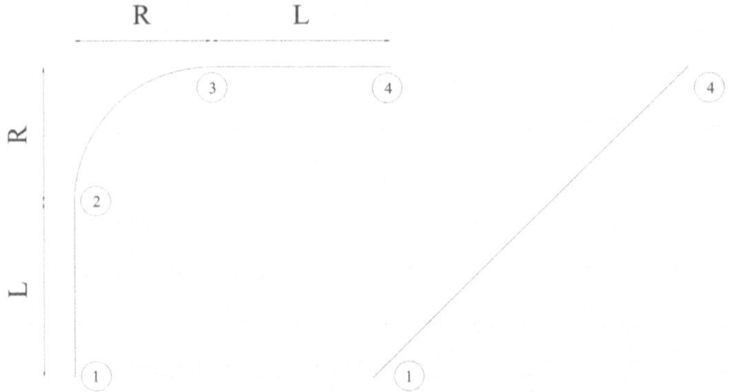

Fig. 2. Example of the equivalent elbow.

Table 2. Properties of pipes and equivalent elbow.

Properties	Original pipe	Equivalent elbow
Thickness (mm)	8.18	6.61
Moment of inertia (mm^4)	3.03×10^7	2.49×10^7

2.2 Seismic Input

The case study tank-pipeline systems were subjected to 3D nonlinear time-history analysis using the FEMA P695 far-field ground motion set [6] scaled to three increasing intensities. This set is composed of 22 pairs of horizontal ground motion records from historical strong seismic events around the world. The scaling process was conducted by matching the 5% damped median absolute acceleration response spectrum of the 44 records with the ASCE 7–22 [10] design response spectrum at a period equal to 1.0 s. The design spectrum was calculated for a seismic design category D_{max} per ASCE 7–22 [10], with a design earthquake (DE) spectral acceleration of 0.6 g at 1.0 s ($Sa_{T=1s}$ = 0.6 g) and a maximum considered earthquake (MCE) spectral acceleration of 0.9 g at 1.0 s ($Sa_{T=1s}$ = 0.9 g). The other intensity corresponds to 0.5 times the design earthquake (0.5DE) and is characterized by a spectral acceleration of 0.3 g at 1.0 s ($Sa_{T=1s}$ = 0.9 g). Figure 3 shows the median and envelope acceleration response spectra of the FEMA P695 far-field ground motion set scaled to the three intensity levels along with the corresponding target ASCE 7–22 spectra.

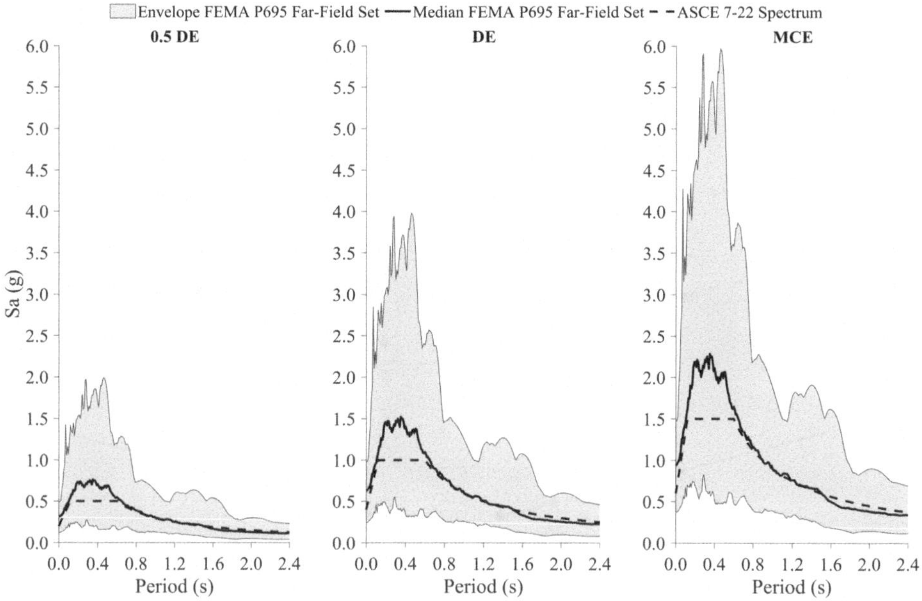

Fig. 3. Median and envelope acceleration response spectra of the FEMA P695 far-field ground motion set scaled to the three intensity levels and the corresponding ASCE 7–22 spectra.

3 Nonlinear Time-History Analysis Results

The results of the nonlinear time-history analyses were assessed in terms of median peak absolute axial forces and median peak absolute moments for each pipe element at three sections: both ends and the middle point. As shown in Fig. 1, configuration A is composed of two pipe elements, configuration B is composed of four pipe elements, and configuration C is composed of six pipe elements. Figures 4, 5, 6, 7, 8 to 9 show the median peak axial force and median peak moment with their respective plus minus one standard deviation of the three case study tank-pipeline configurations. The median peak absolute values show that the geometry of the pipeline layout, as expected, can have a significant impact on the seismic demand, especially on the bending moments. Pipe elements closer to changes in direction exhibit larger bending moments; moreover, it can be observed that these elements presented peak moments in both directions as the middle span section tended to show rather smaller peak moment values compared to both ends of the pipe element. In addition, due to the highly rigid response of both tanks (i.e. impulsive periods smaller than 0.01 s), there were no significant differences in the seismic responses among the different analyzed intensities as their response is dominated by the peak ground acceleration rather than spectral acceleration.

Finally, regarding the axial force demand, the pipes connected to the tanks and those located where there was a change in the direction layout tended to exhibit larger axial force demands.

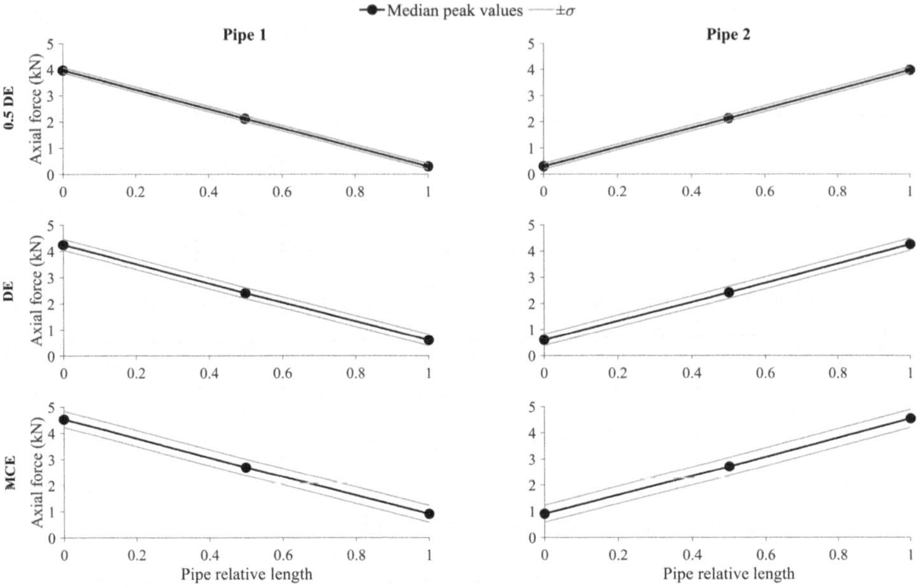

Fig. 4. Median peak axial forces of the pipelines of configuration A.

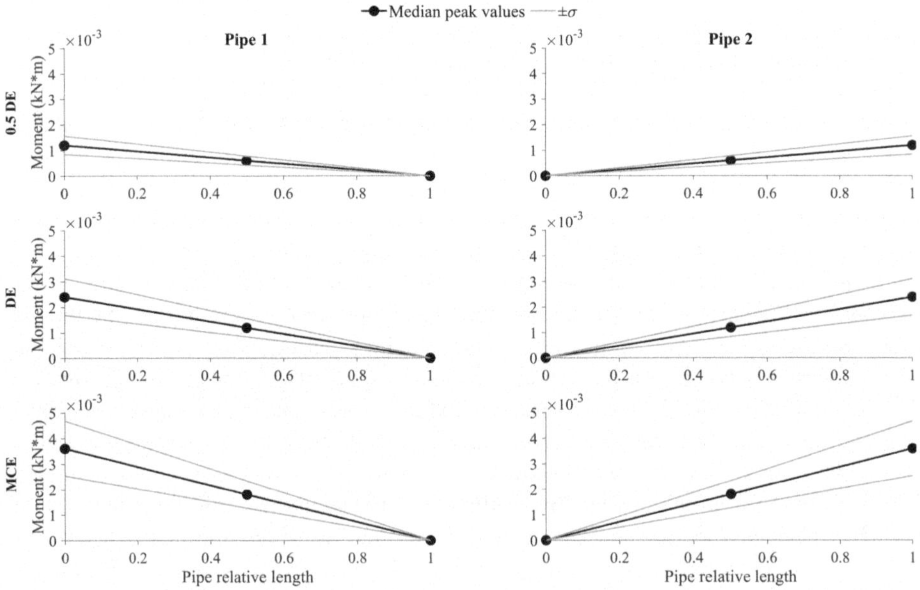

Fig. 5. Median peak moments of the pipelines of configuration A.

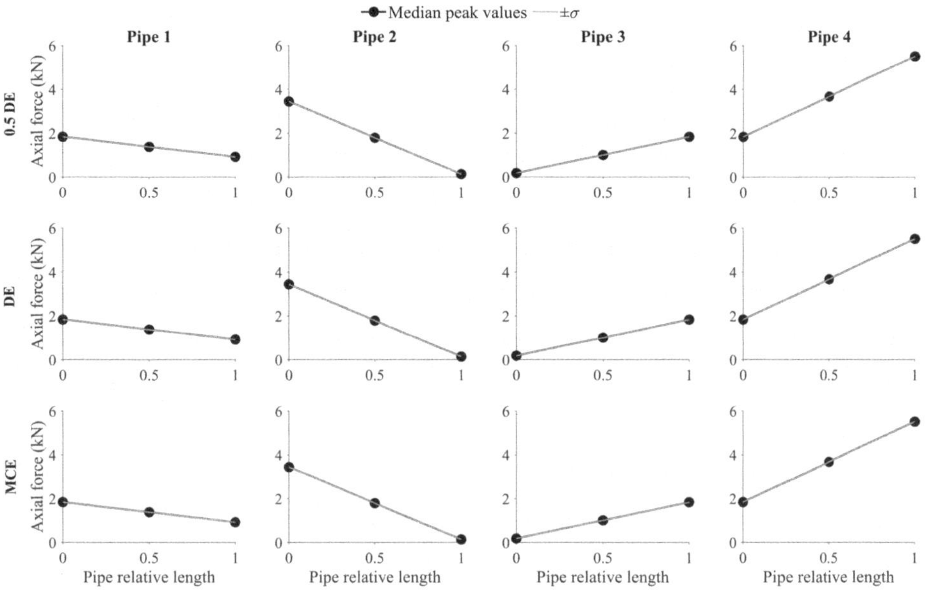

Fig. 6. Median peak axial forces of the pipelines of configuration B.

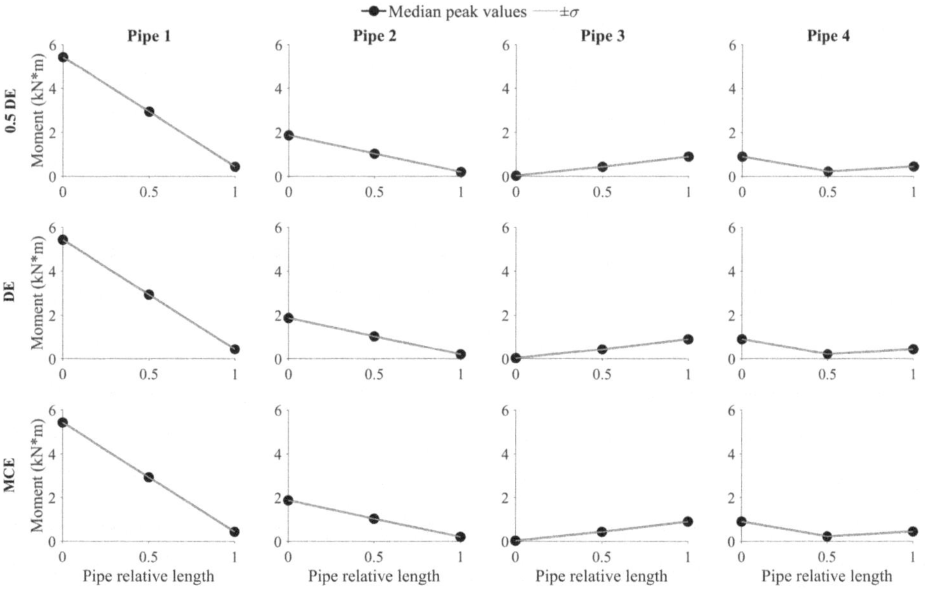

Fig. 7. Median peak moments of the pipelines of configuration B.

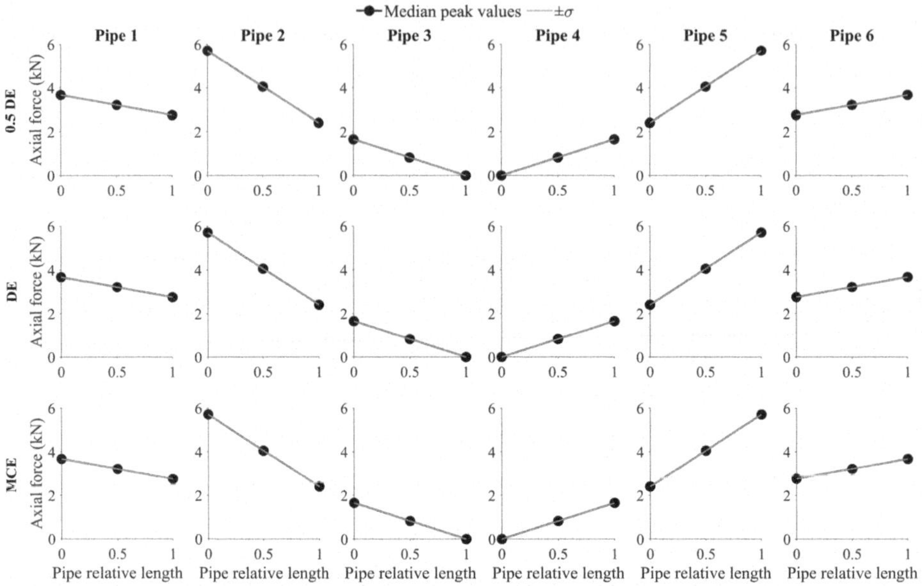

Fig. 8. Median peak axial forces of the pipelines of configuration C.

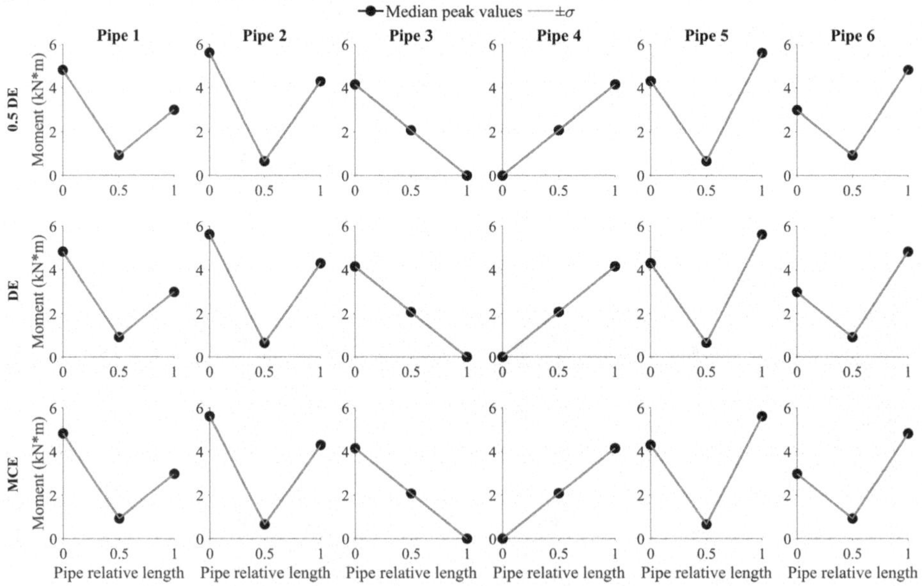

Fig. 9. Median peak moments of the pipelines of configuration C.

4 Conclusions

Na-Tech disasters derived from seismic events can significantly affect the resilience capacity of society and communities. In this regard, industrial plants, specifically steel liquid storage tanks and their corresponding pipeline connections and layouts, are key components for accurate and manageable modelling procedures due to their seismic vulnerability and complexity. This study explored the implementation of several modelling simplifications for the seismic assessment of tank-pipeline interactions through the evaluation of three case study setups subjected to nonlinear time-history analyses under three intensity levels. The results showed that the geometrical layout configuration can significantly impact the seismic demand, especially in terms of bending moment on the pipe elements. It was also observed that the displacements of both tanks were rather small. Although it can be explained by the small period that governs the impulsive mass of the selected tanks, further evaluation of the inclusion of the convective behaviour should be carried out to improve the accuracy of the results.

Acknowledgments. The authors gratefully acknowledge the National Institute for Insurance against Accidents at Work (INAIL) for their financial contributions to the MITPLANT project through the "Bando Bric 2022" framework.

References

1. Van Loenhout, J., Below, R.: Human cost of Disasters: An Overview of the Last 20 years 2000–2019, Centre of Research on the Epidemiology of Disasters – United Nations Office for Disaster Risk Reduction. Belgium, Brussels (2021)
2. Bevere, L., Ewald, M., Wunderlich, S.A.: decade of major earthquakes: lessons for business, Swiss Re Institute, Zurich, Switzerland (2019)
3. Campedel, M., Analysis of major industrial accidents triggered by natural events reported in the principal available chemical accident databases. JRC Sci. Tech. Reports, JRC42281 (2008)
4. Miglietta, C., Perrone, D., Gabbianelli, G., Micelli, F., Ciucci, M.: Release of hazardous substance due to seismic loads: influence of pipe to tank connection in fragility estimation. Chem. Eng. Trans. **111**, 127–132 (2024). https://doi.org/10.3303/CET24111022
5. Grajales-Ortiz, C., Melissianos, V.E., Bakalis, K., Kohrangi, M., Vamvatsikos, D., Bazzurro, P. Relationships between earthquake-induced damage and material release for liquid storage tanks. In: Proceedings of the 18th World Conference on Earthquake Engineering. Presented at the 18th World Conference on Earthquake Engineering. Milan, Italy (2024)
6. FEMA. Quantification of building seismic performance factors FEMA P695. Federal Emergency Management Agency. Washington D.C., United States (2009)
7. Calvi, G.M., Nascimbene, R.: Seismic Design and Analysis of Tanks. Wiley (2023)
8. Bursi, O.S., Reza, M.S., Abbiati, G., Paolacci, F.: Performance-based earthquake evaluation of a full-scale petrochemical piping system. J. Loss Prev. Process Ind. **33**, 10–22 (2015). https://doi.org/10.1016/j.jlp.2014.11.004
9. EN 13480–3. Metallic Industrial Piping e Part 3: Design and Calculation (2002)
10. ASCE. Minimum design loads and associate criteria for buildings and other structures ASCE/SEI 7–22. American Society of Civil Engineers. United States (2021)

Fragility Curves of Deformed Ground-Supported Liquid Storage Tanks

Raihan Rahmat Rabi(✉) and Giorgio Monti

Sapienza University of Rome, Rome, Italy
raihan.rahmatrabi@uniroma1.it

Abstract. Liquid storage tanks are crucial in industrial infrastructure, but they face significant risks during seismic events. This paper presents a framework to assess the seismic vulnerability of ground-supported tanks by developing fragility curves using Incremental Dynamic Analysis (IDA). The approach combines finite element modelling with a simplified Double Degree of Freedom (2DoF) spring system in OpenSeesPy to capture the tank's dynamic response. Fragility curves are generated for four key failure modes: liquid spillage, base shear-induced buckling, overturning, and elephant foot buckling. These curves offer insights into failure probabilities across seismic intensities, supporting improved resilience measures for tanks in earthquake-prone areas.

Keywords: cylindrical tanks · vulnerability · fragility curves · seismic analysis

1 Introduction

Liquid storage tanks are essential to modern infrastructure, holding water, oil, and hazardous materials for uses ranging from firefighting to industrial processing. Given their global distribution, many are situated in seismically active regions and require research into earthquake-resistant designs and assessments to evaluate tank vulnerability [1–9].

Past seismic events have highlighted the vulnerability of storage tanks, especially to buckling and support failure under strong ground shaking. Studies [6, 10] have documented damage to small storage tanks in Italy (2012) and Chile (2010), identifying shell buckling and anchor failure as key issues. Additionally, Sezen and Whittaker [11] observed damage to larger storage tanks in the United States, Turkey, and Japan, where foundation instability and liquid sloshing also contributed to damage, causing structural harm and spillage.

The dynamic behavior of storage tanks under seismic loading has been a research focus for over 50 years. Foundational studies by Jacobsen [12] and Housner [13] introduced models to analyze tanks' dynamic responses using La Place's equation to maintain mass and momentum in the tank-liquid system. These studies proposed a two-component motion model—impulsive for the tank's lower part and convective for surface waves—now commonly used in seismic design codes, including API 650 [14], ASCE/SEI 7-10 [15], and NZSEE [16].

X. Lu et al. (Eds.): EOS 2024, LNCE 638, pp. 152–166, 2025.
https://doi.org/10.1007/978-3-031-90690-9_14

Fragility curves, which predict the probability of structural damage under different seismic intensities, are widely used in assessing seismic vulnerability. However, while fragility curves for buildings and bridges are well-developed [17–19], they remain less common for industrial structures like tanks. Early tank fragility assessments relied on observed earthquake damages, with ALA [20, 21] and HAZUS [22] providing some initial empirical fragility curves.

Recent studies have refined these models. For example, Iervolino et al. [23] developed fragility curves for unanchored tanks, highlighting the influence of tank geometry. Berahman and Behnamfar [24] used Bayesian methods to evaluate elephant foot buckling as a failure mode, while Phan et al. [25] incorporated uncertainty factors into fragility models through Gaussian process regression. These works reflect growing knowledge of tank fragility and highlight the importance of design improvements in seismic areas.

Many studies have traditionally assumed idealized tank geometries, overlooking the effects of prior seismic deformations, which can alter tank stiffness, mass distribution, and natural periods, impacting seismic responses. To address this, this study proposes a framework integrating a finite element model (FEM) of the deformed tank shell with a simplified double-degree-of-freedom (2DOF) system. This approach provides a more realistic vulnerability assessment by capturing modified dynamic properties based on the tank's actual geometry, thus offering valuable insights for seismic risk mitigation. Figure 1 shows the proposed framework consisting of three phases for developing fragility curves of ground-supported cylindrical tanks with pre-existing deformations.

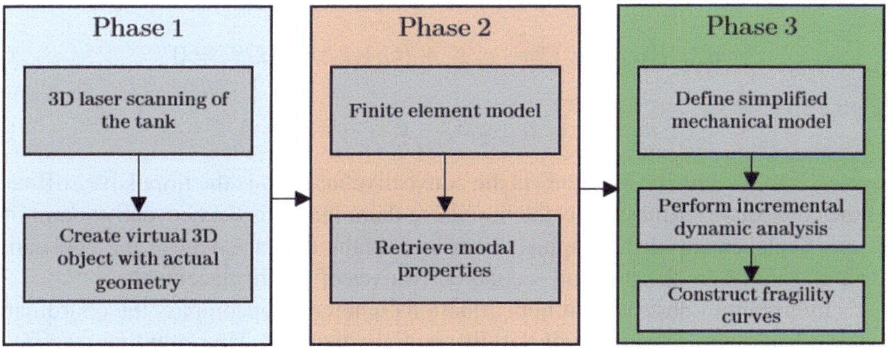

Fig. 1. Proposed framework for developing fragility curves of the fluid storage tank

2 Seismic Analysis of Liquid Storage Tanks

2.1 Mechanical Model

Fluid-structure interaction introduces considerable computational challenges, especially when using finite element models to explicitly simulate fluid confinement and tank shell contact dynamics. Despite advances in computational power, these simulations are resource-intensive, particularly for performance-based engineering (PBEE) frameworks, which typically require more analyses than conventional code-based approaches like CEN [26]. While code-based methods may involve limited nonlinear response history analyses, PBEE often demands many more, highlighting the need for efficient surrogate models.

For fragility curve development, a simplified mass-spring model is used in place of a 3D finite element model, as shown in Fig. 2. This stick-model, consisting of two springs and two masses, represents the impulsive and convective components of the tank's dynamic response efficiently.

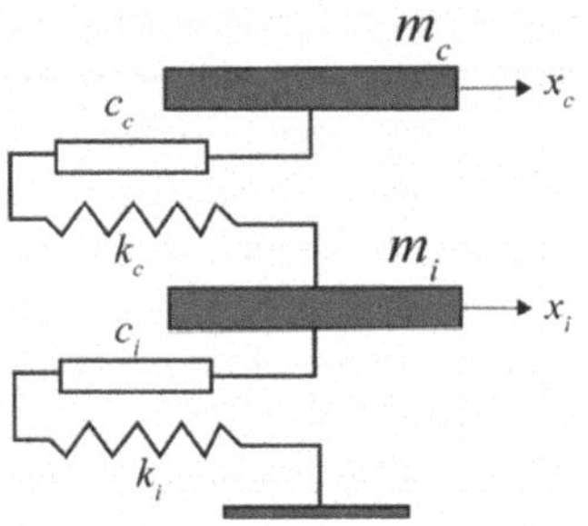

Fig. 2. Schematic of the two degree of freedom system (2-DOFs) representing impulsive and convective components of the tank

The equation of motion of the 2-DOFs system is written as:

$$m_i\ddot{x}_i + (c_i + c_c)\,\dot{x}_i - c_c\,\dot{x}_c + (k_i + k_c)x_i - k_c x_c = 0 \tag{1}$$

$$m_c\,\ddot{x}_c - c_c\,\dot{x}_i + c_c\dot{x}_c - k_c x_i - k_c x_c = 0 \tag{2}$$

where m_i is the impulsive mass, m_c is the convective mass, k_i is the impulsive stiffness, k_c is the convective stiffness, c_i is the impulsive damping, c_c is the convective damping, x_i is the displacement of the impulsive mass, x_c is the displacement of the convective mass and, $\dot{x}$ and $\ddot{x}$ are the first and second derivatives of the displacements.

It's important to observe that both equations inherently encompass the coordinates x_i and x_c, leading to a coupling of the differential equations. This coupling signifies a significant interdependence between the movements of these masses. Conceptually, this implies that the motion of one mass substantially influences the motion of the other. To facilitate a more structured representation, Eq. (1) and (2) are written into a matrix format:

$$[m]\{\ddot{x}(t)\} + [c]\{\dot{x}(t)\} + [k]\{x(t)\} = 0 \tag{3}$$

where:

$$[m] = \begin{bmatrix} m_i & 0 \\ 0 & m_c \end{bmatrix}$$

$$[c] = \begin{bmatrix} c_i + c_c & -c_c \\ -c_c & c_c \end{bmatrix}$$

$$[k] = \begin{bmatrix} k_i + k_c & -k_c \\ -k_c & k_c \end{bmatrix}$$

$$\{x\} = \begin{Bmatrix} x_i \\ x_c \end{Bmatrix}$$

The mass, damping, and stiffness matrices consistently maintain a square structure, possessing dimensions aligned with the number of degrees of freedom.

The equation of motion of the 2d of system given in Eq. (3) represents the fluid tank behavior.

2.2 Hydrodynamic Analysis

The seismic analysis of liquid storage tanks is complex due to fluid-structure interactions, where seismic energy transfers from the ground through the tank to the liquid. This process causes a dual response: one portion of the liquid moves with the tank walls, while the upper portion exhibits sloshing, characterized by minimal lateral movement along the tank wall [27].

The hydrodynamic response can be divided into two main components: impulsive and convective motions. In impulsive motion, the liquid moves rigidly with the tank, following wall acceleration. In contrast, convective motion involves long-period oscillations, creating vertical movement on the fluid's surface [28]. This separation aids in lateral tank analysis, as impulsive pressures stem from wall inertia forces, while convective pressures arise from fluid oscillations, influenced by impulsive effects.

The fluid pressure p is modelled as sum of the three components: hydrostatic pressure p_h, and hydrodynamic pressure made up of impulsive p_i and convective pressure p_c.

$$p = p_h + p_i + p_c \tag{4}$$

The estimation of the convective component of hydrodynamic pressure employs a cylindrical coordinate system, integrating non-dimensional coordinates (see Fig. 3):

$$\xi = r/R \tag{5}$$

$$\varsigma = z/H \tag{6}$$

This approach, encapsulated by the expressions below, enables a comprehensive representation of the convective pressure's characteristics [29]:

$$p_c(\xi, \varsigma, \theta) = C_{cn}(\xi, \varsigma)\rho R\ddot{u}_{cn}cos\theta \tag{7}$$

$$C_{cn}(\xi, \varsigma) = \sum_{n=1}^{\infty} \frac{2\cosh(\lambda_n\gamma\varsigma)J_1(\lambda_n\xi)}{(\lambda_n^2 - 1)J_1(\lambda_n)\cosh(\lambda_n\gamma)} \tag{8}$$

where J_1 is the first order Bessel function, γ is the ratio of filling fluid height to the tank's radius (H/R), λ_n is the *n-th* zeros of the Bessel function of the first kind and first

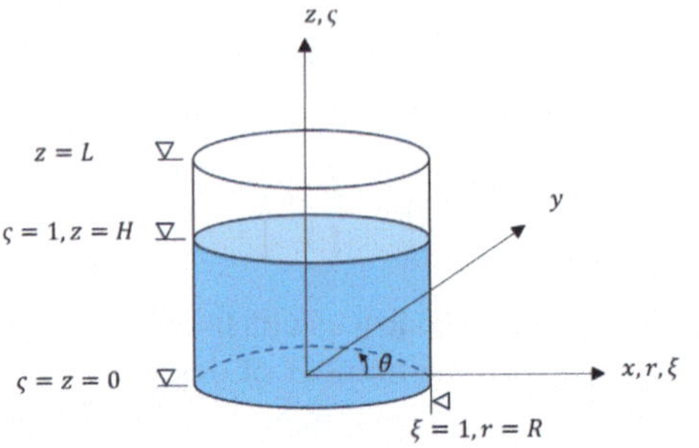

Fig. 3. Coordinate system of cylindrical tanks

order. The first three positives of roots are $\lambda_1 = 1.841$, $\lambda_2 = 5.331$ and $\lambda_3 = 8.536$. $\ddot{u}_{cn}$ is the time history of the response of convective mass mode. Furthermore, the impulsive flexible distribution of pressure on the tank wall could be obtained as follows [30]:

$$p_i(z, \theta, t) = c_0'(z)\rho H \ddot{u}_i cos\theta \tag{9}$$

where $c_0'(z)$ is a dimensionless function that defines the height-wise distribution of the impulsive hydrodynamic component, and $\ddot{u}_i$ is the time history of the response of impulsive mass mode. According to Veletsos and Yang [30], $c_0'(z)$ depends on ξ, ς the assumed deflection configuration function and the ratio of the tank mass to the contained liquid mass. The calculation method is detailed in Veletsos and Yang [30].

2.3 Failure Modes

Elastic Buckling
Elastic buckling typically occurs either near the base or at some distance above the bottom of a tank. When elastic buckling takes place, the affected region tends to bend inward, creating multiple wrinkles along the circumference of the tank wall. This type of buckling is primarily caused by significant axial compression forces, which result from the overturning moment due to seismic actions, as well as the self-weight of both the tank wall and the roof. Euler's critical buckling stress for cylinders subjected to axial compression at the elastic level is given by the following expression [26, 31]:

$$\sigma_e = \frac{E_s t_w}{R \cdot \sqrt{3(1 - \upsilon^2)}} \tag{10}$$

where E_s is the modulus of elasticity, t_w is the thickness of the tank wall, R is the radius and υ is the Poisson's ratio.

The critical overturning moment (OM_{cr}) and the structural base shear (Fs_{cr}) associated with the elastic buckling of the tank wall are expressed as follows [31]:

$$OM_{cr} = \sigma_e \pi R^2 t_w \tag{11}$$

$$FS_{cr} = \tau_e \pi R t_w \tag{12}$$

where τ_e is elastic shear buckling stress and it can be obtained using the following equation given in [31]:

$$\tau_e = 0.07708 \frac{\pi^2 E_s}{\left(1 - \upsilon^2\right)^{\frac{5}{8}} \sqrt{\frac{H}{R}}} \left(\frac{R}{t_w}\right)^{-5/4} \tag{13}$$

Elephant Foot Buckling

Elephant's foot buckling, also referred to as elastic-plastic buckling, is among the most common failure modes in circular steel storage tanks. This phenomenon occurs due to the high concentration of compressive meridian stress that arises in the tank shell when the base lifts off the ground support. During seismic events, hydrostatic and hydrodynamic pressures acting on the tank walls increase the internal pressure, contributing to local yielding and elastic-plastic buckling (*elephant's foot buckling*), as illustrated in Fig. 4. Furthermore, this type of buckling extends around the tank perimeter near the lower section due to the critical combination of vertical compressive stress, hoop tensile stress, and high shear stress during seismic loading.

Fig. 4. EFB of storage tank during Northridge earthquake in California [32].

The critical buckling limit related to elastic–plastic buckling is defined as [26, 33]:

$$\sigma_{cr} = \frac{\sigma_e}{r+1}\left[1 - \left(\frac{pR}{t_w f_y}\right)^2\right]\left(1 - \frac{1}{1.12 + r^{1.5}}\right)\left(r + \frac{f_y}{250}\right) \tag{14}$$

where σ_e is the Euler's critical buckling stress at elastic level in Eq. (10), f_y is the yield stress of the tank wall, $r = R/(400t_w)$ and p is the total internal pressure acting on the tank wall in Eq. (4).

The elephant foot buckling occurs when the meridian stress (σ_m) in tank wall exceeds the critical buckling limit given in Eq. (14). The meridian stress (σ_m) is calculated using the following equation:

$$\sigma_m = \frac{OM}{\pi R^2 t_w} + \frac{W}{2\pi R t_w} \tag{15}$$

where OM is the overturning moment and W is the total weight of the tank.

Sloshing

Sloshing motion is triggered by the vertical displacement of the liquid mass during seismic activity. In tanks with inadequate freeboard, sloshing can exert upward pressure on the roof and may rupture the connection between the roof and tank walls, leading to potential spillage of the tank's contents. The freeboard height (H_f) is the distance between the height of the liquid surface (H) and the roof of the tank (L) (see Fig. 3):

$$H_f = L - H \tag{16}$$

According to API 650 [14] the height of the sloshing wave (δ_s) can be calculated as:

$$\delta_s = R \cdot \frac{S_a(T_c)}{g} \tag{17}$$

where R is the radius of the tank, $g = 9.82 m/sec^2$ is the acceleration of gravity and $S_a(T_c)$ is the spectral acceleration for the site hazard at the convective period, T_c.

3 Fragility Curves and the Damage States

A fragility curve (FC) describes the probability of exceeding a given threshold representing a damage state (DS) in terms of engineering demand parameter (EDP) for a given intensity measure (IM):

$$P[EDP \geq DS|IM] = \frac{N_{fail}}{N_{sim}} \tag{18}$$

where, if expressed in statistical terms, N_{fail} is the total number of times demand exceeded capacity, and N_{sim} is the total number of simulations.

Fragility curves can be represented using a lognormal cumulative distribution function. Hence, the probability of the damage exceeding a Damage State for a given intensity measure (IM) has been assumed as:

$$P(D \geq DS|IM) = \Phi\left(\frac{lnX - \mu_{lnX}}{\beta_{lnX}}\right) \tag{19}$$

where Φ is the standard cumulative distribution function, lnX is the natural logarithm of the variable X (spectral acceleration or PGA for which the Damage State (DS) is attained) and μ_{lnX} and β_{lnX} are the mean and the standard deviation of the natural logarithms of the distribution of X respectively.

Table 1. Description of the Damage States

Damage State	Description	Capacity	Demand	Fragility function
DS_{SL}	Liquid overspill	H_f in Eq. (16)	δ_s in Eq. (17)	$P(\delta_s > H_f \mid IM)$
DS_{BS}	Exceedance of critical base shear	FS_{cr} in Eq. (12)	base shear (FS)	$P(FS > FS_{cr} \mid IM)$
DS_{OM}	Exceedance of critical overturning moment	OM_{cr} in Eq. (11)	Overturning moment (OM)	$P(OM > OM_{cr} \mid IM)$
DS_{EFB}	Exceedance of critical buckling limit	σ_{cr} in Eq. (14)	σ_m in Eq. (15)	$P(\sigma_m > \sigma_{cr} \mid IM)$

The fragility curves for the liquid storage tanks are developed based on the selected demand, capacity, and intensity measures, and are described in Table 1.

Incremental dynamic analysis (*IDA*) is a parametric analysis method that has recently emerged in several different forms to estimate more thoroughly structural performance under seismic loads. It involves subjecting a structural model to ground motion record, each scaled to multiple levels of intensity, thus producing parameterized response curves versus intensity level. The flow chart of the procedure to obtain FCs using *IDA* is shown in Fig. 5.

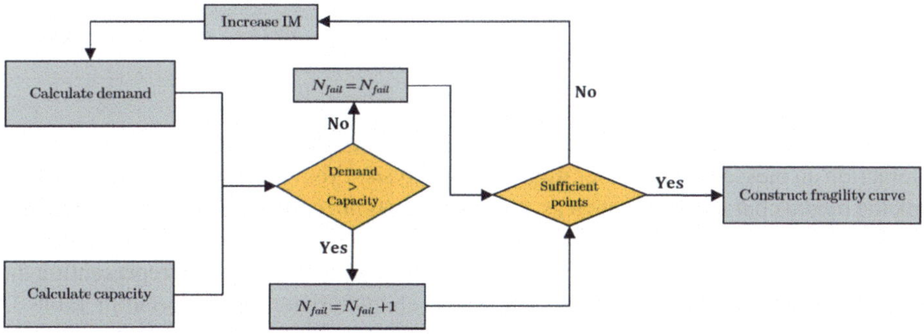

Fig. 5. Fragility curve development procedure using Incremental Dynamic Analysis (*IDA*)

4 Case Study

4.1 Finite Element Modelling

A cylindrical, ground-supported fluid storage tank was selected as a representative case study, reflecting typical design parameters used in petroleum depots. The geometrical and mechanical properties of the tank are detailed in Table 2. The tank has a diameter of 54.9 m and a height of 14.6 m. The thickness of the tank wall varies from 21 mm at the base to 9 mm at the top.

In this study, a detailed finite element (*FE*) model of the liquid storage tank was developed using ABAQUS software [34], as illustrated in Fig. 6. The tank was modeled in ABAQUS [34] using S4R shell elements, a four-node type with six degrees of freedom per node, allowing for both bending and membrane behavior. The liquid inside was represented with potential-based fluid elements, enabling automatic fluid-structure coupling. Figure 6 shows details of the finite element model, covering key structural parts like the roof, bottom, and cylindrical walls, excluding nonstructural elements such as nozzles, piping, and stairs.

A 3D laser scan was conducted to capture the tank's actual shell shape, identifying deformations for accurate finite element modeling. This virtual model supports the analysis of the tank's dynamic response, with Fig. 6 displaying wall deformations obtained from the scan.

Table 2. Mechanical and geometrical properties of case study cylindrical tank

Diameter $D(m)$	Height $L(m)$	Fluid height $H(m)$	Wall thickness $t(mm)$	Fluid density $\rho(kg/m^3)$	Steel strength $f_y(MPa)$	Elastic modulus $E_s(MPa)$
54.9	14.6	13.5	21–9	980	235	210000

The tank is assumed to be located on a rock foundation, rendering soil-structure interaction an inessential consideration in the analysis. From the dynamic analysis of the tank the modal characteristics such as, impulsive mass m_i, convective mass m_c, impulsive stiffness k_i, convective stiffness k_c, impulsive frequency f_i and convective frequency f_c are retrieved, which are used to define the two degree of freedom system representing the simplified mechanical model of the tank. These characteristics are reported in Table 3.

4.2 Selection of Ground Motion Records

A critical step in assessing seismic vulnerability is selecting appropriate earthquake ground motions for incremental dynamic analysis, considering both the type and quantity of records. Pulse-like ground motions, with characteristics like long-period velocity pulses and permanent displacement, exert greater demands on structures than non-pulse-like motions due to impulsive effects typical in near-fault regions.

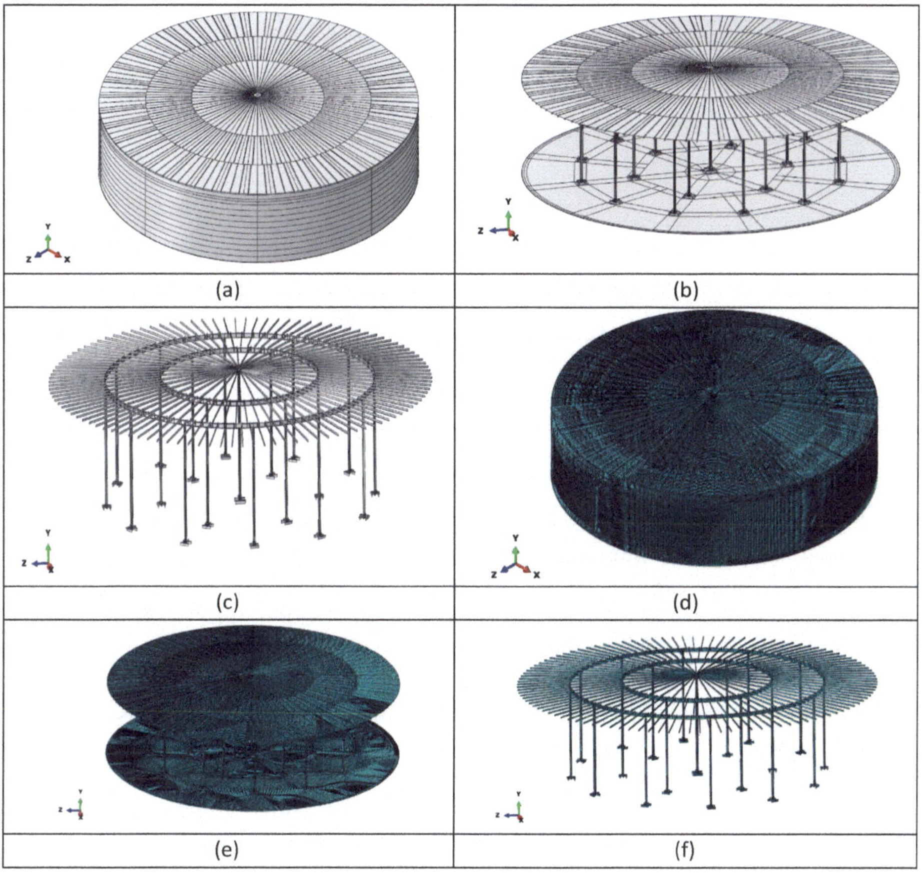

Fig. 6. Deformation modeling of the existing tank; a) color map of the deformed tank shell; b) top view; c) profile

Table 3. Summary of the impulsive and convective component characteristics

$f_i(HZ)$	$f_c(HZ)$	$m_i(ton)$	$m_c(ton)$	$k_i(kN/m)$	$k_c(kN/m)$
6.39	0.11	9391	22181	2159800	10500

Fragility function accuracy improves with the number and type of ground motions. For mid-rise buildings, 10–20 records have been shown to yield reliable results [35]. Following this approach, this study uses 28 records from the NGA-West2 [36] database, scaled in increments of 0.05 g from 0.005 g to 2.0 g PGA. Figure 7 presents the pseudo-acceleration response spectra and their average at 2% damping for these motions.

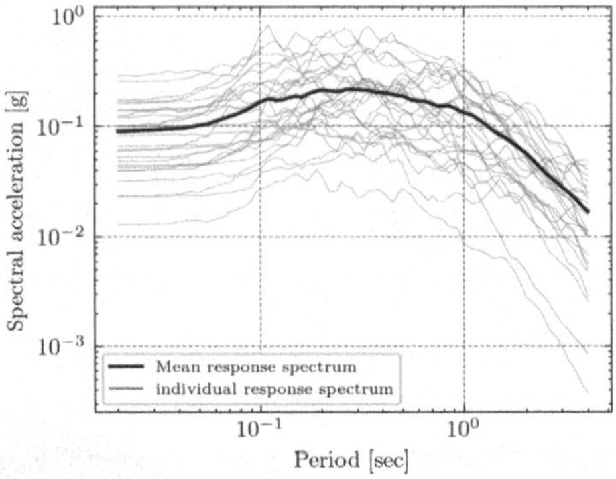

Fig. 7. Response spectra of selected ground motions.

4.3 Results and Discussion

A double-degree-of-freedom model capturing impulsive and convective properties from FEM modal analysis was implemented in OpenSeesPy [37] for Incremental Dynamic Analysis (IDA) under various ground motions. Each ground motion was incrementally scaled from a PGA of 0.005 g to 2.0 g in 0.05 g steps, resulting in 11,200 simulations. IDA with 28 accelerograms evaluated critical response parameters, including sloshing height, base shear, overturning moment, and meridional stress.

The results, as shown in Fig. 8, illustrate how different failure mechanisms emerge under increasing IM levels. Specifically, the analysis reveals that the elastic buckling of the tank wall, caused by base shear, and the so-called "elephant foot" buckling, are the primary governing failure modes for liquid storage tanks. These failure mechanisms are critical for understanding how tanks behave during earthquakes and for ensuring their structural integrity. Fragility curves (Fig. 9) were generated for the tank's Engineering Demand Parameters (EDPs), providing a probabilistic assessment of failure risk. The analysis shows elastic wall buckling, induced by shear forces, as the most likely failure mode, with "elephant foot" buckling as the second. The steep fragility curve for wall buckling indicates high sensitivity to seismic intensity increases, where even a slight PGA rise sharply raises failure probability.

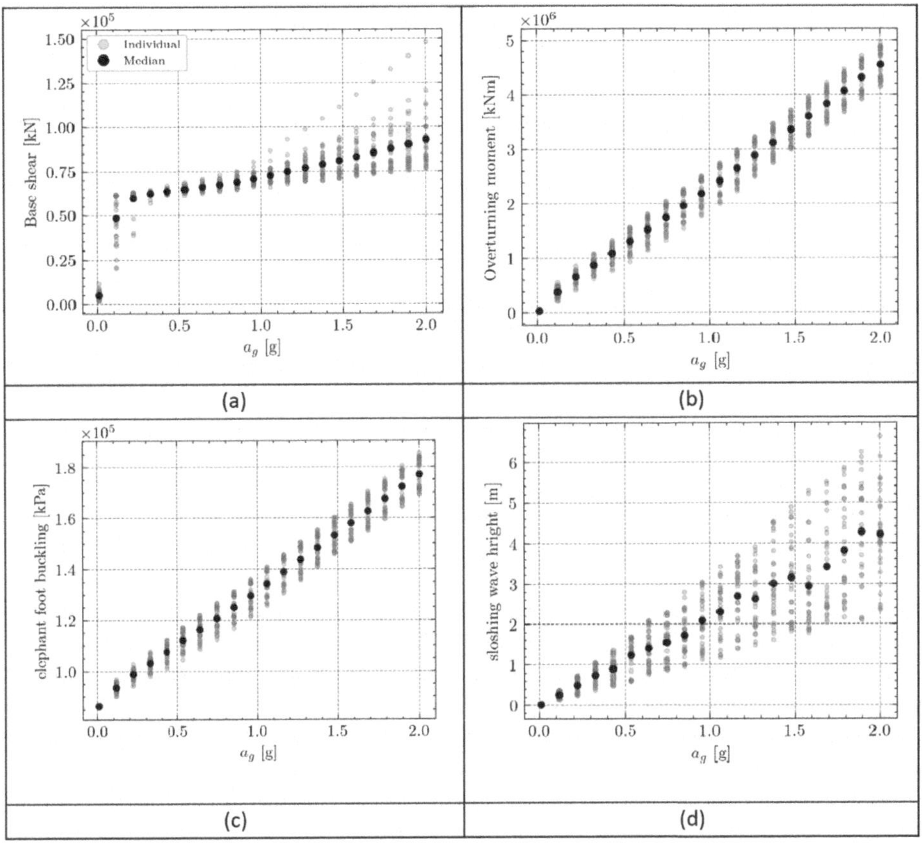

Fig. 8. Results of incremental dynamic analysis from 28 accelerograms: a) base shear; b) overturning moment; c) elephant foot buckling; d) sloshing wave height

Table 4 summarizes the median (μ) and standard deviation (β) of fragility curves, essential for understanding the tank's response distribution under seismic demand. The median represents a 50% failure probability intensity, while β indicates performance variability, aiding in risk mitigation planning for earthquake resilience.

Table 4. Mean (μ) and standard deviation (β) of the Fragility Curves

DS_{SL}		DS_{FS}		DS_{OM}		DS_{EFB}	
μ	β	μ	β	μ	β	μ	β
0.461	0.149	0.118	0.056	1.141	0.126	0.249	0.058

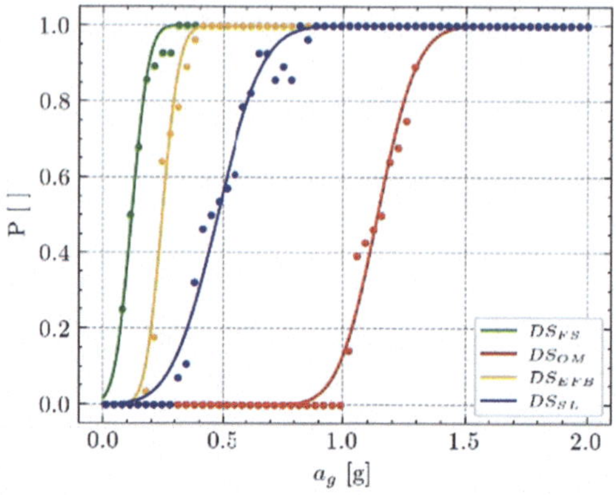

Fig. 9. Resulting fragility curves for the case study cylindrical tank

5 Conclusions

This study presents a refined framework for assessing the seismic vulnerability of cylindrical liquid storage tanks with pre-existing deformations, combining finite element analysis (FEA) with a simplified mechanical model. This approach significantly reduces computational costs while maintaining accuracy, facilitating the derivation of fragility curves based on incremental dynamic analysis (IDA). Applied to an existing tank, the framework produced fragility curves for four critical failure modes: liquid overflow, elastic buckling due to base shear, overturning moment failure, and elephant foot buckling.

The results highlight elastic buckling due to base shear and elephant foot buckling as the primary failure risks. Elastic buckling results from lateral seismic forces deforming the tank shell, while elephant foot buckling occurs at the tank's lower portion where stress concentrates. The risk of liquid overflow is minimal due to sufficient freeboard, and overturning failure is unlikely, as the tank's height-to-radius ratio below 1 classifies it as a stable "squat" tank, less susceptible to overturning.

This framework provides an efficient method for evaluating and retrofitting deformed liquid storage tanks, in the direction of improving seismic resilience.

References

1. De Angelis, M., Giannini, R., Paolacci, F.: Experimental Investigation on the seismic response of a steel liquid storage tank equipped with floating roof by shaking table tests. Earthq. Eng. Struct. Dyn. **39**, 377–396 (2010). https://doi.org/10.1002/EQE.945
2. Phan, H.N., Paolacci, F., Di Filippo, R., Bursi, O.S.: Seismic vulnerability of above-ground storage tanks with unanchored support conditions for na-tech risks based on gaussian process regression. Bull. Earthq. Eng. **18**, 6883–6906 (2020). https://doi.org/10.1007/S10518-020-00960-7/TABLES/7
3. Alessandri, S., Caputo, A.C., Corritore, D., Giannini, R., Paolacci, F., Phan, H.N.: Probabilistic risk analysis of process plants under seismic loading based on monte carlo simulations. J. Loss Prev. Process Ind. **53**, 136–148 (2018). https://doi.org/10.1016/J.JLP.2017.12.013

4. Bakalis, K., Vamvatsikos, D., Fragiadakis, M.: Seismic risk assessment of liquid storage tanks via a nonlinear surrogate model. Earthq. Eng. Struct. Dyn. **46**, 2851–2868 (2017). https://doi.org/10.1002/EQE.2939
5. Bakalis, K., Kohrangi, M., Vamvatsikos, D.: Seismic intensity measures for above-ground liquid storage tanks. Earthq. Eng. Struct. Dyn. **47**, 1844–1863 (2018). https://doi.org/10.1002/EQE.3043
6. Colombo, J.I., Almazán, J.L.: Experimental investigation on the seismic isolation for a legged wine storage tank. J. Constr. Steel Res. **133**, 167–180 (2017). https://doi.org/10.1016/J.JCSR.2017.02.013
7. Merino Vela, R.J., Brunesi, E., Nascimbene, R.: Derivation of floor acceleration spectra for an industrial liquid tank supporting structure with braced frame systems. Eng. Struct. **171**, 105–122 (2018). https://doi.org/10.1016/J.ENGSTRUCT.2018.05.053
8. Sezen, H., Livaoglu, R., Dogangun, A.: Dynamic analysis and seismic performance evaluation of above-ground liquid-containing tanks. Eng. Struct. **30**, 794–803 (2008). https://doi.org/10.1016/J.ENGSTRUCT.2007.05.002
9. Campedel, M., Cozzani, V., Garcia-Agreda, A., Salzano, E.: Extending the quantitative assessment of industrial risks to earthquake effects. Risk Anal. **28**, 1231–1246 (2008). https://doi.org/10.1111/J.1539-6924.2008.01092.X
10. Brunesi, E., Nascimbene, R., Pagani, M., Beilic, D.: Seismic performance of storage steel tanks during the may 2012 Emilia, Italy, earthquakes. J. Perform. Constr. Facil. **29**, 04014137 (2014). https://doi.org/10.1061/(ASCE)CF.1943-5509.0000628
11. Sezen, H. Whittaker, A.S.: Seismic performance of industrial facilities affected by the 1999 Turkey earthquake. **20**(1), 28–36 (2006). https://doi.org/10.1061/(ASCE)0887-3828(2006)20:1(28)
12. Jacobsen, L.S.: Impulsive hydrodynamics of fluid inside a cylindrical tank and of fluid surrounding a cylindrical pier. Bull. Seismol. Soc. Am. **39**, 189–204 (1949). https://doi.org/10.1785/BSSA0390030189
13. Housner, G.W.: Dynamic pressures on accelerated fluid containers. Bull. Seismol. Soc. Am. **47**, 15–35 (1957). https://doi.org/10.1785/BSSA0470010015
14. API 650 Welded Tanks for Oil Storage. American Petroleum Institute. Washington, DC. (2010)
15. ASCE ASCE/SEI 7–16 Minimum Design Loads and Associated Criteria for Buildings and Other Structures. American Society of Civil Engineers (2016)
16. NZSEE Recommendations for Seismic Design of Storage Tanks. New Zealand Society for Earthquake Engineering (2009)
17. Vamvatsikos, D., Cornell, C.A.: Applied incremental dynamic analysis. **20**(2), 523–553 (2004). https://doi.org/10.1193/1.1737737
18. Alavi, B., Krawinkler, H.: Behavior of moment-resisting frame structures subjected to near-fault ground motions. **33**(6), 687–706 (2004). https://doi.org/10.1002/eqe.369
19. Choi, E., DesRoches, R., Nielson, B.: Seismic fragility of typical bridges in moderate seismic zones. Eng. Struct. **26**, 187–199 (2004). https://doi.org/10.1016/J.ENGSTRUCT.2003.09.006
20. Alliance AL. Seismic Fragility Formulation for Water Systems: Part 2- Appendices. ASCE (2001)
21. Alliance AL. Seismic Fragility Formulation for Water Systems: Part 1-Guideline. ASCE (2001)
22. Agency, F.E.M.: Multi-Hazard Loss Estimation Methodology, Earthquake Model. HAZUS-MH MR4 Technical Manual. Washington, D. C (2003)
23. Iervolino, I., Fabbrocino, G., Manfredi, G.: Fragility of standard industrial structures by a response surface based method. J. Earthquake Eng. **8**, 927–945 (2004). https://doi.org/10.1142/S1363246904001717

24. Berahman, F., Behnamfar, F.: Probabilistic seismic demand model and fragility estimates for critical failure modes of un-anchored steel storage tanks in petroleum complexes. Probab. Eng. Mech. **24**, 527–536 (2009). https://doi.org/10.1016/J.PROBENGMECH.2009.03.005
25. Phan, H.N., Paolacci, F., Alessandri, S.: Enhanced seismic fragility analysis of unanchored steel storage tanks accounting for uncertain modeling parameters. J. Pressure Vessel Technol. Trans. ASME **141** (2019). https://doi.org/10.1115/1.4039635/366261
26. CEN Eurocode 8: Design of Structures for Earthquake Resistance – Part 4: Silos, Tanks and Pipelines. ENV-1998–4. Comité Européen de Normalisation. Brussels, Belgium. (2006)
27. Seleemah, A.A., El-Sharkawy, M.: Seismic response of base isolated liquid storage ground tanks. Ain Shams Eng. J. **2**, 33–42 (2011). https://doi.org/10.1016/J.ASEJ.2011.05.001
28. Moslemi, M., Kianoush, M.R., Pogorzelski, W.: Seismic response of liquid-filled elevated tanks. Eng. Struct. **33**, 2074–2084 (2011). https://doi.org/10.1016/J.ENGSTRUCT.2011.02.048
29. Standardization, E.: Eurocode 8-Design of Structures for Earthquake Resistance-Part 4: Silos, Tanks and Pipelines. Specific principles and application rules for tanks (2006)
30. Veletsos, A.S., Auyang, J.: Earthquake response of liquid storage tanks. Adv. Civil Eng. Through Eng. Mech. (1977)
31. Okada, J., Iwata, K., Tsukimori, K., Nagata, T.: An evaluation method for elastic-plastic buckling of cylindrical shells under shear forces. Nucl. Eng. Des. **157**, 65–79 (1995). https://doi.org/10.1016/0029-5493(95)00985-L
32. Malhotra, P. Practical Nonlinear Seismic Analysis of Tanks. **16**, 473–492 (2000). https://doi.org/10.1193/1.1586122
33. Rotter, J.M.: Local collapse of axially compressed pressurized thin steel cylinders. J. Struct. Eng. (United States) **116**, 1955–1970 (1990). https://doi.org/10.1061/(ASCE)0733-9445(1990)116:7(1955)
34. Dassault Systemes Simulia ABAQUS (2023)
35. Shome, N.: Probabilistic Seismic Demand Analysis of Nonlinear Structures; Stanford University (1999)
36. Pacific Earthquake Engineering Research Center PEER NGA Database. https://www.ngawest2.berkeley.edu/
37. Zhu, M., McKenna, F., Scott, M.H.: OpenSeesPy: python library for the opensees finite element framework. SoftwareX **7**, 6–11 (2018). https://doi.org/10.1016/J.SOFTX.2017.10.009

Masonry Structures and Geotechnical

A Specialized OSPy Tool for Territorial Seismic Risk Analysis of Masonry Building Clusters Using Typological-Mechanical Fragility Curves

Marco Vailati[1], Raihan Rahmat Rabi[2(✉)], and Giorgio Monti[2]

[1] University of L'Aquila, L'Aquila, Italy
[2] Sapienza University of Rome, Rome, Italy
raihan.rahmatrabi@uniroma1.it

Abstract. In historic Mediterranean towns, masonry building clusters (MBCs) are formed over centuries through gradual construction. These clusters reflect numerous deferred interventions, often with different materials and techniques, leading to the unique and complex structures we observe today. Recognizing this complexity and the absence of standardized regulations, the 2008 Italian code introduced guidelines for structural assessment of MBCs. These guidelines allow engineers to divide MBCs into separate Structural Units (SUs) and perform simplified nonlinear analyses on each unit to assess both individual and collective seismic performance. A few years later, Technical Note DT 212, issued by the National Research Council (CNR), expanded on these guidelines to assess the reliability of existing buildings, specifically addressing MBCs. This study builds on these frameworks, introducing a specialized tool within the OpenSeesPy platform for assessing MBCs' seismic reliability through fragility curves. Developed for efficient, large-scale seismic risk analysis, this tool integrates powerful Python libraries like Pandas and Matplotlib, enabling users to visualize territorial maps of input variables (*e.g.*, SU structural types) and damage scenarios. Validated through numerous studies, it provides an accessible, robust solution for MBC seismic risk assessment at territorial scale.

Keywords: Masonry Building Clusters (MBCs) · Seismic reliability · Structural Units (SUs) · Nonlinear assessment procedure · OpenSeesPy framework

1 Introduction

Masonry Building Clusters (MBCs), also referred to as masonry building aggregates, are defining elements of historical urban landscapes, especially in Mediterranean regions, shaped by centuries of gradual construction practices. These clusters exhibit a complex architectural patchwork due to subsequent interventions, each introducing varied construction techniques and materials, which complicate seismic risk assessment.

In 2008, the Italian regulatory framework introduced guidelines to address the seismic vulnerability of MBCs, marking a pivotal step in assessing these complex structures. These guidelines subdivide MBCs into distinct Structural Units (SUs) to allow

X. Lu et al. (Eds.): EOS 2024, LNCE 638, pp. 169–182, 2025.
https://doi.org/10.1007/978-3-031-90690-9_15

for individual nonlinear analyses, thereby enabling professionals to evaluate the seismic performance of each SU and provide a comprehensive assessment of the MBC as a whole [1–3]. While these guidelines offer a foundation, significant gaps remain in the standardized assessment process, particularly given the heterogeneous nature of MBCs. Over the past 15 years, numerous studies have explored various factors influencing the seismic response of Sus [4–8], including the development of specialized software with graphical interfaces [9]. However, the vulnerability assessment of MBCs is still impacted by high uncertainties due to limited data, and many practical scenarios require input parameters with considerable dispersion.

In the years following, the Italian National Research Council (CNR) issued Technical Note DT 212, introducing a framework for reliability assessment of existing buildings and specifically addressing MBCs. This note underscores the importance of a probabilistic approach to seismic risk evaluation, recognizing the sensitivity of certain response parameters that strongly influence seismic performance [10, 11] and advocating for the use of fragility curves in risk assessment [12–17]. Recent advancements, such as an archetype-based automated procedure for deriving global-local seismic fragility for masonry building aggregates [18], and the integration of multi-resolution data sources [19, 20], have further enriched the methodological toolkit available for practitioners, expanding the accuracy and applicability of seismic assessments.

This study seeks to advance understanding of MBC seismic reliability by developing a nonlinear procedure tailored to the unique complexities of these structures. Employing a typological-mechanical approach, the methodology begins with creating a representative abacus of typical openings for masonry buildings in an Italian historic center. The approach then incorporates Monte Carlo simulations, extending previous nonlinear methodologies into a probabilistic framework to derive fragility curves in closed form.

The entire procedure has been implemented in OSPy [21], a Python interface for OpenSees, proposed here to support user-friendly seismic risk assessments on a territorial scale. Leveraging powerful Python libraries like Pandas and Matplotlib, OSPy allows users to visualize input variables, such as the structural types of SUs, and generate damage scenario maps. This visualization is essential for comprehending the spatial distribution of seismic risk, aligning with recent studies that highlight the role of Geographical Information Systems (GIS) in risk analysis [22, 23].

Numerous publications [24, 25] highlight the capabilities of OpenSeesPy, which empowers engineers and researchers to analyze the seismic behavior of structures efficiently through Python, enabling advanced simulations of complex models.

Beyond facilitating territorial-scale seismic risk analysis, the developed tool also supports fragility curve generation for small clusters of existing buildings. This allows professionals to derive global fragility curves for individual Structural Units (SUs) and for systems of interconnected SUs by combining the responses of individual masonry walls and including in the model also the openings layout.

In summary, as urban landscapes evolve and seismic threats persist, tools like OSPy are critical for effective risk assessment and management. This study not only adds to the discourse on seismic risk in masonry buildings but also establishes a practical framework for future research and application in the field.

2 Basic Methodology for Seismic Assessment of MBCs

The seismic assessment of masonry building clusters (MBCs) relies on developing a capacity curve for each structural unit (SU) within the cluster, using a systematic step-by-step approach. This process begins with defining the capacity curve for individual masonry walls, which are modeled as a series system consisting of two primary resistance mechanisms. Each wall's capacity curve is then used to develop the capacity curve for each storey, conceptualized as a parallel system of walls that share the load-bearing requirements across the structure. Finally, by combining the storey-level capacity curves, the overall capacity curve for the structural unit is derived, represented through a simplified "stick" model, where the structural unit is seen as a series system of interconnected storeys.

The strength of this methodology lies in its ability to unify multiple analytical expressions within a coherent framework. Some expressions are drawn from the Italian building code, with additional refinements from recent studies, offering a clear understanding of the role played by different key parameters in seismic performance. One of the significant advantages of this approach is its reliance on an analytical, data-light framework that allows for effective modeling of masonry clusters even under low Level of Detail (LOD) conditions. By treating the structure as an assemblage of series and parallel subsystems, this approach provides a robust yet simplified model of the MBC's structural behavior, making it especially suited for large-scale, territorial studies. The model enables rapid modal analysis and efficient generation of fragility curves with minimal computational requirements, achievable even in simple spreadsheet programs.

As discussed further in the text, many of these complexities, such as SU interactions and torsional effects at the boundaries, are seldom addressed in low-LOD and large-scale studies. The proposed methodology, therefore, offers a pragmatic solution by focusing on the primary behaviors of MBCs and providing an efficient approach for seismic assessment that balances accuracy with computational efficiency.

3 Typological-Mechanical Approach to Derive Fragility Curves Using Closed-Form Equations

The following method outlines a systematic approach for deriving the seismic capacity curve of masonry structural units, progressing from individual wall capacities to an integrated model of the entire structural unit.

3.1 Wall Capacity Curve Including Openings

The capacity curve for a masonry wall is derived by treating it as a series system comprising two resisting mechanisms: shear and bending. The overall stiffness is then:

$$k_w^{-1} = \beta^{-1}\left(k_s^{-1} + k_b^{-1}\right) \tag{1}$$

where k_s and k_b represent the stiffnesses for shear and bending, defined as:

Table 1. Opening configurations on a masonry panel and variables used to compute the effective height H_{eff}. The width l and the partial height $h\prime$ refer to the grey-shaded portion of the wall.

<table>
<tr><th>Case</th><th>Configuration</th><th>Variables</th></tr>
<tr><td>1</td><td></td><td>
</td></tr>
<tr><td>2</td><td></td><td>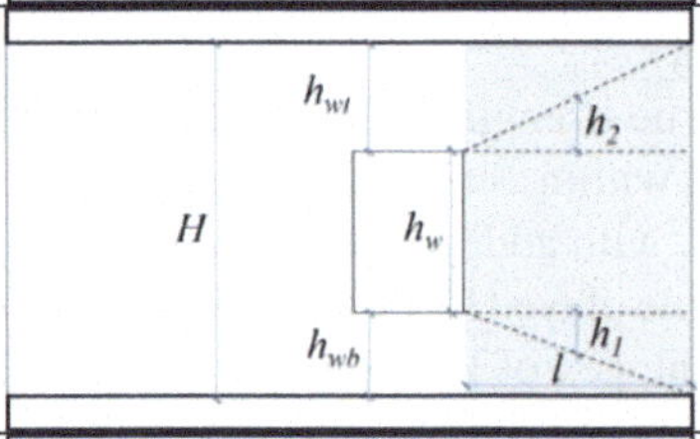
</td></tr>
<tr><td>3</td><td></td><td>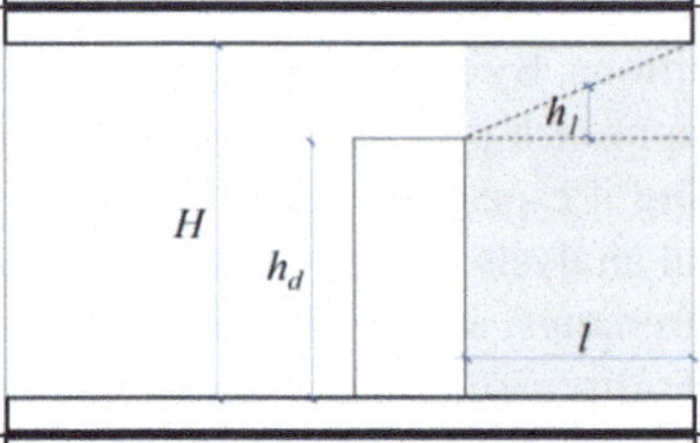
</td></tr>
<tr><td>4</td><td></td><td>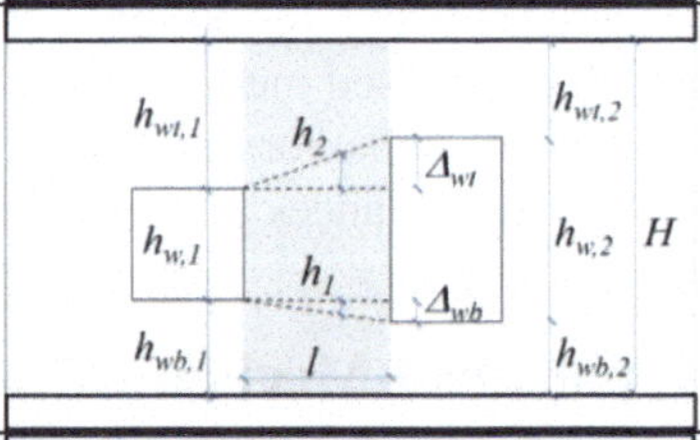
</td></tr>
<tr><td>5</td><td>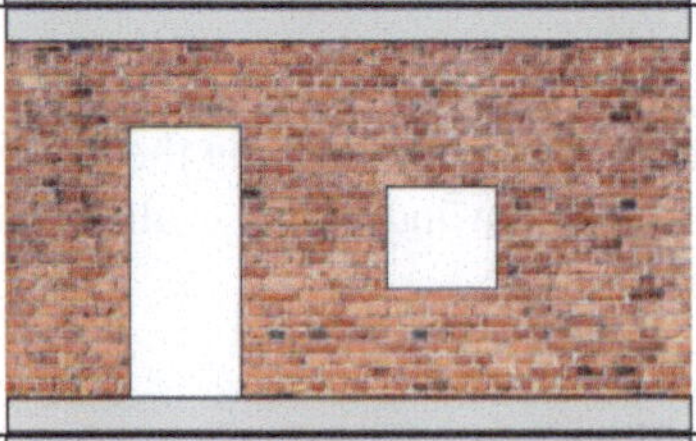</td><td>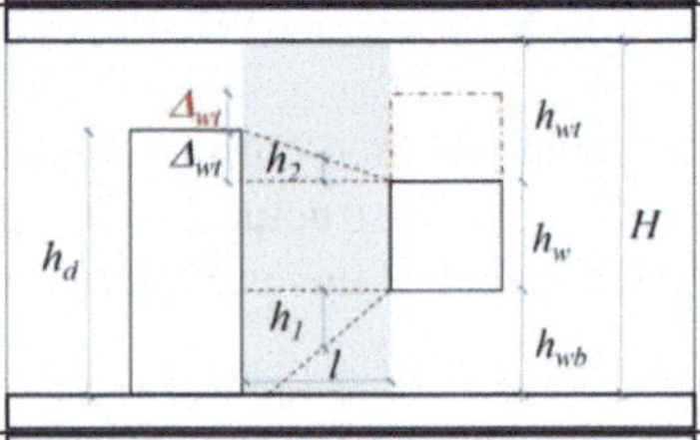
</td></tr>
<tr><td>6</td><td>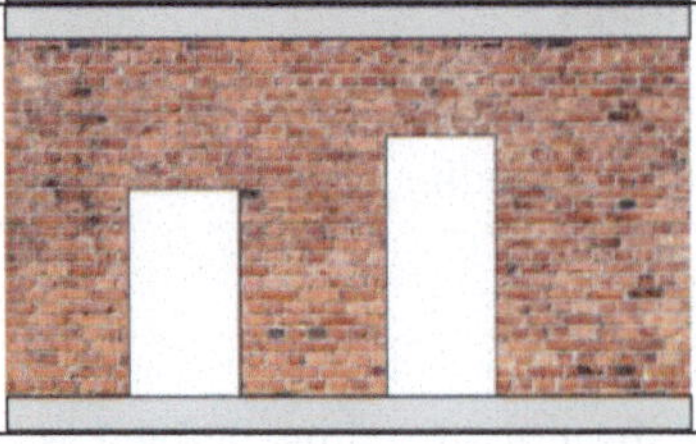</td><td>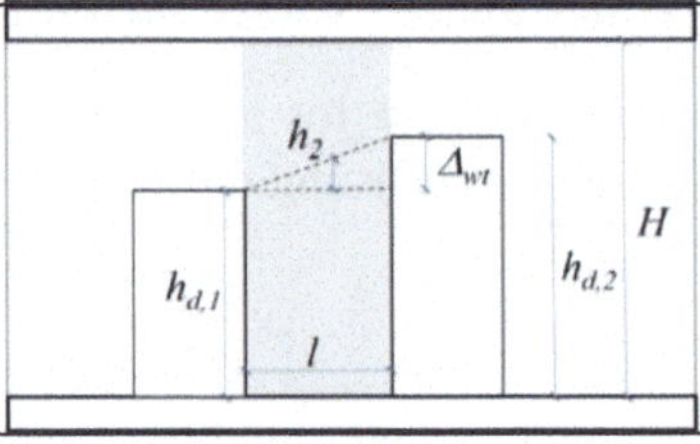
</td></tr>
</table>

Table 2. Determination of partial height $h\prime$ used to compute H_{eff} for the 6 cases in Table 1.

Case	Sub-case	$h\prime =$
1		H
2		$h_w + h_1 + h_2$
3		$h_d + h_1$
4	$\Delta_{wt}, \Delta_{wb} > 0$	$h_{w,1} + h_1 + h_2$
	$\Delta_{wt} = 0$and $\Delta_{wb} > 0$	$h_{w,1} + h_1$
	Δ_{wb} or $\Delta_{\mathrm{wt}} > \max(h_{w,i})$	$\min(h_w, i)$
	$\Delta_{wt} > 0$and $\Delta_{\mathrm{wt}} = 0$	$h_{w,1} + h_2$
	$\Delta_{wt} = \Delta_{wb} = 0$	h_w
5	$\Delta_{wt} = 0$	$h_w + h_1$
	$(\Delta_{wt} > 0 or < 0 and h_{wb} < h_d$	$h_w + h_1 + h_2$
	$h_{wb} > h_d$	h_w
6	$\Delta_{wt} = 0 aligned openings$	$h_{d,1} = h_{d,2}$
	$\Delta_{wt} > 0 non - aligned opening$	$\min(h_{d,i}) + h_2$

$$k_s = \beta \frac{GA_w}{1.2H_{eff}} \text{and} k_b = \beta \frac{10EI_w}{\alpha H_{eff}^3} \tag{2}$$

where β is a stiffness reduction factor, G and E are the shear and elastic moduli, A_w and I_w are the shear area and moment of inertia of the wall section, α depends on the wall end-constraints and H_{eff}

$$H_{eff} = \frac{h' + l(H - h')}{3h'} \leq H \tag{3}$$

where H is the inter-story height, and l is the wall length between openings or between the opening side and the wall end (corners). The most important parameter is $h\prime$, which accounts for openings, evaluated as shown in Table 1 and Table 2 for 6 common cases.

The force capacity of the wall is the minimum between the capacities for shear and bending:

$$f_w = \min(f_s, f_b) \tag{4}$$

The yield displacement is determined as:

$$d_y = \frac{f_w}{k_w} \tag{5}$$

The ultimate displacement depends on the weakest resisting mechanism:

$$d_u = \begin{cases} 0.5\% H_{eff} + f_s/k_b & \text{if } f_w = f_s \\ 1.0\% H_{eff} & \text{if } f_w = f_b \end{cases} \tag{6}$$

3.2 Storey Capacity Curve

For a given storey, modeled as a parallel system, the force-displacement curve is obtained by summing the individual wall curves:

$$F(d) = \sum_i f_i(d) \quad (7)$$

The storey capacity curve can be represented in piecewise form using specific points corresponding to limit states, such as Operative (OLS), Damage (DLS), Life Safety (LLS), and Collapse (CLS), with interstorey forces at each limit state as:

$$F_{OLS} = K_s d_{OLS} \quad (8)$$

$$F_{DLS} = F_s - \sum\nolimits_i k_i d_{y,i} - d_{DLS} \quad (9)$$

$$F_{LLS} = F_s - \sum\nolimits_i f_{w,i} \cdot H\left(d_{LLS} - d_{u,i}\right) \quad (10)$$

$$F_{CLS} = F_s - \sum\nolimits_i f_{w,i} \cdot H\left(d_{CLS} - d_{u,i}\right) \quad (11)$$

where K_s and F_s represent the total stiffness and force, and $\langle\cdot\rangle$ denote McAuley brackets.

A typical capacity curve of a storey is shown in Fig. 1, along with its corresponding piecewise curve as secant segments between consecutive limit states.

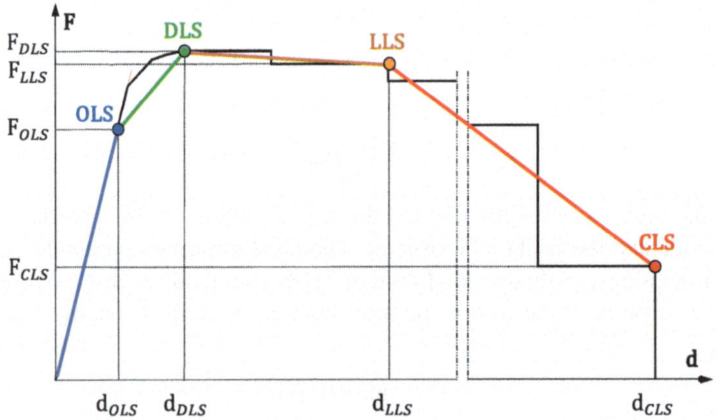

Fig. 1. Typical storey capacity curve and corresponding piecewise curve as secant segments between consecutive limit states.

3.3 Structural Unit Capacity Curve

The structural unit (SU) is modeled as a series system ("stick-model") of storeys, with each mass located at height z_i (Fig. 2). Seismic forces are applied horizontally with shape ψ_i, which can be either height-proportional or mass-proportional:

$$\psi_i = \frac{m_i z_i}{\sum m_i z_i} \text{or} \psi_i = \frac{m_i}{\sum m_i} \tag{12}$$

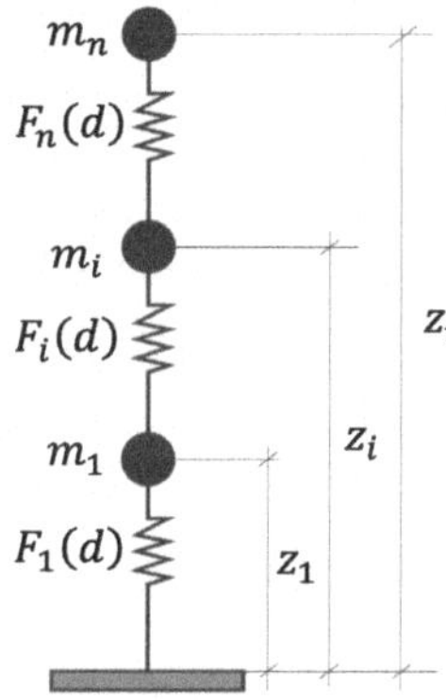

Fig. 2. Structural unit as stick-model series system.

Having defined the force shape ψ_i, the interstorey shear shape τ_i is obtained as:

$$\tau_i = \sum\nolimits_j \psi_j \tag{13}$$

The elastic interstorey displacement shape φ_i at each limit state is found as:

$$\varphi_{LS,i} = \frac{\tau_i}{K_{LS,i}} \left(\sum\nolimits_j \frac{\tau_j}{K_{LS,j}} \right)^{-1} \tag{14}$$

where $K_{LS,i}$ is elastic stiffness of the i-th story at each limit state.

The top displacement $d_{SU,LS}$ and resulting base-shear $V_{SU,LS}$, are derived as:

$$d_{SU,LS} = \min\left(\frac{d_{LS,i}}{\varphi_{LS,i}} \right) \tag{15}$$

$$V_{SU,LS} = F\left(d_{SU,LS} \cdot \varphi_{LS,1}\right) \tag{16}$$

3.4 Equivalent Bilinear SDOF System

The "stick model" can be simplified into a single degree-of-freedom (SDOF) system, with displacements and base-shear given by:

$$d_{LS}^{*} = \frac{d_{SU,LS}}{\Gamma_{LS}} \text{and} V_{LS}^{*} = \frac{V_{SU,LS}}{\Gamma_{LS}} \tag{17}$$

where Γ_{LS} is the participation factor derived from the mass and displacement shapes.

The bilinearized stiffness K^* is obtained at 60% of the peak force:

$$K^{*} = \frac{0.6V_{DLS}^{*}}{d_{DLS}^{*} - \left(1 - \alpha_{y}\right)\frac{V_{DLS}^{*}}{K_{SU,DLS}}} \tag{18}$$

The equivalent yield displacement d_y^* is found by equating the dissipated energy under the bilinear curve with the area A^* under the SU capacity curve:

$$d_{y}^{*} = d_{CLS}^{*} \cdot \left(1 - \sqrt{1 - \frac{2A^{*}}{K^{*} \cdot d_{CLS}^{*2}}}\right) \tag{19}$$

where T^* is the effective period:

$$T^{*} = 2\pi\sqrt{\frac{m_{DLS}^{*}}{K^{*}}} \tag{20}$$

where m_{DLS}^* is the effective mass and K^* is in Eq. (18).

4 Development of Fragility Curves

4.1 Definition of the Fragility Curve

A FC describes the conditional probability that a selected chosen engineering demand parameter (*EDP*), imposed by a given intensity measure *im*, exceeds a capacity threshold $D_{xLS,C}(\mathbf{x})$ identifying a certain Limit State (LS). The capacity is expressed as the function of the system properties vector **x**, which includes geometry variables, material properties, and action (*e.g.*, the axial load). The dispersion of the fragility curves is mainly caused by the materials strength, whose inherent randomness is described in terms of distribution parameters (*e.g.*, median μ and log-standard deviation β). Then, for the LSs considered in this study, *DLS* and *CLS*, , the FC reads (with $x = D, C$):

$$P_{xLS}(im) = P\left[EDP(im, \mathbf{x}) \geqslant D_{xLS,C}(\mathbf{x}) \mid im\right] \tag{21}$$

Analytical fragility assessments involve two main steps: (a) establishing the $EDP(im, \mathbf{x})$ relationship, and (b) determining capacity thresholds $D_{xLS,C}(\mathbf{x})$. Both steps involve uncertainties.

In step (a), the relationship between *EDP* and *im* is formulated. Uncertainties in this step primarily arise from record-to-record variability in ground motion. In non-linear dynamic analyses, this variability is directly integrated into the analysis by using multiple earthquake records. Conversely, in non-linear static analyses or analytical approaches, the variability from record-to-record differences is typically added during the post-analysis phase.

In step (b), capacity thresholds are established to determine the limits at which a structure may fail under seismic loading. In dynamic analyses, relevant uncertainties—primarily due to variations in material properties **x**—are often neglected due to the computational expense associated with including them. In contrast, in static analyses or analytical approaches, these uncertainties can be more easily introduced through Monte Carlo simulations, requiring reduced computational effort.

In this approach, the *EDP* is conveniently chosen as the top displacement while the capacity $D_{xLS,C}(\mathbf{x})$ is determined using the equations provided in previous section, where uncertainties are incorporated in the structural parameters **x** (*i.e.*, geometry, materials, and axial load). The resulting expression of the FC, is then:

$$P_{xLS}(D) = P\big[D \geqslant D_{xLS,C}(\mathbf{x}) \mid D\big] \tag{22}$$

which represents the cumulative distribution function (CDF) of displacement capacity.

4.2 Monte Carlo Analysis

In this study, fragility curves for a single masonry panel are developed under various opening configurations outlined in Table 1. The geometry and structural layout of the masonry panel are illustrated in Fig. 3.

The proposed simplified procedure is employed to derive the fragility curves in Fig. 4 by varying the mechanical properties of the masonry panel as specified in Table 3 and by considering the six cases of opening configurations in Table 1.

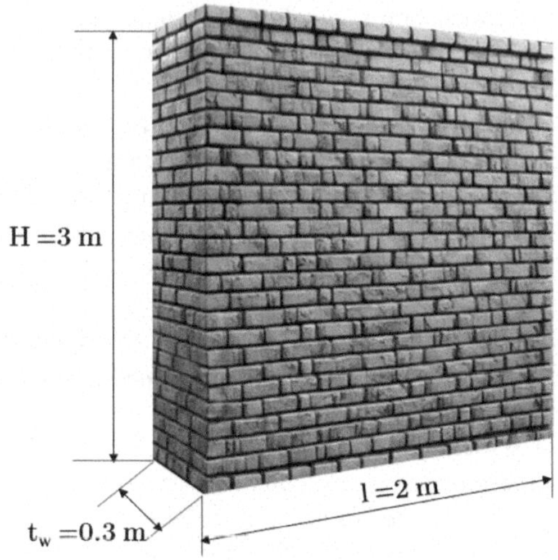

Fig. 3. Geometry of the masonry panel.

Table 3. Mechanical properties of masonry, made from solid bricks and of the constituent blocks. Moduli are taken as mean values, whereas design strengths are obtained by dividing mean values by the partial factor $\gamma_m = 2$.

Parameter	Symbol	Mean value	Design value
Elastic modulus (*MPa*)	E	1500	1500
Shear modulus (*MPa*)	G	500	500
Compression strength (*kPa*)	f_m	3450	1725
Shear strength (*kPa*)	τ_0	90	45
Cohesion strength (*kPa*)	f_{v0}	200	100
Compression strength of blocks (*kPa*)	f_b	4000	2000

From Fig. 4 and Fig. 5 it can be deduced that Case 1, with no openings, exhibits the highest mean value in the fragility curve. This indicates lower vulnerability to damage compared to the other cases. However, when openings are introduced, the mean values of the fragility curves decrease, reflecting a reduction in the panel's overall resistance. The extent of this reduction varies depending on the type, size, and number of openings, as well as their specific positioning within the panel. Each configuration results in a unique influence on structural behaviour, highlighting the sensitivity of the fragility response to such geometric variations. In addition to the mean values, the dispersion of the fragility curves also changes significantly across different cases. Case 1, despite its higher mean, shows greater dispersion, meaning that the absence of openings gives materials uncertainty a stronger influence on the fragility.

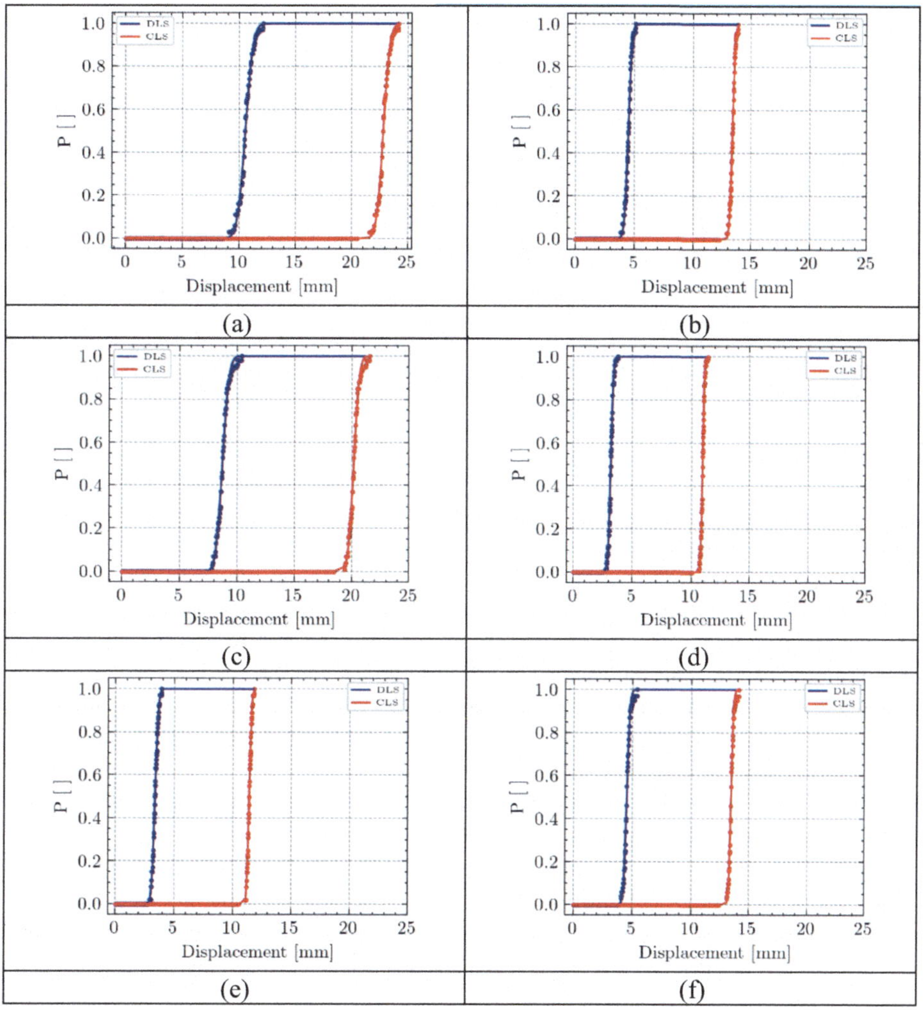

Fig. 4. Fragility curves of the masonry panel with 6 different openings configurations in Table 1: (a) case 1; (b) case 2; (c) case 3; (d) case 4; (e) case 5; (f) case 6.

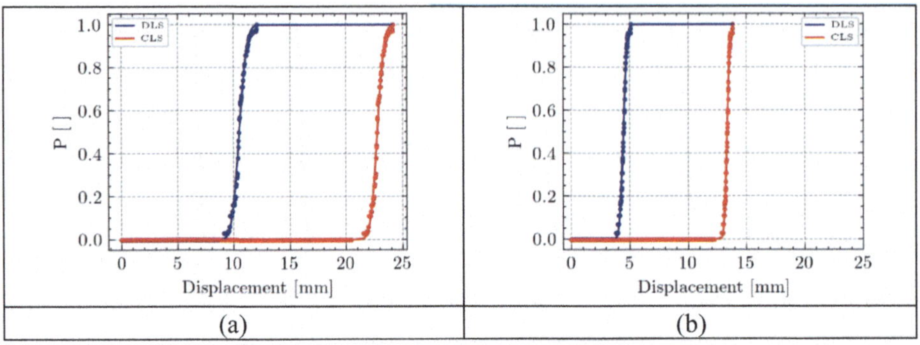

Fig. 5. Sensitivity of the Fragility Curve parameters for the 6 case types in Table 1.

5 Conclusions

This study presents an analytical framework for seismic risk assessment of masonry building clusters (MBCs), addressing their structural complexity and variability. The methodology relies entirely on analytical functions, enabling efficient implementation of Monte Carlo simulations to account for uncertainties in material properties and geometry, including openings. This approach allows generating fragility curves with minimal computational effort, making it well-suited for large-scale applications.

The derived fragility curves highlight the significant influence of wall openings on seismic vulnerability. Configurations with openings demonstrate reduced resistance and increased fragility compared to solid walls, with the extent of this impact varying based on the size, number, and positioning of the openings. These findings underscore the importance of geometric characterization in assessing seismic performance.

By employing a systematic process from wall to structural unit assessments, the framework offers a practical and scalable solution for analyzing the seismic reliability of MBCs. Future work could focus on refining models for structural interactions and exploring more complex boundary conditions to further enhance the accuracy of this methodology.

References

1. Monti, G., Vailati, M.: Procedura Di Analisi Non lineare statica per La valutazione sismica degli edifici in aggregato. In: Proceedings of the XIII ANIDIS Conference, "L'Ingegneria Sismica In Italia". Bologna, Italy (2009)
2. Monti, G., Vailati, M.: Analisi Di Vulnerabilità Sismica Di Edifici in Aggregato: Un Caso Esempio. In: Proceedings of the ANIDIS conference, "L'Ingegneria Sismica In Italia", Bologna, Italy (2009)
3. Formisano, A., Florio, G., Landolfo, R., Mazzolani, F.M.: A method for seismic vulnerability assessment of historical aggregates. In: Proceedings of the Proc of the Workshop Wondermasonry "Design for Rehabilitation of Masonry Structures". Ischia, Italy (2010)
4. Vailati, M.: Effects of Intrinsic and Epistemic Uncertainties on the Seismic Response of Masonry Building Cluster: Development of a Non-Linear Analytical Procedure and Sensitivity Analyses (2011)

5. Vailati, M., Monti, G.: Analisi di aggregati edilizi con solai rigidi e flessibili. In: Proceedings of the Proc. XIV ANIDIS conference, "L'ingegneria Sismica in Italia", vol. I, Bari (2011)
6. Formisano, A., Mazzolani, F.M., Florio, G., Landolfo, R.: A quick methodology for seismic vulnerability assessment of historical masonry aggregates. In: Proceedings of the Proceedings of the COST action C26 Final Conference Urban Habitat Constructions Under Catastrophic Events (2010)
7. Formisano, A., Florio, G., Landolfo, R., Mazzolani, F.M.: Numerical calibration of a simplified procedure for the seismic behaviour assessment of masonry building aggregates. In: Proceedings of the Proceedings 13th International Conference on Civil, Structural and Environmental Engineering Computing, Stirlingshire, Scotland (2011)
8. Di Trapani, F., Villar, S., Di Benedetto, M., Petracca, M., Camata, G.: Seismic response of unreinforced masonry building aggregates: investigation on the "aggregate-effect" based on an elementary building aggregate. Eng. Struct. **316**, 118301 (2024). https://doi.org/10.1016/J.ENGSTRUCT.2024.118301
9. Vailati, M., Monti, G., di Gangi, G.: Seismic assessment of masonry building cluster with a dedicated NLA software. In: Proceedings of the Proc. 2nd International Symposium on Advances in Civil and Infrastructure Engineering, Vietrisul Mare, Italy (2015)
10. Vailati, M., Monti, G., Khazna, M.J., Realfonzo, R.: Probabilistic assessment of masonry building clusters. In: Proceedings of the in Proceedings of the 15th World Conference on Earthquake Engineering, vol. I, 1-9. Lisbona, Portugal (2012)
11. Vailait, M., Monti, G., Realfonzo, R., Khazna, M.J., De Iuliis, M., Valeri, G.: A simplified approach for the seismic assessment of existing masonry structures using few analyses. In: Proceedings of the In 4th International Workshop on Archeology, Cryptoportici, Hypogea, Geology, Geotechnics, Geophysics. Rome, Italy
12. Rapone, D., Brando, G., Spacone, E., De Matteis, G.: Seismic vulnerability assessment of historic centers: description of a predictive method and application to the case study of Scanno (Abruzzi, Italy). Int. J. Architectural Heritage **12**, 1171–1195 (2018). https://doi.org/10.1080/15583058.2018.1503373
13. Basaglia, A., Aprile, A., Spacone, E., Pilla, F.: Performance-based seismic risk assessment of urban systems. Int. J. Architectural Heritage **12**, 1131–1149 (2018). https://doi.org/10.1080/15583058.2018.1503371
14. Sandoli, A., Calderoni, B., Lignola, G.P., Prota, A.: Seismic vulnerability assessment of minor Italian urban centres: development of urban fragility curves. Bullet. Earthq. Eng. **20**, 5017–5046 (2022). https://doi.org/10.1007/S10518-022-01385-0/FIGURES/17
15. Brando, G., De Matteis, G., Spacone, E.: Predictive model for the seismic vulnerability assessment of small historic centres: application to the inner Abruzzi region in Italy. Eng. Struct. **153**, 81–96 (2017). https://doi.org/10.1016/J.ENGSTRUCT.2017.10.013
16. Brando, G., Cianchino, G., Rapone, D., Spacone, E., Biondi, S.: A CARTIS-based method for the rapid seismic vulnerability assessment of minor Italian historical centres. Int. J. Disaster Risk Reduction **63**, 102478 (2021). https://doi.org/10.1016/J.IJDRR.2021.102478
17. Vailati, M., Monti, G.: Low-LOD fragility curves of structural units in masonry building clusters for territorial risk analysis. Eng. Struct. **286**, 116143 (2023). https://doi.org/10.1016/J.ENGSTRUCT.2023.116143
18. Ruggieri, S., et al.: An archetype-based automated procedure to derive global-local seismic fragility of masonry building aggregates: META-FORMA-XL. Int. J. Disaster Risk Reduct. **95**, 103903 (2023). https://doi.org/10.1016/J.IJDRR.2023.103903
19. Polese, M., Di Ludovico, M., Gaetani d'Aragona, M., Prota, A., Manfredi, G.: Regional vulnerability and risk assessment accounting for local building typologies. Int. J. Disaster Risk Reduct. **43**, 101400 (2020). https://doi.org/10.1016/J.IJDRR.2019.101400

20. Tosto, C., Leggieri, V., Ruggieri, S., Uva, G.: Investigation of architectural typological parameters influencing seismic vulnerability of masonry buildings in historical centres: the case of Puglia. Procedia Struct. Integrity **44**, 2036–2043 (2023). https://doi.org/10.1016/J.PROSTR.2023.01.260
21. Zhu, M., McKenna, F., Scott, M.H.: OpenSeesPy: python library for the opensees finite element framework. SoftwareX **7**, 6–11 (2018). https://doi.org/10.1016/J.SOFTX.2017.10.009
22. Zuccaro, G., Dolce, M., Perelli, F.L., De Gregorio, D., Speranza, E.: CARTIS: a method for the typological-structural characterization of Italian ordinary buildings in urban areas. Front Built Environ. **9**, 1129176 (2023). https://doi.org/10.3389/FBUIL.2023.1129176/BIBTEX
23. Coïsson, E., Ferretti, D., Lenticchia, E., Zanazzi, E.: GIS methodologies for the management of seismic risk and the damage prevention on masonry-built heritage. RILEM Bookseries **46**, 1169–1180 (2024). https://doi.org/10.1007/978-3-031-39450-8_95
24. Guo, J., Ye, A., Wang, X., Guan, Z.: OpenSeesPyView: python programming-based visualization and post-processing tool for OpenSeesPy. SoftwareX **21**, 101278 (2023). https://doi.org/10.1016/J.SOFTX.2022.101278
25. Martins, L., Silva, V., Crowley, H., Cavalieri, F.: Vulnerability modellers toolkit, an open-source platform for vulnerability analysis. Bullet. Earthquake Eng. **19**, 5691–5709 (2021). https://doi.org/10.1007/S10518-021-01187-W/FIGURES/16

Seismic Response of Masonry Aggregate Buildings by Explicit Modelling Different Interconnection Degrees

Sofia Villar[1](✉), Marilisa Di Benedetto[1], Antonio Pio Sberna[1], Massimo Petracca[2], Guido Camata[3], and Fabio Di Trapani[1]

[1] Dipartimento di Ingegneria Strutturale, Edile e Geotecnica, Politecnico di Torino, Turin, Italy
sofia.villar@polito.it

[2] ASDEA Software Technology, Via Alide Breviglieri 8, Pescara, Italy

[3] Università degli studi "G. D'Annunzio di Chieti-Pescara, Pescara, Italy

Abstract. Unreinforced aggregate masonry buildings constitute a significant portion of Mediterranean heritage and are characterized by construction processes that evolved alongside urban development. This evolution often resulted in the expansion and transformation of original structures into large buildings composed of multiple units with heterogeneous characteristics in materials, geometry, and boundary conditions. Moreover, phased construction processes introduced uncertainty regarding the degree of inter-unit connection. These variations critically influence their structural behavior during earthquakes, increasing their vulnerability. The assessment of these structures presents two primary challenges derived from the large scale of the problem: achieving accurate behavior representation through modeling assumptions and optimizing analysis speed without compromising accuracy. This study aims to present modelling strategies based on the OpenSees framework, oriented to balance these challenges. The main guidelines focus on material properties definition, and the modeling of interactions between units, slab deformability, and lintel behavior. The strategies are based on the use of 2D shell elements and the homogenized material approach, applied to an elementary aggregate building designed to study the "aggregate effect." Validation of these strategies and studies on different boundary conditions using static and dynamic non-linear analyses will be discussed, focusing on model performance.

Keywords: Masonry · Unreinforced Masonry Buildings · Modelling Strategies · FEM

1 Introduction

Recent and past seismic events have highlighted the vulnerability of unreinforced masonry building aggregates, a common feature in the historical centers of Italy and other Mediterranean countries. These aggregates emerge through gradual expansion over decades or centuries, as new structural units are added to pre-existing ones. This evolution results in complex structural configurations that frequently diverge from the

X. Lu et al. (Eds.): EOS 2024, LNCE 638, pp. 183–192, 2025.
https://doi.org/10.1007/978-3-031-90690-9_16

original design intent. Depending on the timeline of their construction, these aggregates often exhibit a variety of masonry types and construction techniques. The assessment of such buildings is particularly challenging due to geometric uncertainties, including incomplete knowledge of their configurations, slab flexibility, and slab-wall connection quality. Material uncertainties, driven by the orthotropic behavior of masonry, further complicate the evaluation process, as they significantly influence the seismic response [1, 2]. Moreover, structural units may share boundary walls or remain independent, with the degree of connection between units often uncertain, which directly influences the distribution of compressive forces across units. Its presence, absence, or progressive degradation over time affects building performance, making this variable both case-specific and difficult to model accurately, often requiring additional assumptions and simplifications. These factors make detailed modeling complex and computationally demanding. Among modeling approaches, micro-modeling—where masonry units and mortar joints are modeled separately—is highly accurate (e.g., [3–8]). However, dynamic analysis of large, non-linear masonry structures like aggregates is computationally intensive due to long dynamic inputs, small time steps needed to secure convergence, and material non-linearity. To balance efficiency and accuracy, simplified methods such as homogenized masonry models are often employed. These models treat masonry as a continuum, drastically reducing computational demand by minimizing the number of elements. Examples include 3D and 2D macro-elements and the Equivalent Frame Method (EFM), which simplifies analysis by modeling piers and spandrels as 1D elements. EFM presents computational advantages and practical applicability, particularly when openings are regular [9]. However, EFM is less reliable for irregular layouts. In such cases, 3D and 2D continuum models offer a better alternative, as they account for distributed contact properties between walls, effectively capturing interactions even in complex configurations. This study aims to recommend modeling strategies that enhance analysis efficiency while maintaining sufficient detail to capture multiple mechanisms triggered by progressive damage during dynamic analysis. A validation of the modelling approach is presented and several aggregate building models with different slab flexibility and inter-unit connection strength are tested under static and dynamic input.

2 Modelling Strategies for Aggregate Buildings in Opensees

2.1 Modelling Walls and Wall Connectivity

The proposed modeling strategies rely on the homogenized masonry approach to model walls, where there is no distinction between the individual components of masonry (brick and mortar). Instead, a single material with equivalent properties that reflect masonry behavior is considered (Fig. 1). Homogenized material properties can be derived through experimental test calibration, available literature (e.g., building codes), or empirical formulations such as [10]. For modeling material behavior, the approach utilizes the *ASDConcrete3D* constitutive model [11]. This model employs a tension-compression damage framework incorporating two scalar damage indices and distinct failure criteria to differentiate behavior under tension and compression. Additionally, it incorporates auto-regularization of fracture energy to reduce mesh-size dependency and employs an implicit/explicit (IMPL-EX) integration scheme [12] to address issues related to strong non-linearities. These features significantly reduce computational burden. In terms of geometry, the use of 2D elements is facilitated by advanced thick shell formulations, such as *ASDShellQ4* (4-node element) and *ASDShellT3* (3-node element) [13]. These elements are particularly effective at integrating out-of-plane shear behavior with membrane and bending behavior, enabling accurate modeling of combined failure mechanisms. This capability is further enhanced through the use of a *LayeredShell* fiber section, which incorporates multiple integration points along the thickness, allowing for the assignment of varying material properties [14]. In aggregate buildings, two levels of wall connectivity are distinguished: (i) connections between walls within the same structural unit (SU) and (ii) connections between different adjacent SUs. Lack of connection can be effectively modeled using a contact condition with zero-length elements governed by friction laws, such as the *zerolengthContactASDimplex* [15]. This element can be oriented arbitrarily and accommodates nodes with different degrees of freedom (DOFs), allowing seamless connection between different element types. Partial connections with degrading cohesive behavior can be modeled by introducing a continuum material as a separate geometry with a non-linear constitutive model. In the current study, the latter approach is employed and the *DamageTC3D* material is used for this purpose. The interface material is connected to adjacent walls through an *equalDOF* condition, which also enables the use of varying mesh sizes (Fig. 1).

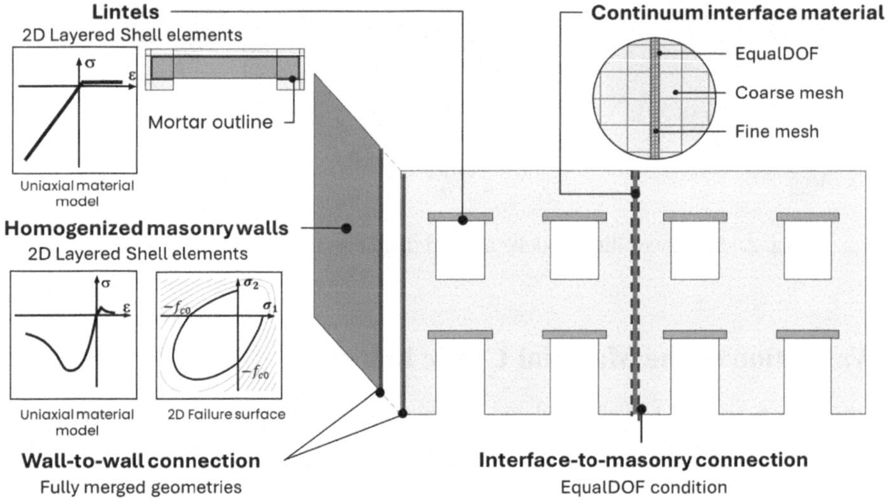

Fig. 1. Masonry aggregate buildings model assembly strategies

Modeling additional elements, such as lintels, is highly recommended as they influence the response of spandrels by providing additional flexural resistance. To avoid overestimating this resistance, their sliding behavior can be modeled by introducing a thin layer of mortar with low tensile strength. The constitutive law for this layer is designed to exhibit compressive elastic behavior and reduced tensile strength, characterized by elastic-perfectly plastic behavior (Fig. 1). This is achieved using the *ASDConcrete3D* model in its user-defined mode with no plastic behavior (damage vectors equal to 1).

2.2 Modelling Slab Flexibility

One of the most influential variables in the dynamic behavior of masonry buildings is slab flexibility. Careful consideration is required to accurately replicate the various mechanisms it triggers. The simplest assumption is that of a rigid diaphragm, which results in a stiffness-dependent distribution of seismic forces and can be easily modeled using a *rigidDiaphragm* condition. This approach introduces a master node that is fixed to allow horizontal displacements and in-plane rotation, with all nodes belonging to the same structural unit (SU) connected to it (Fig. 2.a). Flexibility, on the other hand, can be achieved by explicitly modeling the beams and wooden planks that constitute the slab (Fig. 2.b). The slab's flexibility can be controlled by modeling the wood as a shell element and adjusting its thickness and in-plane shear modulus. Orthotropic behavior can be modeled using the *elasticOrthotropic* material or the *Orthotropic Material Wrapper* on top of the material constitutive model. Connections between beams and walls can be modeled either as rigid, using an embedded condition, or by employing *equalDOF* constraints to simulate simply supported conditions. Additionally, pounding or degrading connections can be simulated by introducing zero-length elements with a specific behavior.

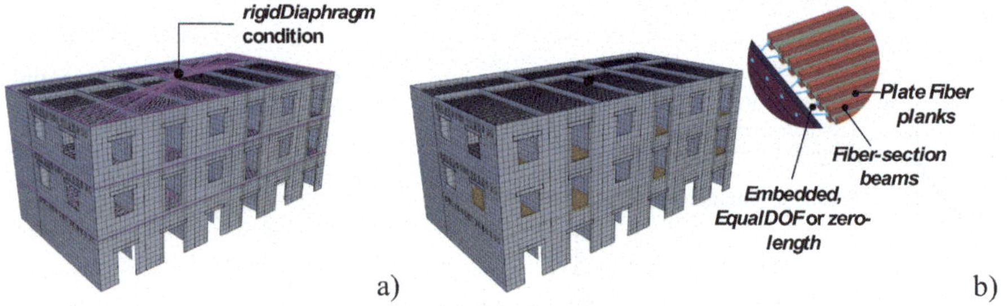

Fig. 2. Slab flexibility models: a) rigid diaphragm; b) flexible diaphragm

3 Validation of the Material Cyclic Behavior

The modelling approach was validated against a quasi-static cyclic experimental test carried on a two-story masonry building. The structure is composed of 4 orthogonal walls with a thickness of 250 m each, measuring 6 m × 4.4 m in plan with a height of 6.4 m. All walls are rigidly connected to each other except for the "Door Wall", which is connected to its opposite wall (named "Window Wall") by a series of steel beams (IPE 140) creating a flexible diaphragm. A full description of the test is provided in [16]. The main objective of the validation was to calibrate the homogenized masonry material parameters for cyclic behavior. To do so, the test was replicated numerically using the compressive strength (6.2 MPa) and Young's modulus (1490 MPa) obtained in previous experimental campaigns [17] and adjusting the tensile strength and tensile fracture energy to match the experimental curve. Table 1 summarizes the material properties derived from the calibration. Results of the validation are exhibited in terms of force-displacement curves (Fig. 3) and cumulated damage (Fig. 4).

Table 1. Masonry material properties

Compressive strength f_m (MPa)	Tensile strength f_t (MPa)	Young Modulus E_m (MPa)	Fracture energy in compression Gc (N/mm)	Fracture energy in tension G_t (N/mm)
6.2	0.16	1490	20.0	0.12

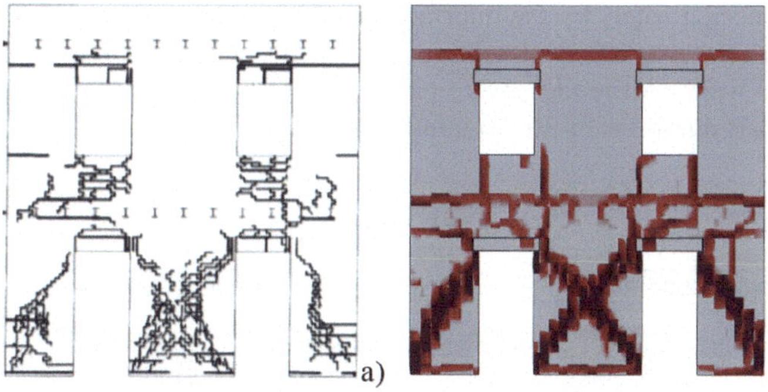

Fig. 3. Damage comparison at the end of the test: a) experimental; b) numerical

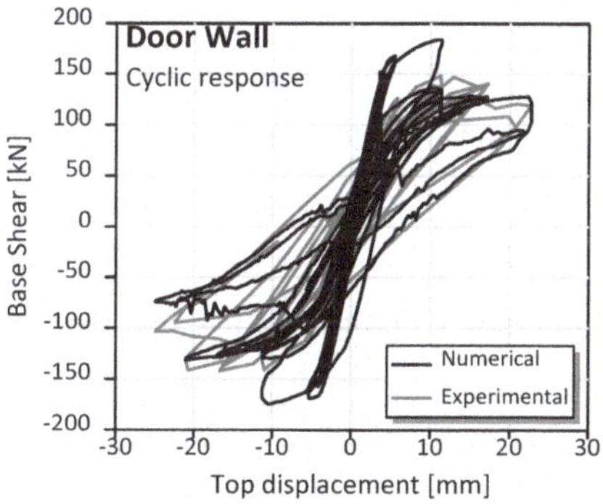

Fig. 4. Cyclic response comparison.

4 Numerical Response of Aggregate Buildings Under Different Slab Flexibility and Unit-Interconnection Conditions

The modeling strategies were applied to a regular aggregate building composed of three similar units, purposely designed to study the influence of different variables on the response of each individual structural unit (SU). The structure covers a total area of 24 × 12.6 m in plan and reaches a height of 11.20 m, a full description of the model is provided in a previous study [18]. Six different models were developed in the STKO [19] platform for OpenSees [20], incorporating two distinct slab conditions: a rigid diaphragm and a flexible diaphragm. The flexible diaphragm was modeled using 3 cm thick wooden planks represented as a *plateFiber* element with an *elasticOrthotropic* material and an in-plane shear modulus (*G*) of 300 MPa. For both slab typologies, the units composing the aggregate were modeled as follows: (a) rigidly connected to each other (AB), (b) partially connected with an interface material exhibiting non-linear behavior (SCAB), and (c) isolated (SU). The units were labeled U1, U2, and U3 from left to right. A summary of all cases is shown in Fig. 5. Both non-linear static analysis (NLSA) and non-linear dynamic analyses (NLDA) were performed to test the models in the longitudinal direction. Since the SCAB model was designed as three individual SU models connected solely by the interface material to ensure compatibility with AB and SU models, the extraction of force-displacement curves focused exclusively on the longitudinal walls. All models were partitioned to take advantage of *openseesMP* achieving analysis durations of 30–40 min for NLSA and 3–4 h for NLDA.

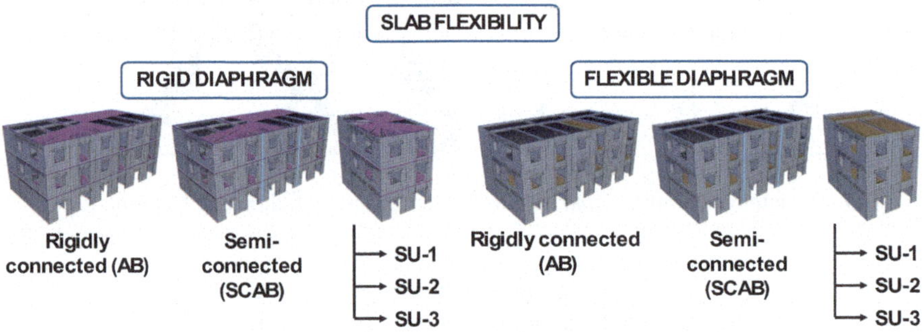

Fig. 5. Numerical models developed in STKO

4.1 Non-linear Static Analysis: Considerations and Results

Performing pushover analysis on masonry structures involves two key complexities in modeling horizontal forces. First, the direct application of loads to walls differs by diaphragm type. For rigid diaphragms, forces are easily applied to a master node and distributed through the diaphragm. In flexible diaphragms, where slabs connect to walls only at beam ends, load transfer becomes challenging, as slabs cannot effectively distribute forces. Directly applying nodal forces at beam ends is discouraged due to masonry's low tensile resistance, which can cause local damage and hinder load transfer. This issue can be mitigated by creating rigid regions around beams using *rigidLink* elements, but this solution is limited to beams perpendicular to the load direction. Alternatively, distributed horizontal forces along wall lengths can be applied. The second complexity lies in quantifying the load distribution each wall receives. This study addresses this by proposing tributary mass-proportional forces distributed at the top of walls for each floor. Numerical comparisons between AB and SU models are presented for Unit 3 in both rigid (Fig. 6.a) and flexible diaphragm cases (Fig. 6.b).

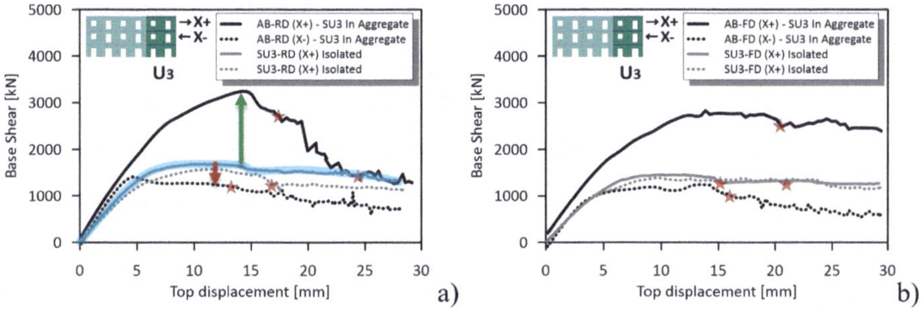

Fig. 6. Pushover comparison between AB and SU models of Unit 3 in a) rigid diaphragm condition, b) flexible diaphragm condition.

Results exhibited that inner units experience increased resistance due to the aggregate condition, with minimal influence from the polarity of seismic forces, while external units exhibit significant resistance variations depending on seismic force direction, driven by pronounced changes in compression stresses at their base. The flexibility of the slab affected the magnitude of these variations. SCAB models were not assessed using pushover analysis because the degrading interface material causes the external units to detach from the central unit (which contains the control node), leading to measurement errors for these units.

4.2 Non-linear Dynamic Analysis: Considerations and Results

All six models were dynamically tested using a single ground motion input (NS component of AQG - Colle Grilli, L'Aquila, Italy, recorded on April 6th, 2009). For RD models, the analysis was discretized to align with the signal's sampling rate (0.005 s), while for FD models, the time step was halved to better capture non-linear behavior and enhance convergence. Rayleigh damping with a 3% damping ratio ($\xi = 0.03$) was applied. The frequency range was constrained between the first and fifth modal frequencies, determined from prior eigenvalue analyses of the models. Results for U3 in all configurations are presented in terms of force-displacement curves (Figs. 7 and 8) and damage (Fig. 9).

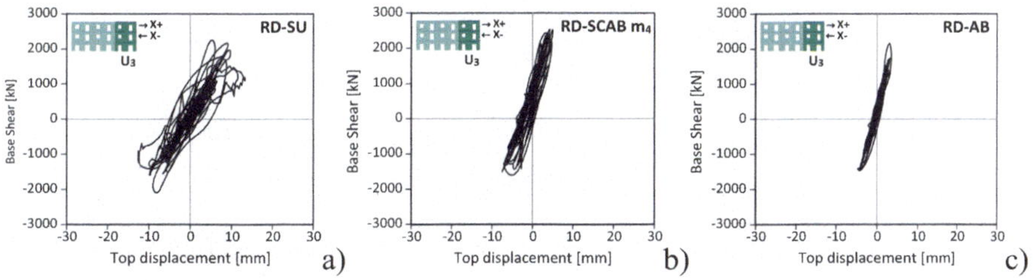

Fig. 7. NLDA comparison of the Unit 3 of a) SU model b) SCAB model and c) AB model for Rigid Diaphragm condition

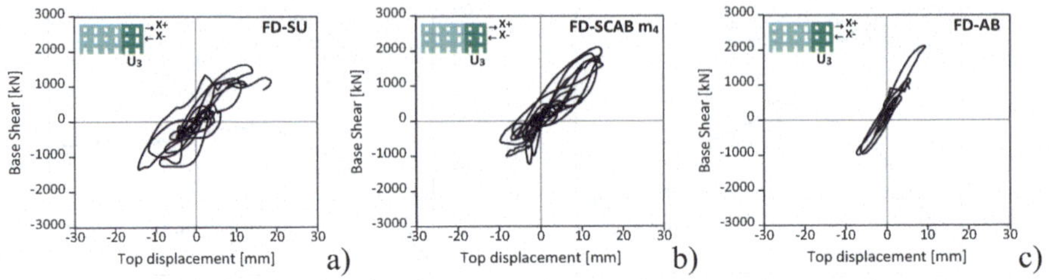

Fig. 8. NLDA comparison of the Unit 3 of a) SU model b) SCAB model and c) AB model for Flexible Diaphragm condition

The results for the Rigid Diaphragm case exhibit a clear asymmetry in the maximum base reaction values depending on the direction of the load when analyzed in SCAB and AB configurations. The compression experienced in these cases, when the action occurs in the X+ direction, enhances performance as previously observed in the NLSA. In contrast, the SU model does not exhibit this asymmetry. Additionally, the degree of connection directly impacts the maximum displacement reached and the dissipated energy (derived from accumulated damage) experienced by the unit, with the rigidly connected structure representing the best-case scenario by remaining in the elastic phase. For the Flexible Diaphragm, the results still indicate the previously described asymmetry, although with lower resistance variations compared to the isolated (SU) case. The variation in the degree of connectivity between units, while evident when compared to the AB case, is not as pronounced. Furthermore, independent behavior of the walls was observed during the analysis, raising questions about the representativeness of the barycenter as a control node for measuring responses. The barycenter's response reflects only the behavior of the wall to which it is attached, potentially missing collapses in more flexible, distant walls. This is evident in the AB case, which appears to remain in the elastic phase; however, upon closer inspection, the front wall collapses due to

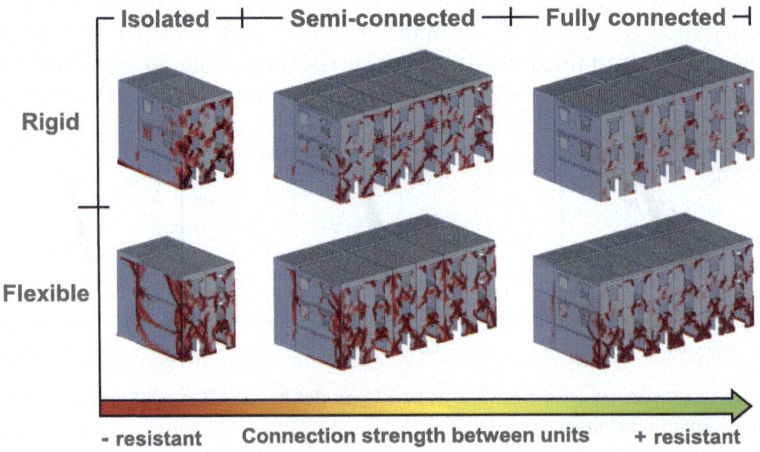

Fig. 9. Damage comparison between different degrees of unit-interconnectivity and slab typology

elevated shear forces at the base piers (see Fig. 9). A comparison of the final damage states across all six cases is presented in Fig. 9, highlighting the different mechanisms activated by changes in slab flexibility and the reduced level of damage achieved as unit interconnectivity improves.

5 Conclusions

The study focuses on developing modeling strategies for masonry aggregate buildings, aiming to improve the balance between cost-effective and detailed models by utilizing numerous advanced material and element formulations available in OpenSees. A homogenized masonry approach using shell elements was employed for this purpose. The model was validated by replicating a cyclic experimental test on a full-scale masonry building to obtain suitable values for tensile strength and fracture energy. These strategies were applied to six different aggregate models with varying boundary conditions in terms of slab flexibility and unit interconnectivity. The models were tested using non-linear static (NLSA) and dynamic (NLDA) analyses to assess the influence of multiple variables on the behavior of individual units. Results showed that resistance increased asymmetrically depending on the unit's position within the aggregate: inner units benefited the most from the aggregate condition, showing increased resistance with minimal influence from seismic force polarity, while external units exhibited the largest resistance variations due to significant base compression stress changes caused by the direction of seismic forces. Rigid diaphragms were found to reduce unit displacements and damage levels as interconnectivity improved, especially in regular building configurations. Conversely, flexible diaphragms caused independent behavior of the walls, raising concerns about the representativeness of the geometrical barycenter for measuring structural response. While improved connectivity generally limited damage and displacement demands, this observation cannot be generalized for structures with irregular geometrical configurations. Additional research is required to expand the database of geometrical configurations and enhance the applicability of these findings.

Acknowledgements. This paper was supported by DPC-ReLuis 2024–2026, WP10, Subtask 10.2 – "Analysis of complex masonry systems".

References

1. Penna, A., Morandi, P., Rota, M., Manzini, C.F., da Porto, F., Magenes, G.: Performance of masonry buildings during the Emilia 2012 earthquake. Bull. Earthq. Eng. **12**, 2255–2273 (2012)
2. Sberna, A.P., Demartino, C., Vanzi, I., Marano, G.C., Di Trapani, F.: Cost-effective topology optimization of masonry structure reinforcements by a linear static analysis-based GA framework. Bull. Earthq. Eng. **22**(8), 4143–4167 (2024)
3. Petracca, M., Pelà, L., Rossi, R., Zaghi, S., Camata, G., Spacone, E.: Microscale continuous and discrete numerical models for nonlinear analysis of masonry shear walls. Constr. Build. Mater. **149**, 296–314 (2017)

4. Di Trapani, F., Giordano, L., Mancini, G.: Progressive collapse response of reinforced concrete frame structures with masonry infills. J. Eng. Mech. **146**(3), 04020002 (2020)
5. Di Trapani, F., Tomaselli, G., Cavaleri, L., Bertagnoli, G.: Macroelement model for the progressive-collapse analysis of infilled frames. J. Struct. Eng. **147**(6), 04021079 (2021)
6. Di Trapani, F., Vizzino, A., Tomaselli, G., Sberna, A.P., Bertagnoli, G.: A new empirical formulation for the out-of-plane resistance of masonry infills in reinforced concrete frames. Eng. Struct. **266**, 114422 (2022)
7. Di Trapani, F., Oddo, M.C., Sberna, A.P., Mendola, L.: Structural health monitoring of masonry structures using stress sensors: experimental induced damage tests and proposed approach for real-time monitoring. Constr. Build. Mater. **449**, 138077 (2024)
8. Di Trapani, F., Di Benedetto, M., Petracca, M., Camata, G.: Local infill-frame interaction under seismic loads: investigation through refined micro-modeling. Eng. Struct. **315**, 141–296 (2024)
9. Tomić, I., Beyer, K.: Shake-table test on a historical masonry aggregate: prediction and postdiction using an equivalent-frame model. Bull. Earthq. Eng. **12**, 6225–6258 (2023)
10. European Committee for Standardization. Eurocode 6: Design of masonry structures – Part 1-1: General rules for reinforced and unreinforced masonry structures (EN 1996-1-1). Brussels, Belgium: CEN (2005)
11. Petracca, M., Camata, G., Spacone, E., Pelà, L.: Efficient constitutive model for continuous micro-modeling of masonry structures. Int. J. Architectural Heritage **17**(1), 134–146 (2023)
12. Oliver, J., Huespe, A.E., Cante, J.C.: An implicit/explicit integration scheme to increase computability of non-linear material and contact/friction problems. Comput. Methods Appl. Mech. Eng. **197**, 1865–1889 (2008)
13. ASDEA Software Technology. ASDShellQ4 Element (2020). https://opensees.github.io/OpenSeesDocumentation/user/manual/model/elements/ASDShellQ4.html
14. Lu, X., Xie, L., Guan, H., Huang, Y., Lu, X.: A shear wall element for nonlinear seismic analysis of super-tall buildings using OpenSees. Eng. Struct. **61**, 195–208 (2014)
15. ASDEA Software Technology. zeroLengthContactASDimplex Element https://opensees.github.io/OpenSeesDocumentation/user/manual/model/elements/zeroLengthContactASDimplex.html
16. Magenes, G., Calvi, G.M., Kingsley, G.R.: Seismic Testing of a Full-Scale, Two-Story Masonry Building: Test Procedure and Measured Experimental Response. Experimental and Numerical Investigation on a brick Masonry Building Prototype - Numerical Prediction of the Experiment, Report 3.0 - G.N.D.T, Pavia, Italy (1995)
17. Anthoine, A., Maganotte, G., Magenes, G.: Shear-compression testing and analysis of brick masonry walls. In: Proceedings of the 10th European Conference on Earthquake Engineering, Vienna, Austria (1994)
18. Di Trapani, F., Villar, S., Di Benedetto, M., Petracca, M., Camata, G.: Seismic response of unreinforced masonry building aggregates: Investigation on the "aggregate-effect" based on an elementary building aggregate. Eng. Struct. **316**, 118301 (2024)
19. Petracca, M., Candeloro, F., Camata, G.: ASDEA Software STKO user manual (2017)
20. McKenna, F., Fenves, G.L., Scott, M.H.: Open System for Earthquake Engineering Simulation. University of California, Berkeley, CA (2000)

An Insight into the Sensitivity of Seismic Basin Effects to Boundary Conditions

Alireza Duzandeh[1](✉), Davide Noè Gorini[2], Marco Uzielli[1], Andrea De Luca[1], Andrea Geppetti[1], Johann Facciorusso[1], and Claudia Madiai[1]

[1] Department of Civil and Environmental Engineering, University of Florence, Via di S. Marta 3, Florence, Italy
alireza.duzandeh@unifi.it

[2] Department of Civil, Environmental and Mechanical Engineering, University of Trento, via Mesiano, 77, Trento, Italy

Abstract. Under complex stratigraphic and morphological conditions, such as the ones typically occurring in basins, the realistic assessment of seismic site effects requires a refined, continuum-based modeling of the entire soil deposit. The effectiveness of the assessment is highly influenced by the assignment of boundary conditions. In this context, this paper presents the preliminary steps of a broader research path aimed at developing practice-oriented standards for basin effects under seismic loading. To this end, site response analyses are carried out on an example two-dimensional basin model using OpenSees and comparing the outputs of multiple boundary conditions techniques. Specifically, the following methods are employed along the lateral boundaries: kinematic constraints, Lysmer and Kuhlemeyer viscous absorbers, massive boundary columns and free-field boundary columns. The propagation of the seismic waves from the outcrop to the lower boundary is instead analyzed through a deconvolution procedure. Although preliminary, these results highlight some potentially relevant factors influencing the seismic performance of boundary conditions and suggest some practical indications about their applicability in boundary value problems.

Keywords: Basin effects · Site response analysis · Boundary conditions · Parametric study · OpenSees

1 Introduction

Site response analysis (SRA) is a fundamental approach for the assessment of the seismic behavior of the subsoil. Given the complexity of the phenomena involved in seismic wave propagation, SRA can entail advanced numerical modeling in lieu of simplified approaches like those typically included in technical provisions.

A critical aspect of advanced numerical modeling is the accurate definition of boundary conditions (BCs), which can influence significantly the reliability of the predictions as they define how the model interacts with the lateral and underlying subsoil not represented numerically. A less realistic definition of BCs can lead to under- or overestimating site effects, thereby hindering the quality of seismic hazard estimation at site-scale.

X. Lu et al. (Eds.): EOS 2024, LNCE 638, pp. 193–201, 2025.
https://doi.org/10.1007/978-3-031-90690-9_17

This paper illustrates the results of a study included within a broader research project aimed at developing seismic code-compatible approaches for quantifying basin effects [1, 2]. While previous studies focused on the influence of BCs on specific SRA-related problems (e.g., [3, 4]), the ongoing research effort will involve an extensive parametric analysis and the subsequent systematic, data-driven assessment of these effects. In this view, this paper provides an initial insight into the comparative assessment of BC modeling effects on SRA by analyzing a single basin model in a seismic scenario. SRA is conducted by using distinct BC techniques, and the resulting outcomes are critically compared and interpreted.

2 Numerical Modeling

By virtue of its versatility and open-source policy, the Python version of the finite element analysis framework OpenSees [5] was used to implement a continuum-based numerical representation of the example basin, along with the definition of the seismic input, as detailed in the following sections [6].

2.1 Geometry

The reference geometry for the example numerical study is shown in Fig. 1. The basin is composed of a double layer of surficial deposits resting on a sigmoid-shaped bedrock. The geometry of the reference basin scheme is defined by the following dimensions (in meters): $H = 200\,\text{m}$, $H_1 = 100\,\text{m}$, $H_2 = 100\,\text{m}$, $L_i = 150\,\text{m}$, $L_{ol} = 200\,\text{m}$, $L_{or} = 300\,\text{m}$, $L_{el} = L_{ol}$, $L_{er} = L_{or}$, and $H_e = H$.

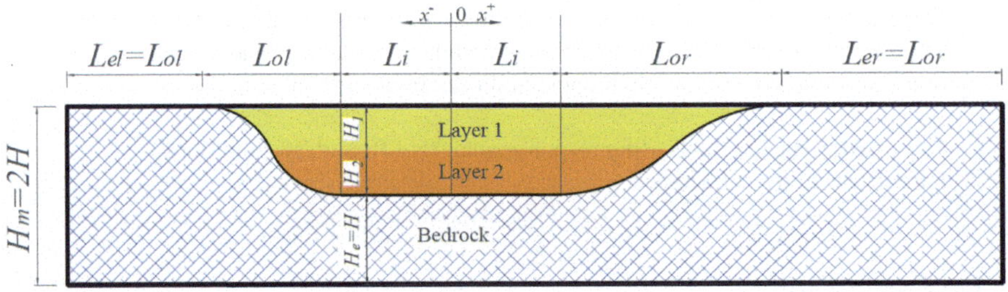

Fig. 1. Reference basin layout

2.2 Model

The cyclic behavior of the layers constituting the basin was described by the pressure-dependent multi-surface constitutive law (PDMY) developed by [7]. The PDMY provides a multi-linear response with kinematic hardening describing the nonlinear cyclic behavior of coarse-grained soils. The bedrock was instead modeled as a linear-elastic isotropic material (named EI hereinafter). The entire deposit is assumed to be dry.

The constitutive models were calibrated considering first the suggested values provided on the relative OpenSeesWiki webpage [8]. The subsequent assignment of material parameters was aimed at obtaining representative values of the stiffness moduli for different soil categories prescribed by European provisions (EC8, [9]). As per the latter, typical power laws were assumed for simulating the in-depth variation of the shear wave velocity.

2.3 Seismic Input Motion

Given the preliminary scope of this study, the seismic performance of the reference basin layout was investigated by considering a single seismic scenario. Specifically, the latter is represented by the horizontal EW component of 1980 Irpinia (Italy) earthquake recorded at the Bisaccia station. This time history is part of a suite of records selected for the broader purpose of the present research path, spanning over EC8-compatible target response spectra. Figure 2 shows the acceleration time history, and the corresponding 5%-damped elastic acceleration response spectrum of the considered seismic motion.

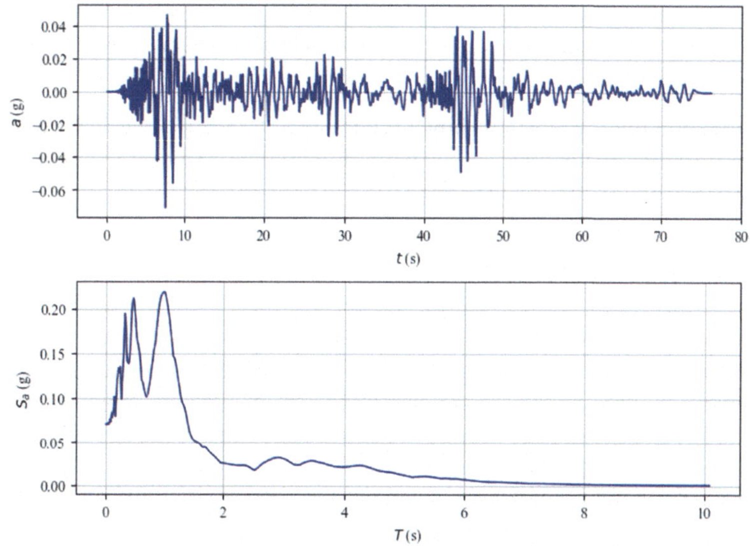

Fig. 2. Acceleration time history and response spectrum of the seismic input motion

2.4 Gravity Analysis

The OpenSees numerical model of the reference layout (Fig. 1) was implemented according to a staged analysis procedure, aimed at reproducing the effective stress state of the soil under lithostatic conditions, as a fundamental point to get reasonable assessment of the dynamic performance. To this end, the adopted staged analysis simulated the progressive deposition of the superficial deposits composing the basin.

In the light of the above, gravity analysis was implemented as the succession of the stages shown in Fig. 3. Stage 1 was devoted to the initialization of the stress state in the bedrock only, restraining all nodes corresponding to basin layers. In Stage 2, the restraints to the nodes of Layer 2 in the basin (see Fig. 1) were removed and a new static analysis was carried out to let the system settle. Finally, Stage 3 reproduced the further modification of the stress state in the domain induced by the activation of the superficial basin layer, No. 1.

Boundary Conditions

This study focuses on a comparative investigation on the seismic response of the considered case employing different modelling of lateral boundary conditions, namely:

I) EqualDOF: nodes along opposite boundaries at the same elevation undergo the same displacements [10].
II) L&K dashpots: viscous dampers are connected to the boundary nodes [11].
III) Massive columns: massive soil columns, having a thickness much greater than the soil elements along the boundary are introduced at both edges and rigidly connected to the inner domain in the horizontal direction (see, for instance, [12].
IV) Free-field columns: the edge columns in Point III are no longer directly connected to the nodes of the inner domain as viscous dashpots are interposed between them (e.g., see [12, 13]).

The corresponding layouts are shown schematically in Fig. 4. A displacement time history was imposed on the nodes at the base of the model, as the convoluted motion of the reference seismic record (Sect. 2.3).

3 Discussion of the Results

The seismic performance of the basin was assessed in terms of horizontal displacements and spectral accelerations, the latter taken in correspondence to the fundamental period of the deposit ($T_1 = 4.82\,\mathrm{s}$) as a concise indicator of dynamic amplification effects typically of interest for assessments of the built environment. For the considered seismic scenario, these quantities were computed for a set of seven scrutiny nodes, shown in Fig. 5, spaced along the ground level to capture the spatial variability of site effects and the relative influence of the adopted BCs.

The ability of the considered BCs to simulate realistically the displacement demand at the ground level was quantified in terms of the following concise indicator:

$$\frac{d_{\max}(\mathrm{ref})}{d_{\max}(\infty)} \tag{1}$$

where $d_{\max}(\mathrm{ref})$, is the maximum horizontal displacement computed with the considered BCs models (see Cases I-IV in Sect. 2.4) and $d_{\max}(\infty)$ is the one relating to an "enlarged model" of the basin, i.e., a model having a horizontal extension of the bedrock 10 times greater the one of the basin, making boundary effects in the latter negligible. The relative results are represented in Fig. 6.

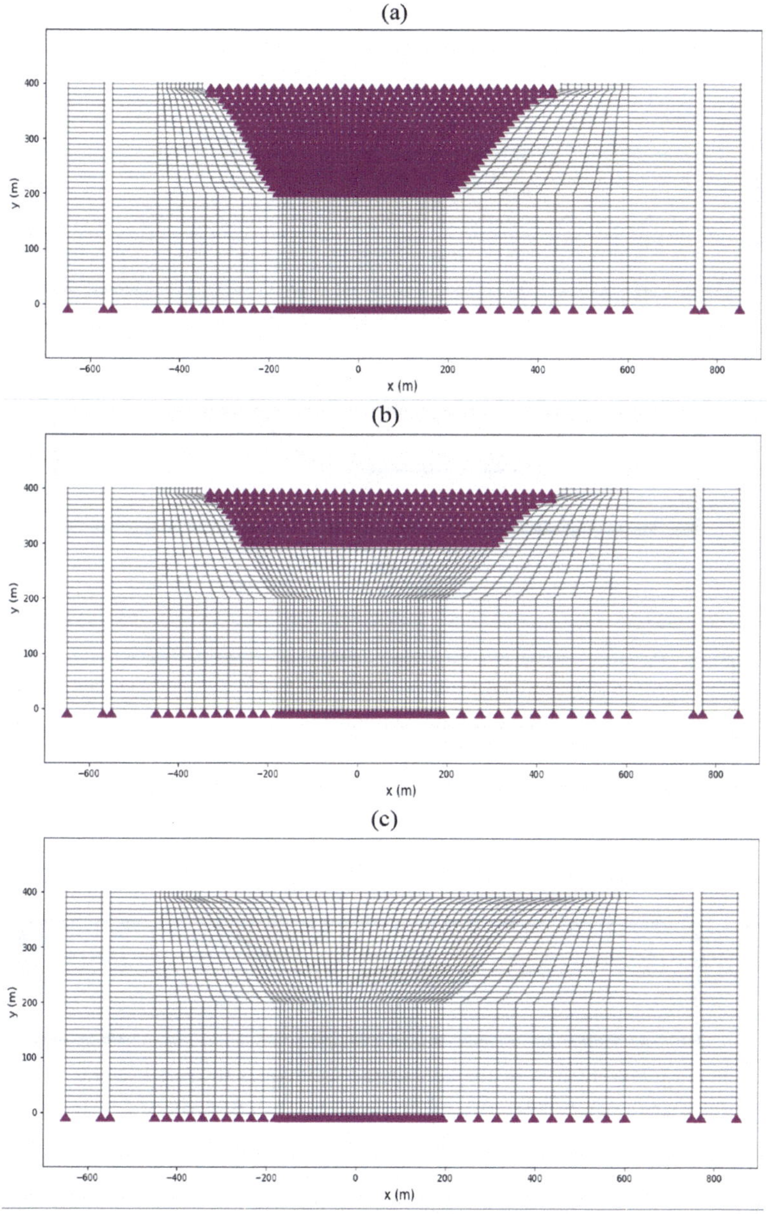

Fig. 3. Staged analysis procedure for gravity analysis: (a) realization of the bedrock domain; progressive introduction of the (b) first and (c) second superficial layers of the basin (restraints represented by pink triangles)

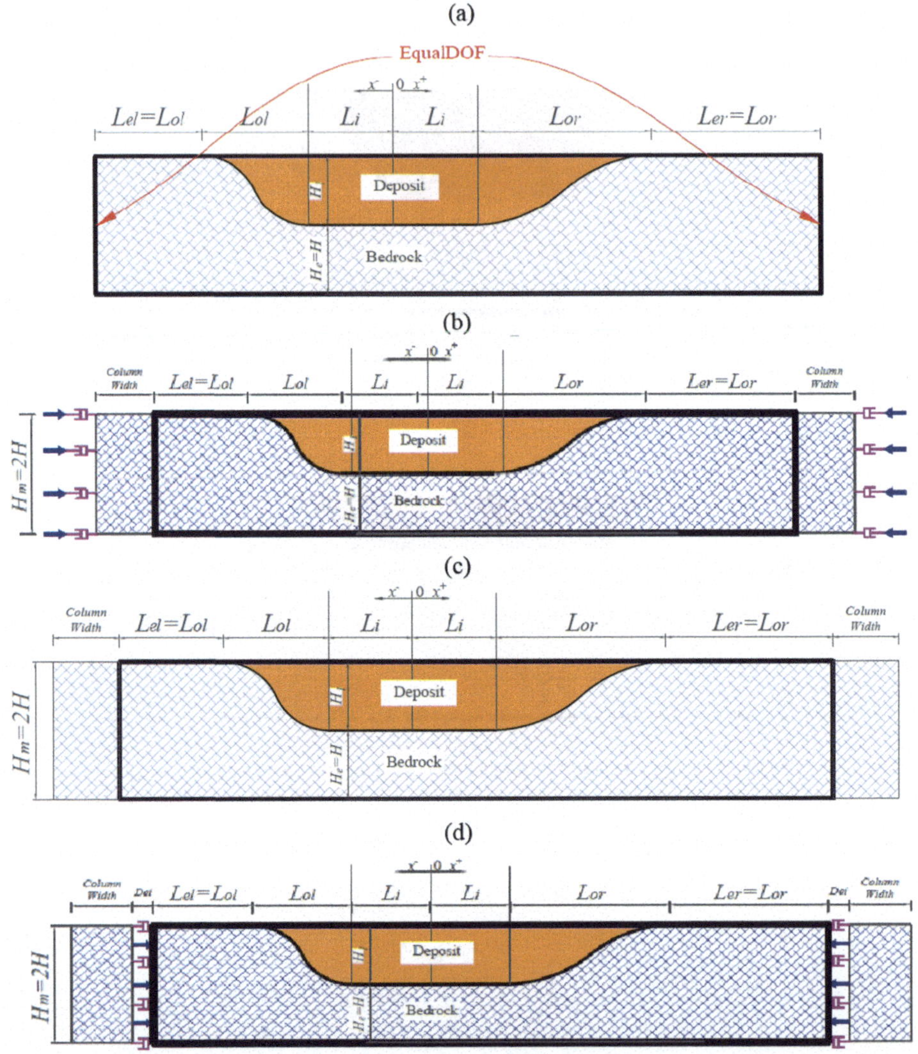

Fig. 4. Schematic representation of the considered techniques to model lateral boundary conditions: (a) kinematic constraints; (b) L&K dashpots; (c) massive columns; (d) free-field columns

Similarly, a spectral acceleration ratio was defined as:

$$\frac{S_{\mathrm{a,T_1}}(\mathrm{ref})}{S_{\mathrm{a,T_1}}(\infty)} \tag{2}$$

with $S_{\mathrm{a,T_1}}$ (ref) the reference spectral acceleration of the considered model and $S_{\mathrm{a,T_1}}(\infty)$ the one of the enlarged model. The results related to the analyzed cases are shown in Fig. 7.

The results highlight the strong sensitivity of the basin response to the considered BCs methods. Notably, the use of L&K dashpots tends to underestimate the dynamic response because of the too much energy dissipated along the lateral boundary. Therefore, the use of this technique would imply a much greater horizontal extension of the model. By contrast, the use of EqualDOFs improves the basin response, with a slightly more pronounced amplification of the motion in proximity to the lateral boundaries due to the full reflection of the seismic energy reaching the boundary. Still, the free-field and massive columns provide a more consistent performance, more closely aligned to the one of the reference enlarged domain. A further element of interest lies in the asymmetry of the outputs, observed both in terms of displacement and spectral acceleration ratios, stemming from the non-symmetric geometry of the basin.

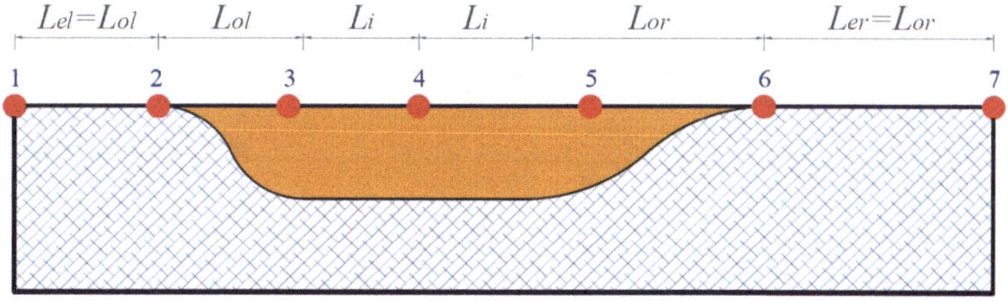

Fig. 5. Scrutiny nodes

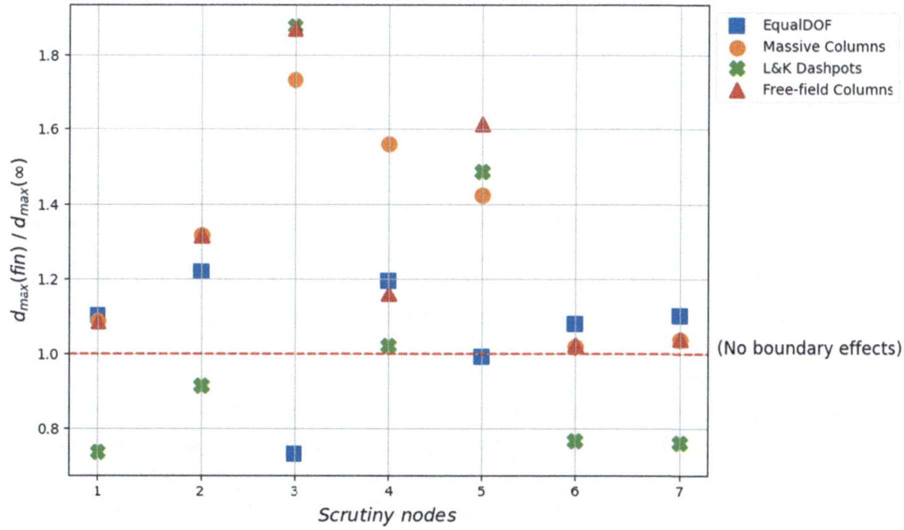

Fig. 6. Displacement ratio plotted as a function of the scrutiny nodes in the basin.

Interestingly, displacement ratios are lower on the right side of the basin (where the extent of the basin flank is larger) with respect to the left side, while spectral acceleration shows the opposite behavior. This is in agreement with the outcome of previous research (e.g., [1]), which highlighted the complex dependency of basin effects (both in terms of patterns and quantitative descriptive parameters) from numerous factors. Hence, the single modeling instance addressed in this paper only allows preliminary inferences regarding causal relationships.

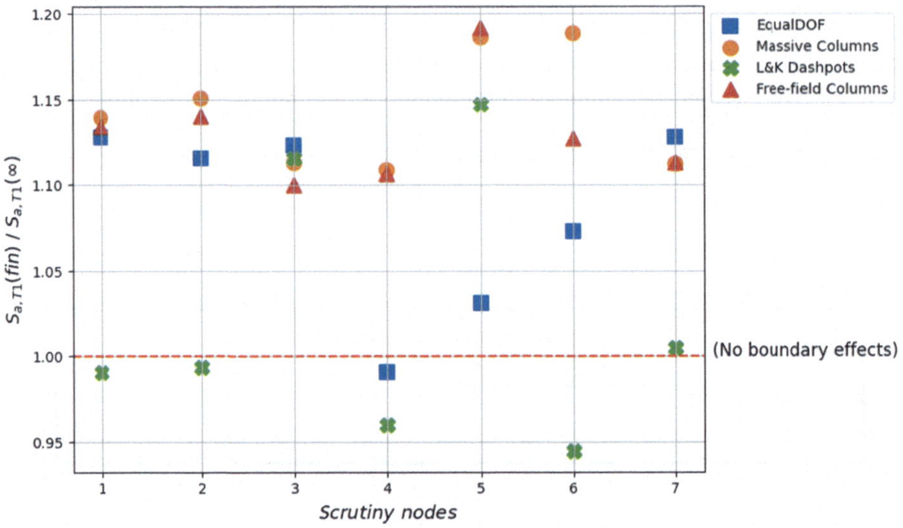

Fig. 7. Spectra acceleration ratio computed for the scrutiny nodes in the basin.

4 Concluding Remarks

This paper presents preliminary results to be further developed and assessed in the context of a broader research project aimed at quantifying the impact of boundary conditions on basin seismic performance, particularly under complex topographic and stratigraphic conditions. The outputs of the considered modeling scenario highlight the sensitivity of basin performance metrics, such as displacement and significant spectral accelerations, to the choice of modeling strategies for boundary conditions. The comparison among different boundary conditions provides initial insights into the potential over- or underestimation of the response associated with each technique. The spatial variability of these effects is also highlighted, emphasizing the importance of a rational selection of boundary modeling techniques. In fact, this comparative study confirms the higher potential of free-field and massive columns to provide more reliable predictions of site effects under complex conditions, compared to L&K absorbers and kinematic constraints requiring a significant greater extension of the soil domain.

Acknowledgements. This study was partially supported by DPC-ReLUIS Research Project 2022–2024.

References

1. Uzielli, M., Facciorusso, J., Madiai, C.: Statistical calibration of two-dimensional seismic aggravation effects for homogeneous basins. Soil Dyn. Earthq. Eng. **163**, 107533 (2022)
2. Madiai, C., et al.: Preliminary investigations into two-dimensional aggravation phenomena of seismic ground response in two-layered shallow sedimentary basins. Bullet. Geophys. Oceanogr. **63**(4), 575–596 (2022)
3. Sánchez-Sesma, F.J., Rivero, M.: Boundary Conditions for site response analysis in nonlinear soil models. Soil Dyn. Earthq. Eng. **24**(4) (2004)
4. Boulanger, R.W., Ziotopoulou, K.: Dynamic site response analysis with equivalent linear and nonlinear methods. J. Geotech. Geoenviron. Eng. **139**(10) (2013)
5. McKenna, F., Scott, M.H., Fenves, G.L.: Nonlinear finite-element analysis software architecture using object composition. J. Comput. Civ. Eng. **24**(1), 95–107 (2010)
6. Zhu, M., McKenna, F., Scott, M.H.: OpenSeesPy: Python library for the OpenSees finite element framework. SoftwareX **7**, 6–11 (2018)
7. Yang, Z., Elgamal, A., Parra, E.: Computational model for cyclic mobility and associated shear deformation. J. Geotech. Geoenvironmental Eng. **129**(12), 1119–1127 (2003)
8. OpenSeesWiki. https://opensees.ist.berkeley.edu/wiki/index.php?title=Main_Page
9. Code, P.: Eurocode 8: Design of Structures for Earthquake Resistance-Part 1: General Rules, Seismic Actions and Rules for Buildings, p. 10. European Committee for Standardization, Brussels (2005)
10. Gorini, D.N., Marrazzo, P., Nastri E., and Montuori, R.: Effectiveness of an innovative seismic resilient superelevation in an archetype, existing soil-structure system. Bullet. Earthq. Eng. (2024)
11. Lysmer, J., Kuhlemeyer, R.L.: Finite dynamic model for infinite media. J .Eng. Mech. Div. **95**(4), 859–877 (1969)
12. Gorini, D.N.A.R.F.: Seismic performance of slopes: do the dynamic features of the landslide matter?. In: Proceedings of the 10th European Conference on Numerical Methods in Geotechnical Engineering - NUMGE (2023), 26–28 June 2023. London, UK (2023)
13. Yeganeh, N., Fatahi, B.: Effects of choice of soil constitutive model on seismic performance of moment-resisting frames experiencing foundation rocking subjected to near-field earthquakes. Soil Dyn. Earthq. Eng. **121**, 442–459 (2019)

A Case Study on Nonlinear and Equivalent Linear Behavior of Site Response Analysis Using OpenSees, Plaxis3D and Deepsoil

Shima Sdeghzadeh[1,2(✉)], Atilla Ansal[1], and Cristoforo Demartino[2]

[1] Department of Civil Engineering, Özyegin University, Nisantepe, Orman Sk. No: 13, 34794 Istanbul, Türkiye
Shima.sadeghzadeh@uniroma3.it, Atilla.ansal@ozyegin.edu.tr
[2] Department of Architecture, Roma Tre University, Largo G. B. Marzi 10, 00153 Rome, Italy
Cristoforo.demartino@me.com

Abstract. According to seismic design guidelines, detailed consideration of soil properties is essential for major structures at specific locations to ensure accurate incorporation into the seismic design process. Site response analysis plays a crucial role in evaluating soil behavior and ground motion characteristics at the surface level. Consequently, design spectra and acceleration time histories are modified to reflect the influence of site effects on the overall structural design. In this study, three-directional numerical modeling and simulation of a soil profile in Istanbul were carried out using the OpenSees finite element program, with results validated against the PLAXIS finite element software. The analyses employed a hazard-compatible set of acceleration time histories alongside locally recorded data from the Istanbul Rapid Response Network's vertical array. The acceleration response spectra and surface-level acceleration time histories obtained from the simulations were subsequently compared with results from one-directional site response analyses and the recorded data from the vertical array.

Keywords: Site response analysis · seismic design · three-directional numerical simulation · Istanbul Rapid Response Network

1 Introduction

Earthquakes, with their unpredictable and multi-directional ground motion, present significant challenges to the resilience and safety of structures. Ensuring structural integrity during seismic events necessitates the consideration of ground motion from multiple directions, as the dynamic behavior of buildings and other structures is profoundly influenced by their interaction with the underlying soil-foundation system. Focusing exclusively on a single direction of ground motion, while neglecting the potential effects of multi-directional seismic forces, can lead to incomplete assessments of structural safety. Consequently, a comprehensive approach that accounts for the multi-directional nature of earthquake-induced ground motion is essential for achieving reliable seismic performance. This approach contributes to the development of structures capable of

X. Lu et al. (Eds.): EOS 2024, LNCE 638, pp. 202–212, 2025.
https://doi.org/10.1007/978-3-031-90690-9_18

withstanding future earthquakes, thereby enhancing the safety and resilience of the built environment.

Site response analysis (SRA) plays a pivotal role in earthquake engineering, serving as a foundation for creating safer, earthquake-resilient communities. SRA is a critical component of seismic hazard assessment, focusing on the evaluation of site conditions in relation to parameters such as ground motion characteristics, seismic event magnitude, shear wave velocity of the soil, analytical methods employed, and regional fault types. These factors significantly influence seismic motion behavior at the ground surface. However, it has been shown that fault types may not substantially affect ground surface responses under certain conditions (Sadeghzadeh, Ansal & Kurtuluş 2022).

The primary objective of this study is to conduct SRA to gain an in-depth understanding of how local geological and soil characteristics interact with seismic waves originating from an earthquake in multiple directions: East-West (EW), North-South (NS), and Vertical (V). By adopting this multidimensional perspective, the study aims to develop a robust framework for assessing ground surface behavior.

In regions with significant seismic activity, such as Türkiye, the importance of local site response analysis cannot be overstated. Recent seismic events, including the Kahramanmaraş earthquake in 2023 and the Izmir earthquake in 2020, have highlighted the vulnerabilities of structures and infrastructure, underscoring the need for a comprehensive understanding of local geological and geotechnical conditions. These events emphasize the value of employing advanced analytical techniques to enhance the resilience of infrastructure, ultimately reducing the social and economic impacts of seismic disasters.

This research particularly emphasizes the necessity of adopting a three-directional (3D) approach in seismic analysis, as opposed to the traditional one-directional (1D) methods, to capture the multi-directional effects of earthquake ground motion on the surface accurately.

The study focuses on the Zeytinburnu soil profile in Istanbul, utilizing both nonlinear and equivalent linear soil behavior models as key components of seismic response analysis. While equivalent linear methods have been widely used due to their computational efficiency and simplicity, this research highlights the critical need to incorporate more complex nonlinear models to achieve reliable seismic predictions, especially in high-risk seismic zones like Istanbul.

Advanced computational tools, specifically Plaxis3D (Bentley 2022) and OpenSees (Mazzoni et al. 2006), are employed to perform three-directional nonlinear SRA for the Zeytinburnu region. Plaxis3D is particularly effective in modeling soil nonlinearity through elasto-plastic hysteretic models, such as the Hardening Soil model, providing a more accurate representation of soil response to seismic loading. Previous studies (Sadeghzadeh et al. 2023; Ansal et al. 2024) have demonstrated that incorporating three-directional site response analysis in Plaxis3D offers improved predictions of ground surface behavior compared to traditional one-directional equivalent linear methods. This study builds on those findings, aiming to further refine and validate the approach for seismic hazard analysis in Istanbul.

2 Site Condition

The Zeytinburnu Vertical Array Site, located in the southern region of Istanbul, Türkiye, serves as a critical location for the investigation presented in this study (Fig. 1). The site's soil profile, modeled using the DeepSoil program, exhibits a natural frequency of 0.3703 Hz and a natural period of 2.701 s. These parameters provide an initial understanding of the dynamic characteristics of the site, forming the basis for further analysis.

To accurately characterize the subsurface soil properties, particularly the shear wave velocity profile, Suspension PS Velocity Logging tests were performed within fluid-filled boreholes at depths of 30 m, 57 m, and 288 m (Fig. 2). The soil profile's average shear wave velocity was determined to be 382 m/sec. These boreholes were constructed using the rotary mud drilling method, ensuring precision and reliability in the collected data. Observations of drill cuttings during the process revealed significant geological transitions. At approximately 230 m in depth, weathered rock formations were encountered, gradually transitioning into more intact and resilient geological material at around 260 m.

For monitoring and data collection purposes, accelerometers were strategically installed within these boreholes at various depths. This setup enables continuous observation and recording of ground motion, offering valuable insights into the site's seismic response characteristics.

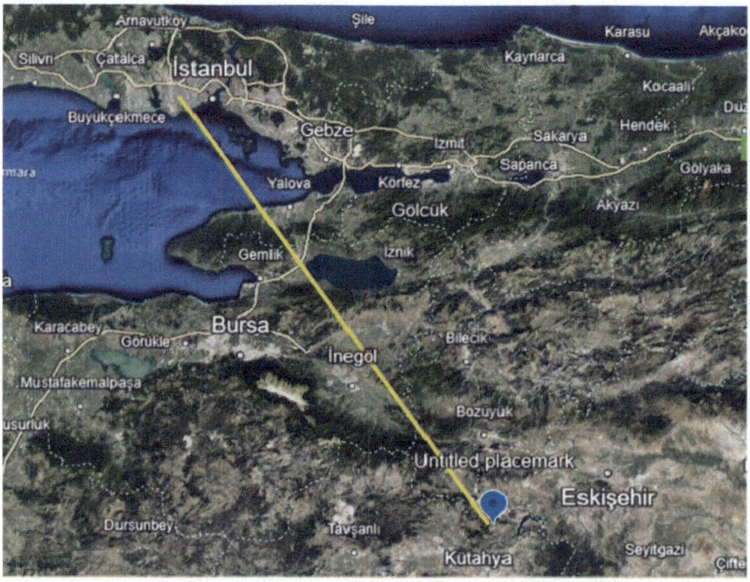

Fig. 1. Distance of study zone to happened earthquake

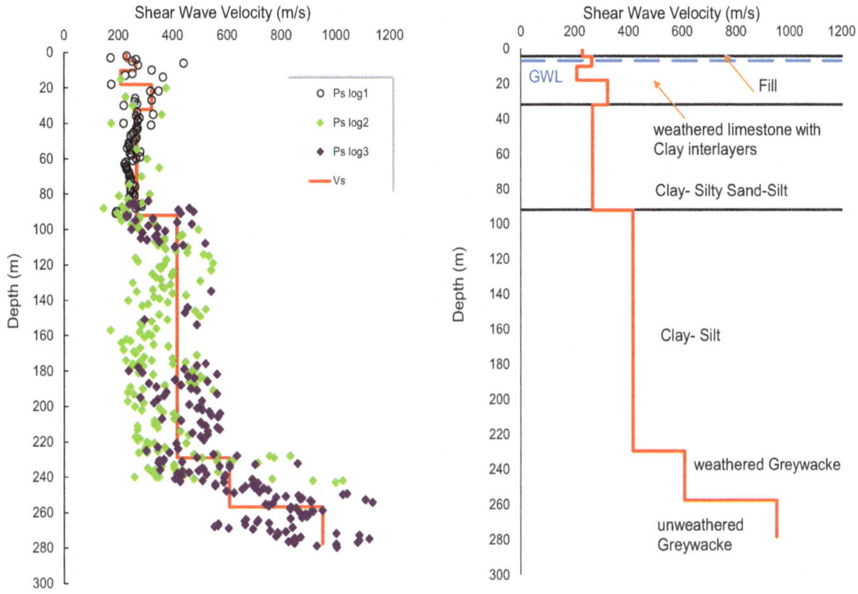

Fig. 2. Variation of shear wave velocity with depth and site stratigraphy

3 Input Seismic Motion

For this parametric study, we analyzed a seismic event recorded in Kütahya, Türkiye, utilizing data from the Istanbul Rapid Response Network and Istanbul Vertical Array stations. This earthquake, which occurred in 2011, had a magnitude of Mw = 5.9, a duration of 134 s, and a distance of approximately 180 km from the studied borehole.

The earthquake's time histories were recorded at depths of 30 m, 57 m, and 288 m, as well as at the surface, using strategically positioned sensors (Table 1). These sensors captured essential seismic data, including acceleration time histories in the East-West (EW), North-South (NS), and Vertical (V) directions. This data forms the basis for understanding the site's behavior during the site response analysis. The ground motion parameters recorded in various directions are summarized in Table 2.

To evaluate the seismic interactions across different depths, site response analyses were conducted using two complementary approaches. First, a three-dimensional (3D) analysis was performed, incorporating the EW, NS, and V components of the Kütahya earthquake simultaneously, employing a nonlinear method. This approach enabled a comprehensive assessment of the multidirectional effects of ground motion. In addition, a one-dimensional (1D) analysis was carried out, focusing on either the EW or NS direction, using an equivalent linear method to estimate ground motion amplification.

The results of these analyses include calculated acceleration time histories and spectral accelerations at the specified depths. These outputs provide critical insights into the site's seismic response and allow for a detailed comparison of the 3D nonlinear and 1D equivalent linear approaches. Table 1 also presents the channel numbers corresponding to the directions of the recorded accelerations, facilitating the organization and interpretation of the seismic data.

Table 1. Channel numbers and corresponding sensors.

No	Sensor	No	Sensor	No	Sensor	No	Sensor
1	288EW	4	Surface EW	7	57V	10	30NS
2	288NS	5	57EW	8	Surface NS	11	30V
3	288V	6	57NS	9	30EW	12	Surface V

Table 2. Ground motions parameters in east-west, north-south and vertical directions.

Parameter	(EW)	(NS)	(V)
Max. Acceleration (g)	0.00062	0.00115	0.00150
Time of Max. Acceleration (sec)	35.8600	35.6550	0.01500
Max. Velocity (cm/sec)	0.12798	0.18761	0.17437
Time of Max. Velocity (sec)	38.9600	36.8050	43.99500
Max. Displacement (cm)	0.09583	0.16544	1.34386
Time of Max. Displacement (sec)	42.1400	47.5200	133.9950
Vmax/Amax: (sec)	0.21137	0.16630	0.11850
Acceleration RMS: (g)	0.00008	0.00012	0.00009
Arias Intensity: (m/sec)	0.00001	0.00003	0.00002
Acceleration Spectrum Intensity (g*sec)	0.00060	0.00105	0.00057
A95 parameter (g)	0.00040	0.00085	0.00046
Predominant Period (sec)	0.54000	0.50000	0.04000
Mean Period (sec)	0.99870	1.07646	1.01040
Damage Index((g)^c)	0.00001	0.00002	0.00001
Number of Effective Cycles	6.89394	2.58277	1.51108

The analysis of recorded values revealed notable amplifications in seismic motions, as illustrated in Fig. 3. These findings highlight the unique seismic response characteristics of the Zeytinburnu soil profile, emphasizing the crucial role of understanding such amplifications in the context of seismic hazard assessment. This investigation underscores the importance of incorporating these insights into the design and planning of infrastructure in this region, contributing to enhanced safety and resilience against seismic events.

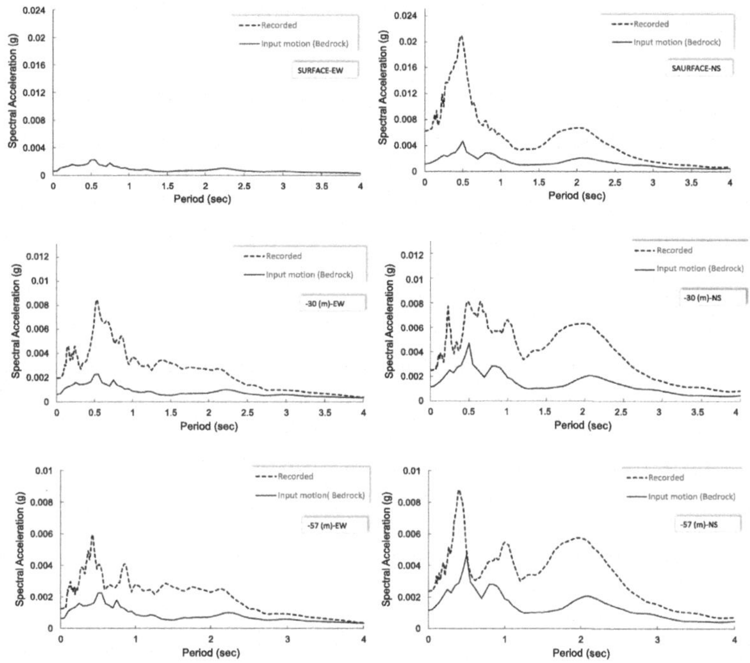

Fig. 3. Bedrock and various recorded motions by accelerometers

4 Method of Analysis

This study employs a comprehensive approach to site response analysis (SRA), utilizing both equivalent linear and nonlinear methods to predict ground surface movements during seismic events. Each method offers distinct advantages, enabling a thorough evaluation of soil behavior under dynamic loading conditions.

4.1 Equivalent Linear Analysis

The equivalent linear analysis, implemented using the DeepSoil program (Hashash et al. 2016), is based on the premise that soil behavior under cyclic loading during earthquakes can be effectively approximated through simplified models. This approach, originally proposed by Seed and Idriss (Idriss & Seed 1970), has gained widespread recognition due to its balance between computational efficiency and accuracy.

In this methodology, the modulus reduction and hysteretic damping characteristics of the soil are iteratively adjusted based on the shear strain levels in each soil layer. This iterative process refines the soil's shear modulus and damping values to match strain-specific behaviors, bridging the gap between purely linear and fully nonlinear analysis. By incorporating soil-specific modulus reduction and damping curves, this method provides a more accurate representation of soil response compared to purely linear methods, making it a valuable tool for characterizing dynamic soil behavior.

4.2 Nonlinear Analysis

Given the inherently complex, multi-phase nature of soil, nonlinear analysis is considered a robust method for capturing soil behavior under dynamic loading conditions. Nonlinear analysis integrates elasto-plastic properties to provide a more realistic depiction of soil responses to varying stress levels. For this study, fully nonlinear site response analyses were performed using time-domain methods, which solve the equation of motion incrementally over small-time steps.

4.3 Plaxis3D Nonlinear Model

In Plaxis3D, the Hardening Soil model (Schanz, Vermeer & Bonnier, 1999) was employed as the nonlinear material model. This model simulates key soil behaviors, including stress history and plastic yielding, using three critical stiffness parameters:

$\mathbf{E_{50}}$: Secant stiffness from a standard drained triaxial test.

$\mathbf{E_{ur}}$: Unloading/reloading stiffness.

$\mathbf{E_{oed}}$: Tangent stiffness for primary oedometer loading.

Following industry best practices, default ratios from the Plaxis3D library were applied, where $\mathbf{E_{50}} = \mathbf{E_{oed}}$ and $\mathbf{E_{ur}} = \mathbf{3E_{50}}$. These parameters offer enhanced accuracy in predicting soil deformation magnitudes (Rafaa, Obrzud & Truty, 2018).

Dynamic analyses using the Hardening Soil model incorporate Rayleigh damping to simulate energy dissipation at strain levels exceeding 10^{-4} to 10^{-2}. For this study, damping ratios (ζ_1 and ζ_2) were set at 1% based on recommendations from Seed et al. (1986) and Vucetic & Dobry (1991). The second frequency (f_2) was calculated as $\mathbf{f_2 = nf_1}$, where n represents the nearest odd number of ratios between the fundamental ground motion frequency and the fundamental frequency of the soil profile, as proposed by Hudson (1994).

The soil profile for Zeytinburnu was modeled as a $1 \times 1\ m^2$ cross-section extending to the engineering bedrock. This dimension ensured minimal computational cost without compromising accuracy at the ground surface. A fine mesh was utilized, maintaining an average element size less than 1/8th of the wavelength, adhering to the guidelines of Kuhlemeyer & Lysmer (1973). Absorbing boundary conditions were applied, including compliant base boundaries at the base and free-field boundaries along the sides, to account for wave propagation effects.

4.4 OpenSees Nonlinear Model

In OpenSees (Mazzoni et al. 2006), nonlinear analysis was conducted using SSPBRICK elements to model the soil. The process involved two stages:

Gravity Analysis: Simulating linear elastic material behavior under static gravitational loading.

Dynamic Analysis: Capturing the soil's nonlinear response during seismic events.

Boundary conditions were defined by fixing the base nodes in the vertical direction to simulate the underlying bedrock. Equal degrees of freedom (DOF) were assigned along the sides to replicate simple shear deformation patterns. A Lysmer-Kuhlemeyer dashpot (Kuhlemeyer & Lysmer 1973) was modeled using a viscous-uniaxial material in a zero-length element at the base to simulate energy dissipation during seismic excitation.

5 Site Response Analysis

The primary objective of this research is to perform a detailed comparative analysis of site response for the Zeytinburnu profile, utilizing two distinct methodologies for seismic analysis. The first methodology involves a one-directional site response analysis conducted through the DeepSoil program, which considers ground motion along either the East-West (EW) or North-South (NS) directions. The second methodology employs a three-directional site response analysis using a finite element (FE) program, simultaneously incorporating all three ground motion components: EW, NS, and vertical (V). This study seeks to evaluate the applicability and performance of these contrasting approaches, focusing on one-directional equivalent linear methods versus three-directional nonlinear methods.

The research outcomes include insights into the site-specific seismic response characteristics of the Zeytinburnu profile. Acceleration time histories and spectral accelerations are calculated both at the ground surface and at various depths within the site (Figs. 4 and 5). This multi-depth analysis is crucial for understanding the complex propagation

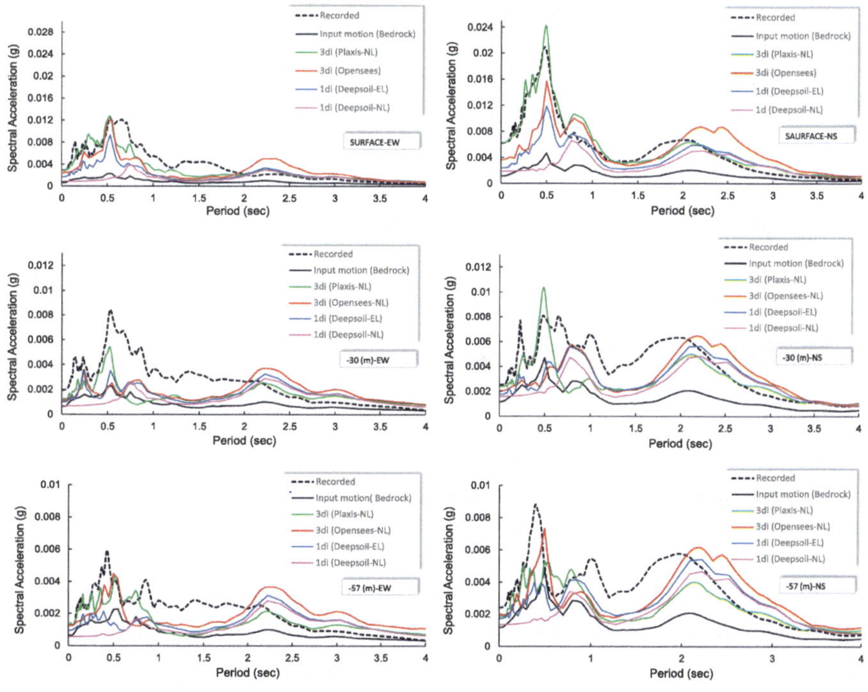

Fig. 4. Spectral acceleration results with respect to ground surface, –30 m and –57 m in East-West and North-South directions.

of seismic waves through the geological layers beneath the surface. By analyzing seismic wave behavior across these depths, the study provides a nuanced perspective on the interaction between seismic motion and soil strata.

Moreover, the data derived from both one-directional and three-directional analyses are compared against recorded ground motions captured by the Vertical Array of Istanbul during the Kütahya earthquake. This comparison serves as a critical benchmark, enhancing the reliability and validity of the analytical results.

In addition to its analytical contributions, this research offers practical insights into the strengths and limitations of one-directional versus three-directional site response analysis approaches, particularly in the unique geological context of the Zeytinburnu profile. These findings have implications for the selection of analysis methods in future seismic hazard assessments and infrastructure design in similar high-risk seismic regions.

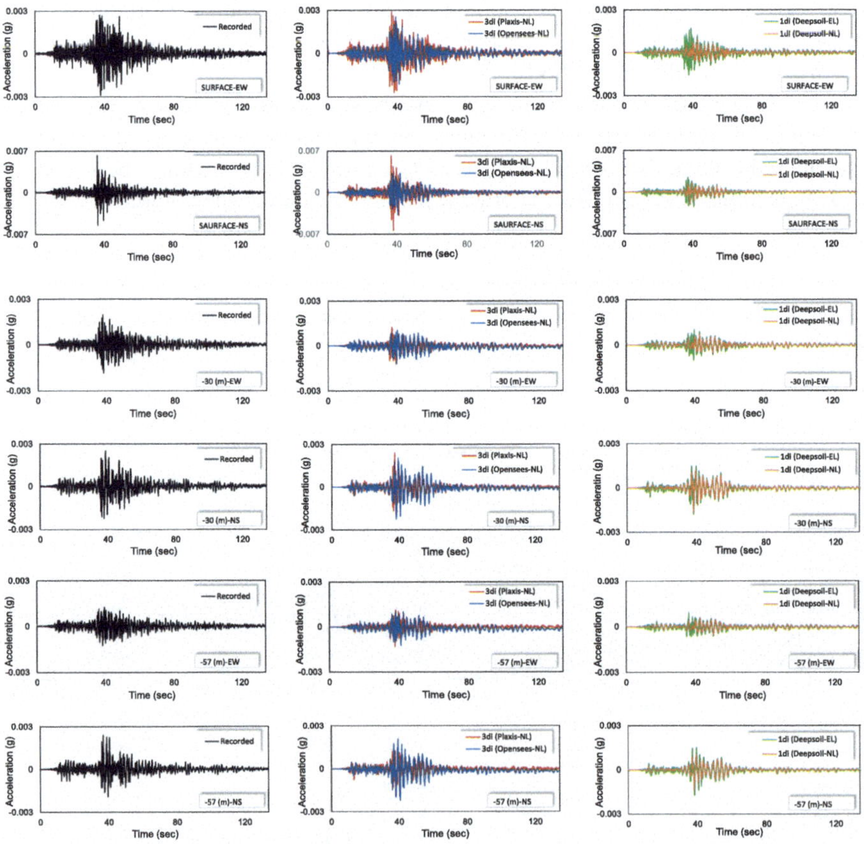

Fig. 5. Acceleration time history results with respect to ground surface, –30 m and –57 m in East-West and North-South directions.

6 Conclusion

This study presented a series of site response analyses for the Zeytinburnu soil profile in Istanbul, Türkiye, under the seismic excitations of the Kütahya earthquake, which had a magnitude of 5.9 and a duration of 134 s. The analyses incorporated simultaneous ground motions in the East-West (EW), North-South (NS), and Vertical (V) directions. The availability of vertical seismometer arrays, including both surface and downhole recordings, provided a valuable opportunity for validating site response models and enhancing the reliability of the findings.

The research emphasizes the importance of considering multi-directional ground motions in site response analysis. The comparative evaluation between three-directional and one-directional approaches revealed several critical trends. When examining spectral acceleration, both the Plaxis3D and OpenSees models demonstrated that the three-directional site response analysis produced results more consistent with accelerometer-recorded ground motions in the EW and NS directions. This finding highlights the efficacy of a comprehensive approach in accurately capturing ground response.

The consistency observed in three-directional analysis was most pronounced at the ground surface, with slight reductions at greater depths. However, the method's effectiveness remained robust across various soil layers, underscoring its reliability in capturing the dynamic behavior of seismic waves. In contrast, the one-directional site response analysis performed using DeepSoil, both with equivalent linear and nonlinear methods, generally underestimated ground motions compared to the three-directional approaches. These results underline the limitations of relying exclusively on a one-directional methodology.

In terms of acceleration time histories, the three-directional site response analysis showed greater agreement with recorded data, outperforming the one-directional method, which tended to yield lower ground peak accelerations (GPAs). This further demonstrates the three-directional method's superiority in accurately predicting ground motions and capturing the complex dynamics of soil behavior.

In summary, this research underscores the critical importance of incorporating multi-directional ground motion in site response analyses, particularly for regions with complex geological conditions and significant seismic activity. The three-directional approach offers a more comprehensive and reliable framework for understanding soil behavior and predicting seismic responses, which is essential for accurate seismic hazard assessments and the development of resilient infrastructure. These findings highlight the need to move beyond traditional one-directional methods to achieve a more nuanced and effective understanding of site-specific seismic risks.

References

Ansal, A., Tönük, G., Sadeghzadeh, S.: Site response analysis in performance based approach. Soil Dyn. Earthq. Eng. **5** (2024)

Bentley, Buy PLAXIS 3D|Geotechnical Analysis Software|Virtuosity (2022)

Obrzud, R.: Andrzej Truty, The Hardening Soil Model-A Practical Guidebook (2018)

Hashash, Y.M.A., Musgrove, M.I., Harmon, J.A., Groholski, D.R., Phillips, C.A., Park, D.: DEEPSOIL 6.1, User Manual. IL, Board of Trustees of University of Illinois at Urbana-Champaign, Urbana (2016)

Martin, H., Idriss, I.M., Beikae, M.: QUAD4M: a computer program to evaluate the seismic response of soil structures using finite element procedures and incorporating a compliant base. Center for Geotechnical Modeling, Department of Civil and Environmental Engineering, University of California, Davis., Hudson, MARTIN (1994)
Idriss, I.M., Seed, H.B.: Seismic response of soil deposits. J. Soil Mech. .Found. Division **96**(2), 631–638 (1970)
Kuhlemeyer, R.L., Lysmer, J.: Finite element method accuracy for wave propagation problems. J. Soil Mech. Found. Div. **99**(5), 421–427 (1973)
Mazzoni, S., McKenna, F., Scott, M.H., Fenves, G.L.: OpenSees command language manual. Pacific Earthq. Eng. Res. (PEER) Center, **264**, 137–158 (2006)
Sadeghzadeh, S., Ansal, A., Kurtuluş, A.: The Effects of Fault Type for the Selection of Input Acceleration Records for Site Response Analysis (2022)
Sadeghzadeh, S., Ansal, A.: A Comparison Study Between 1d And 3d Site Response Analyses Based On Observed Earthquake Acceleration Records… (2023)
Schanz, T., Vermeer, P.A., Bonnier, P.G.: 'The hardening soil model: formulation and verification', beyond 2000 in computational geotechnics. ten years of PLAXIS international. In: Proceedings of the international symposium, Amsterdam, March, pp. 281–296 (1999)
Seed, H.B., Wong, R.T., Idriss, I.M., Tokimatsu, K.: Moduli and damping factors for dynamic analyses of cohesionless soils. J. Geotechnical Eng. **112**(11), 1016–1032 (1986)
Vucetic, B.M., Member, A., Dobry, R.: Effect of Soil Plasticity On Cyclic Response (1991)

Advanced Modeling, Materials, and Optimization

Capacity Assessment of Full and Hollow-Core Rectangular Sections

Raihan Rahmat Rabi(✉) and Giorgio Monti

Sapienza University of Rome, Rome, Italy
raihan.rahmatrabi@uniroma1.it

Abstract. This study develops closed-form equations to calculate yield and ultimate moments and curvatures under combined axial load and uniaxial bending. These equations can be used to determine the force-displacement capacity of bridge piers, accounting for shear capacity and P-delta effects. To verify the accuracy of the proposed equations, a comparison is conducted with results from fiber-section analyses in OpenSeesPy, showing satisfactory agreement. The equations enable efficient multi-parametric and stochastic analyses, allowing for the rapid calculation of fragility curves, which is particularly beneficial for large-scale territorial risk studies. A typical application includes risk analysis of railway and highway networks, where the piers of statically determined bridges are critical components.

Keywords: hollow-core bridge piers · fragility curves · moment-curvature · closed-form equations · sectional analysis

1 Introduction

When conducting seismic risk assessments on territorially distributed railway or highway networks, the availability of efficient tools for generating fragility curves of critical nodes, such as bridges, is crucial. In large-scale networks, it is important to minimize the typically high computational effort required to develop a vast number of fragility curves. Fragility curves express the probability that a structural system will exceed a certain damage state during seismic events of varying intensities and can be constructed using either nonlinear dynamic analysis (NDA) [1] or nonlinear static analysis, also known as pushover analysis [2, 3]. NDA is considered the most accurate and detailed method, as it captures the evolution of structural nonlinear behavior over time, providing deep insights into complex structural responses during earthquakes. However, despite these benefits, NDA demands advanced modeling skills and substantial computational time, making it less practical for large-scale risk studies.

To overcome this limitation and enable more efficient analysis on a larger scale, particularly for the structurally simpler bridge piers, an analytical approach can be adopted. In this method, pushover curves are simplified as bilinear curves that pass through two key points—yield and ultimate—whose coordinates can be derived analytically. Developing closed-form equations for yield and collapse moments and curvatures is the first

X. Lu et al. (Eds.): EOS 2024, LNCE 638, pp. 215–228, 2025.
https://doi.org/10.1007/978-3-031-90690-9_19

step, followed by deriving the corresponding yield and collapse forces and displacements. This approach significantly reduces computational demands while still delivering accurate and representative results. By applying this procedure, analysts can generate fragility curves efficiently from basic input parameters, facilitating large-scale seismic risk studies. Simplified structural assessment techniques like these are frequently used in territorial seismic risk analysis [4–9], which involves evaluating the potential effects of earthquakes on a region's infrastructure.

Pfrang et al. [10] proposed a method to evaluate the relationship between axial load, moment, and curvature, providing a concise overview of the calculation process, though practical application still required computer programming. Carreira and Chu [11] introduced a general method that solves a nonlinear second-degree equation to obtain the moment-curvature relationship, offering flexibility but still involving iterative calculations. Vebo and Ghali [12] developed an analytical approach for reinforced slabs under bidirectional bending moment and torsion, deriving moment-curvature relationships using limit analysis. Kwak and Kim [13] focused on the material nonlinear analysis of RC beams and proposed an approximate trilinear moment-curvature law, using section analysis and iterative methods to find a solution, while satisfying the translation equilibrium condition. They later enhanced this model by incorporating the effect of constant axial force [14], though it still involved iterative calculations. Kim and Lee [15] developed a numerical method to predict the behavior of RC columns under axial force and biaxial bending, validated through experimental tests and finite element analysis. Oller and Barbat [16] proposed a simplified moment-curvature damage evaluation model for RC highway bridge beams, aimed at assessing expected seismic performance, while maintaining equilibrium at the section level through a convergence criterion.

Building on this foundation, this paper contributes to the analysis of RC hollow-core bridge piers by proposing closed-form equations for yield and ultimate moments and curvatures under uniaxial bending and axial force. The formulations provide a robust and efficient method for describing the moment-curvature relationship, enabling accurate predictions of the force-displacement behavior of bridge piers. To validate these equations, an extensive parametric analysis has been performed. The analytical results derived from the proposed equations are compared with numerical results from fiber-section models in OpenSeesPy, demonstrating excellent accuracy. These findings confirm the reliability and effectiveness of the proposed equations in capturing the moment-curvature behavior of RC hollow-core sections.

2 Material Laws

2.1 Steel

A bilinear elastic-plastic stress-strain curve is adopted:

$$\sigma_s(\varepsilon_s) = f_y \min\left(\frac{\varepsilon_s}{\varepsilon_y}, 1\right) = f_y \min(e_s, 1) \tag{1}$$

where σ_s is the steel stress, as function of the steel strain ε_s, ε_y is the steel yield strain (usually 0.002), f_y is the yield strength, and $e_s = \frac{\varepsilon_s}{\varepsilon_y}$ is the normalized strain. Notice that, in the elastic range, $\sigma_s(\varepsilon_s) = f_y e_s = E_s \varepsilon_s$, with E_s the elastic modulus. The steel strain is limited as $\varepsilon_s \leq \varepsilon_{su}$.

In the equilibrium equations of an RC section, the contribution of the web reinforcement distributed along the section depth is expressed by the term $\beta(\varepsilon_s)$, representing the area of the current tension/compression stress block normalized to $f_y \varepsilon_s$:

$$\beta(\varepsilon_s) = \frac{\int_0^{\varepsilon_s} \sigma_s(\varepsilon) d\varepsilon}{f_y \varepsilon_s} \tag{2}$$

An approximate expression for Eq. (2) is:

$$\beta(\varepsilon_s) = \beta_1 \frac{\varepsilon_s}{\varepsilon_{cu}} \tag{3}$$

where $\beta_1 = 0.8$. Notice that the limit strain at the denominator is actually ε_{cu}, the concrete ultimate strain, because it is the maximum compression strain that steel can reach when it adheres to concrete. The coefficient ν, representing the position of the centroid of the stress-strain curve, can be assumed, without significant loss of accuracy, as constant and equal to $\nu = 0.4$.

2.2 Concrete

A triangle-rectangle law is adopted, as in [17]:

$$\sigma_c(\varepsilon_c) = f_c \min\left(\frac{\varepsilon_c}{\varepsilon_{c0}}, 1\right) \tag{4}$$

where σ_c is the concrete stress, as function of the concrete strain ε_c, f_c is the concrete compression strength, ε_{c0} is the strain at peak strength (usually 0.002). The concrete strain is limited as $\varepsilon_c \leq \varepsilon_{cu}$.

In the equilibrium equations of an RC section, the concrete contribution in compression is expressed by the term $\beta(\varepsilon_c)$, representing the area of the current compression stress block normalized to $f_c \varepsilon_c$:

$$\beta(\varepsilon_c) = \frac{\int_0^{\varepsilon_c} \sigma_c(\varepsilon) d\varepsilon}{f_c \varepsilon_c} \tag{5}$$

An approximate expression for Eq. (5) is:

$$\beta(\varepsilon_c) = \beta_1 \frac{\varepsilon_c}{\varepsilon_{cu}} \tag{6}$$

with $\beta_1 = 0.8$. The coefficient ν, representing the position of the centroid of the stress-strain curve, can be assumed, without significant loss of accuracy, as constant and equal to $\nu = 0.4$.

2.3 Confinement Effect

Confinement effect is described by two factors expressing the increase of confined concrete strength $f_{c,c}$ with respect to the unconfined concrete strength f_c and the increase of confined concrete ultimate strain $\varepsilon_{cu,c}$ with respect to the unconfined concrete ultimate strain ε_{cu}, respectively defined as:

$$\eta_f = \frac{f_{c,c}}{f_c} \text{ and } \eta_\varepsilon = \frac{\varepsilon_{cu,c}}{\varepsilon_{cu}} \tag{7}$$

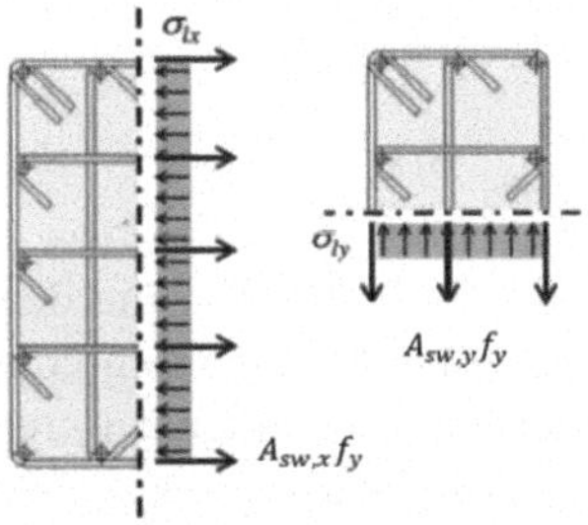

Fig. 1. Estimating the effective lateral confinement pressure $\sigma_2 = \alpha_n \alpha_s \sqrt{\sigma_{lx} \cdot \sigma_{ly}}$.

These two factors can be computed as [18]:

$$\eta_f = 1.000 + 5.0\tilde{\sigma}_2 \text{ for } \tilde{\sigma}_2 \leq 0.05 \tag{8}$$

$$\eta_f = 1.125 + 2.5\tilde{\sigma}_2 \text{ for } \tilde{\sigma}_2 > 0.05$$
$$\eta_\varepsilon = 1 + 0.2\frac{\tilde{\sigma}_2}{\varepsilon_{cu}} \tag{9}$$

where $\tilde{\sigma}_2$ is the normalized effective lateral confinement pressure, effectively computed as:

$$\tilde{\sigma}_2 = \frac{\sigma_2}{f_{ck}} = \frac{1}{3}\alpha_n \alpha_s \omega_w \tag{10}$$

where f_{ck} is the concrete characteristic strength and:

$$\alpha_n = \left(1 - \frac{\sum b_i^2}{6 b_x b_y}\right) \approx 1 - \frac{1}{3 b_x b_y}\left(\frac{b_x^2}{n_{ct,y} + 1} + \frac{b_y^2}{n_{ct,x} + 1}\right) \tag{11}$$

$$\alpha_s = \left(1 - \frac{s}{2b_x}\right)\left(1 - \frac{s}{2b_y}\right) \tag{12}$$

$$\omega_w = \frac{A_{sw,x} b_x + A_{sw,y} b_y}{b_x b_y s} \cdot \frac{f_y}{f_c} \tag{13}$$

where b_i is the spacing between two consecutively tie-restrained or corner bars, b_x and b_y are the sides of the confined concrete core, $n_{ct,x}$ and $n_{ct,y}$ are the number of transverse ties in the two directions, and $A_{sw,x}$ and $A_{sw,y}$ are the stirrups areas in the two directions. The approximation of α_n in Eq. (11). Applies in the common situation of equally spaced longitudinal bars along the core sides.

Following the two definitions above, the concrete stress block area in Eq. (6) is then multiplied by:

$$\beta_{conf} = \eta_f \left(1 - \frac{0.2}{\eta_\varepsilon}\right) \tag{14}$$

while the concrete stress block centroid is at:

$$\nu_{conf} = 0.5 - \frac{0.1}{\eta_\varepsilon} \tag{15}$$

2.4 Section Geometry and Normalized Quantities

All equations proposed in this work are fully parameterized in terms of geometry and material constitutive properties and are formulated for the general case of hollow-core sections. The section has width b, effective depth d, web thickness t_w, flange thickness t_f, and the hole width is $b_w = b - 2t_w$. If needed, equations for full sections are obtained by letting $t_w = b/2$ (i.e., $b_w = 0$) and $t_f/2 = c$, as shown in Fig. 2, where c is the concrete cover.

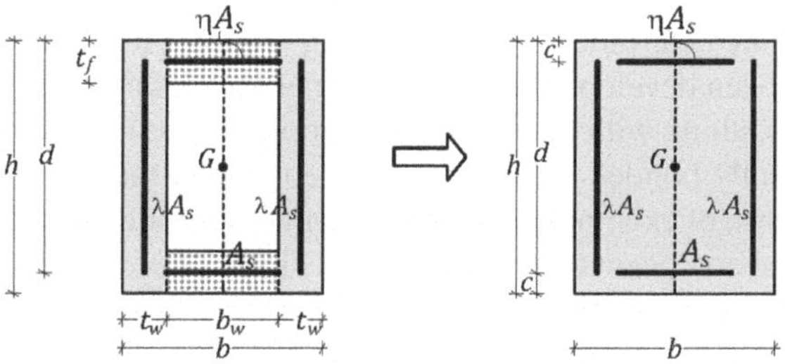

Fig. 2. Hollow and full section geometry.

Reinforcement is placed along four sides. Top and lateral (per side) reinforcements are expressed as ratio of the bottom tensile reinforcement through the factors η and λ, respectively. In hollow sections, top and bottom reinforcement are made of two (outer and inner) layers and are represented by an equivalent area (sum of the two layers) placed at the flange mid-thickness, at a distance $t_f/2$ from the section top. In full sections, top and bottom reinforcement are placed at a distance c from the section top. The lateral reinforcement is considered as smeared over the depth $d - t_f/2$, so that its amount per unit length, including both webs, is $\frac{2\lambda A_s}{d-t_f/2} = \frac{4\lambda A_s}{2d-t_f}$.

The normalized axial load n_S and the longitudinal reinforcement mechanical ratio ω_s are introduced as:

$$n_S = \frac{N_S}{bdf_c} = \frac{\overline{\sigma}_c}{f_c} \tag{16}$$

$$\omega_s = \frac{A_s f_y}{bdf_c} = \rho_s \frac{f_y}{f_c} \tag{17}$$

where σ_c denotes the average concrete stress – an essential parameter at the time most bridge piers were designed – and ρ_s the longitudinal reinforcement geometrical ratio.

In normalized hollow-core rectangular sections, flange and web thicknesses are:

$$\tau_f = \frac{t_f}{d} \text{ and } \tau_w = \frac{t_w}{b} \tag{18}$$

while the hole width is:

$$\beta_w = \frac{b_w}{b} = 1 - 2\tau_w \tag{19}$$

3 Curvature and Moment Capacity

The moment-curvature response is described as a bilinear curve passing through two essential points: the point where yielding takes place in the tension reinforcement, identified as the Damage Limit State (DLS), and the point where either compressed concrete crushes or steel fractures, identified as the Collapse Limit State (CLS). The coordinates of these two points are determined in the following sections. It should be underlined that all equations have been developed by making use of the classical equilibrium and compatibility equations, along with the usual assumptions of section planarity, zero tension concrete, and perfectly bonded rebars. As a side comment, the latter assumption might not be justified in case of existing piers with smooth or corroded reinforcement. In those cases, a loss of adhesion occurs, with consequent slippage of the reinforcing bars. This can also result in a significant rebars strength and stiffness reduction, which, if needed, can be accounted for as explained in [19].

3.1 Damage Limit State (DLS, Tension Reinforcement Yields)

The strain and stress profiles at yield are shown in Fig. 3. The section yields when the bottom tension steel attains the strain $\varepsilon_s = \varepsilon_y$. In this condition, the concrete strain in compression ε_c and the steel strain in compression ε_s' are not known. The lateral steel rebars follow the stress profile shown on the right of Fig. 3. For the sake of simplicity, it is assumed that the lateral rebars are not yielded, which is certainly true for those in tension and a fair assumption for those in compression.

The compatibility equation reads:

$$y_y = d - \frac{\varepsilon_y}{\phi_y} \tag{20}$$

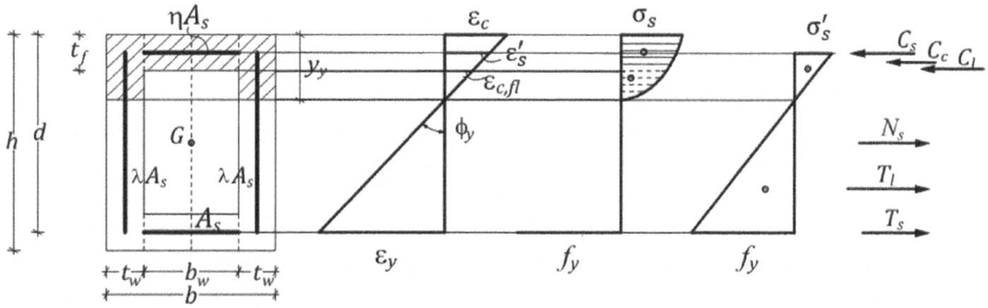

Fig. 3. Cross-section with geometrical and reinforcement quantities, strain profile at yield of bottom steel, stress profiles of concrete, of top/bottom steel, and of lateral steel, and resultants.

where y_y is the distance of the neutral axis from the topmost compressive fiber, ε_y is the steel yield strain and ϕ_y is the yield curvature. In normalized form, Eq. (20) becomes:

$$\xi_y = \frac{y_y}{d} = 1 - \frac{\varepsilon_y}{\phi_y d} \tag{21}$$

Other quantities of interest are:

- the (unknown) strain at the topmost concrete layer:

$$\varepsilon_c = \phi_y y_y = \phi_y d \xi_y = \varepsilon_y \frac{\xi_y}{1-\xi_y} \tag{22}$$

- the (unknown) strain at the flange bottommost concrete layer:

$$\varepsilon_{c,fl} = \varepsilon_c\left(1 - \frac{t_f}{y_y}\right) = \varepsilon_c\left(1 - \frac{\tau_f}{\xi_y}\right) \tag{23}$$

- the (unknown) strain in the top bars:

$$\varepsilon'_s = \phi_y\left(y_y - \frac{t_f}{2}\right) = \phi_y d\left(\xi_y - \frac{\tau_f}{2}\right) \tag{24}$$

which can be further normalized with respect to the steel yield strain as:

$$e'_s = \frac{\varepsilon'_s}{\varepsilon_y} = \frac{\phi_y d}{\varepsilon_y}\left(\xi_y - \frac{\tau_f}{2}\right) = \frac{\xi_y - \frac{\tau_f}{2}}{1-\xi_y} \tag{25}$$

This term serves to express the stress in the top bars as $f_y e'_s$ as in Eq. (1).

Yield Curvature

The yield curvature is found by imposing the translational equilibrium of the forces acting on the cross-section as shown in Fig. 3:

$$\begin{aligned}&\beta(\varepsilon_c) b y_y f_c - \beta(\varepsilon_{c,fl}) b_w (y_y - t_f) f_c + \eta A_s f_y e'_s + \frac{4\lambda A_s}{2d - t_f}\frac{1}{2}\left(y_y - \frac{t_f}{2}\right) f_y e'_s \\ &= N_S + \frac{4\lambda A_s}{2d - t_f}\frac{1}{2}(d - y_y) f_y + A_s f_y\end{aligned} \tag{26}$$

With reference to Fig. 3, the first two terms of the first member represent the resultant concrete compressive force C_c, where the second term (the McAulay brackets $\langle x\rangle$ yield 0 if $x \leq 0$) is the (subtracted) contribution of the hollow core. The third and fourth terms represent the resultant forces of the top C_s and lateral C_l reinforcement in compression, respectively. In the second member, the first term is the axial force N_S, applied at the cross-section centroid, while the second and third terms represent the resultant forces of the tension lateral T_l and bottom T_s reinforcement, respectively.

Using the definition of $\beta(\varepsilon_c)$ given in Eq. (6), and by replacing Eqs. (22) and (23), Eq. (26) becomes:

$$\begin{aligned}&\beta_1\frac{\varepsilon_y}{\varepsilon_{cu}}\frac{\xi_y}{1-\xi_y}by_yf_c-\beta_1\frac{\varepsilon_y}{\varepsilon_{cu}}\frac{\xi_y}{1-\xi_y}\left(1-\frac{\tau_f}{\xi_y}\right)b_wy_y-t_ff_c+\eta A_sf_ye_s'\\&+\frac{4\lambda A_s}{2d-t_f}\frac{1}{2}\left(y_y-\frac{t_f}{2}\right)f_ye_s'=N_S\\&+\frac{4\lambda A_s}{2d-t_f}\frac{1}{2}(d-y_y)f_y+A_sf_y\end{aligned} \tag{27}$$

After dividing by bdf_c, multiplying by $\left(1-\xi_y\right)$ and setting $\overline{\varepsilon}_y=\frac{\varepsilon_y}{\varepsilon_{cu}}$, based on the definitions in Eqs. (20) to (25), and finally using the normalized quantities in Eqs. (16) and (17), Eq. (27) takes on the normalized form:

$$\begin{aligned}&\beta_1\overline{\varepsilon}_y\xi_y^2-\beta_1\overline{\varepsilon}_y\beta_w\xi_y-\tau_f^2+\eta\omega_s\left(\xi_y-\frac{\tau_f}{2}\right)+\frac{2\lambda\omega_s}{2-\tau_f}\left(\xi_y-\frac{\tau_f}{2}\right)^2\\&=n_S\left(1-\xi_y\right)+\frac{2\lambda\omega_s}{2-\tau_f}\left(1-\xi_y\right)^2+\omega_s\left(1-\xi_y\right)\end{aligned} \tag{28}$$

The solution for ξ_y is found by performing a Taylor expansion about $\widehat{\xi}_y=0$. This point produces a simple yet effective approximate solution over the range of interest of the basic parameters, $\omega_s=0.03-0.30$, $\eta=0.5-1.0$, $\lambda=0.0-2.50$, $\tau_f=0.0-0.25$, $\beta_w=0.0-0.75$, $n_S=0.0-0.2$. The expression so obtained is:

$$\xi_y=\begin{cases}\dfrac{n_S+\omega_s\left[(1+\lambda)+\frac{\tau_f}{2}(\eta+\lambda)\right]+\beta_1\beta_w\overline{\varepsilon}_y\tau_f^2}{n_S+\omega_s(1+\eta+2\lambda)+2\beta_1\beta_w\overline{\varepsilon}_y\tau_f} & \geq\tau_f\\[2ex]\dfrac{n_S+\omega_s\left[(1+\lambda)+\frac{\tau_f}{2}(\eta+\lambda)\right]}{n_S+\omega_s(1+\eta+2\lambda)} & \leq\tau_f\end{cases} \tag{29}$$

depending on whether the neutral axis falls below or in the upper flange.

The yield curvature is then obtained from Eq. (21) as:

$$\phi_y=\frac{\varepsilon_y}{d}\frac{1}{1-\xi_y} \tag{30}$$

Yield moment

The yield moment is found from the rotation equilibrium about the section centroid:

$$
\begin{aligned}
M_y = {} & \beta(\varepsilon_c) b y_y f_c \left[\frac{1}{2}\left(d + \frac{t_f}{2}\right) - \nu_1 y_y\right] + \\
& - \beta(\varepsilon_{c,fl}) b_w y_y - t_f f_c \left[\frac{1}{2}\left(d + \frac{t_f}{2}\right) - t_f - \nu_1 (y_y - t_f)\right] \\
& + \eta A_s f_y e'_s \frac{1}{2}\left(d - \frac{t_f}{2}\right) + \frac{4\lambda A_s}{2d - t_f} \frac{1}{2}\left(y_y - \frac{t_f}{2}\right) f_y e'_s \left[\frac{1}{2}\left(d - \frac{t_f}{2}\right) - \frac{1}{3}\left(y_y - \frac{t_f}{2}\right)\right] \\
& + \frac{4\lambda A_s}{2d - t_f} \frac{1}{2}(d - y_y) f_y \left[\frac{1}{2}\left(d - \frac{t_f}{2}\right) - \frac{1}{3}(d - y_y)\right] + A_s f_y \frac{1}{2}\left(d - \frac{t_f}{2}\right)
\end{aligned}
\tag{31}
$$

where each of the six terms at the second member represent, in sequence, the contribution of: concrete in compression, (subtracted) hollow core in compression, compression steel, lateral steel above and below the neutral axis, and tension steel.

Using the definition of $\beta(\varepsilon_c)$ given in Eq. (6), and by replacing Eqs. (22) and (23), Eq. (31). Becomes:

$$
\begin{aligned}
M_y = {} & \beta_1 \frac{\varepsilon_y}{\varepsilon_{cu}} \frac{\xi_y}{1 - \xi_y} b y_y f_c \left[\frac{1}{2}\left(d + \frac{t_f}{2}\right) - \nu_1 y_y\right] + \\
& -\beta_1 \frac{\varepsilon_y}{\varepsilon_{cu}} \frac{\xi_y}{1 - \xi_y}\left(1 - \frac{\tau_f}{\xi_y}\right) b_w y_y - t_f f_c \left[\frac{1}{2}\left(d + \frac{t_f}{2}\right) - t_f - \nu_1 (y_y - t_f)\right] \\
& + \eta A_s f_y e'_s \frac{1}{2}\left(d - \frac{t_f}{2}\right) + \frac{4\lambda A_s}{2d - t_f} \frac{1}{2}\left(y_y - \frac{t_f}{2}\right) f_y e'_s \left[\frac{1}{2}\left(d - \frac{t_f}{2}\right) - \frac{1}{3}\left(y_y - \frac{t_f}{2}\right)\right] \\
& + \frac{4\lambda A_s}{2d - t_f} \frac{1}{2}(d - y_y) f_y \left[\frac{1}{2}\left(d - \frac{t_f}{2}\right) - \frac{1}{3}(d - y_y)\right] + A_s f_y \frac{1}{2}\left(d - \frac{t_f}{2}\right)
\end{aligned}
\tag{32}
$$

After dividing by $bd^2 f_c$, and setting $\bar{\varepsilon}_y = \frac{\varepsilon_y}{\varepsilon_{cu}}$, based on the definitions in Eqs. (20) to (25), and using the normalized quantities in Eqs. (16) and (17), the previous equation takes on the normalized form:

$$
\begin{aligned}
m_y = {} & \beta_1 \bar{\varepsilon}_y \frac{\xi_y^2}{1 - \xi_y} \left[\frac{1}{2}\left(1 + \frac{\tau_f}{2}\right) - \nu_1 \xi_y\right] \\
& -\beta_1 \bar{\varepsilon}_y \frac{1}{1 - \xi_y} \beta_w \xi_y - \tau_f^2 \left[\frac{1}{2}\left(1 + \frac{\tau_f}{2}\right) - \tau_f - \nu_1 (\xi_y - \tau_f)\right] + \\
& + \frac{1}{2}\eta \omega_s \frac{\xi_y - \frac{\tau_f}{2}}{1 - \xi_y}\left(1 - \frac{\tau_f}{2}\right) + \frac{2\lambda \omega_s}{2 - \tau_f}\left(\xi_y - \frac{\tau_f}{2}\right)^2 \frac{1}{1 - \xi_y}\left[\frac{1}{2}\left(1 - \frac{\tau_f}{2}\right) - \frac{1}{3}\left(\xi_y - \frac{\tau_f}{2}\right)\right] \\
& + \frac{2\lambda \omega_s}{2 - \tau_f}(1 - \xi_y)\left[\frac{1}{2}\left(1 - \frac{\tau_f}{2}\right) - \frac{1}{3}(1 - \xi_y)\right] + \frac{1}{2}\omega_s\left(1 - \frac{\tau_f}{2}\right)
\end{aligned}
\tag{33}
$$

3.2 Collapse Limit State (CLS, Concrete Crushing or Steel Fracture)

The strain and stress profiles at collapse are shown in Fig. 4.

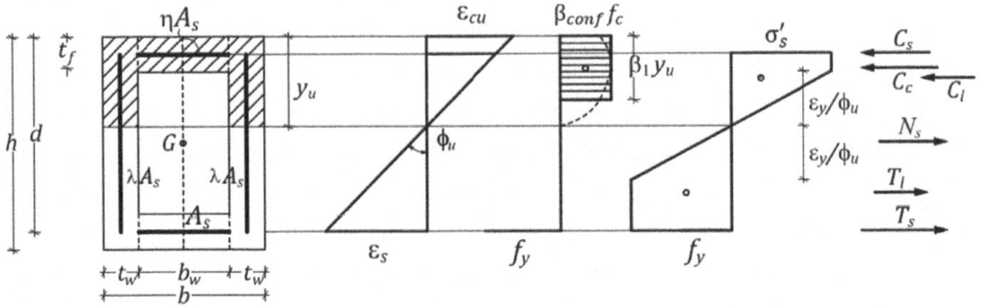

Fig. 4. Cross-section with geometrical and reinforcement quantities, strain profile at crushing of top concrete, stress profiles of concrete and of top/bottom steel, stress profile of lateral steel, and resultants.

The section attains the ultimate state when the topmost concrete fiber attains the strain $\varepsilon_c = \varepsilon_{cu}$ or when the bottommost steel fiber attains the strain $\varepsilon_s = \varepsilon_{su}$. In the former condition, the steel strain in tension ε_s and the steel strain in compression ε_s' are not known. In the latter condition, the concrete strain in compression ε_c and the steel strain in compression ε_s' are not known. The details of the attainment of ultimate moment and curvature capacity are omitted for brevity.

3.3 Summary of Equations

The following Table summarizes the equations to compute the sectional capacity.

4 Equations Verification Through Comparison with Fiber Section

The validation of the equations proposed for the moment-curvature analysis was conducted by comparison with fiber section analysis using OpenSeesPy [20]. A fully parameterized RC section was modeled by employing a zero-length element with section integration by means of fibers. To make the model code-compliant and comparable to the developed equations, the concrete constitutive law was set to follow the zero-tensile strength material model (Concrete01) and steel was modeled as a bilinear material with zero post-yield hardening (Steel01).

The concrete cross-section discretization was subdivided into 50 fiber layers, whereas each rebar was assigned a single fiber. The cross-section was subjected to constant axial force and an increasing curvature. Strain-based failure criteria were defined, such that the yield point corresponds to the attainment of yield strain in the tensile reinforcement, while the ultimate point corresponds to the attainment of ultimate strain in either the topmost compression concrete fiber or the bottommost tension steel fiber.

Table 1. Equations used for computing the sectional capacity at DLS and CLS.

Damage Limitation Limit State (DLS) – Yielding of tension reinforcement	
Neutral axis	$\xi_y = \begin{cases} \frac{n_S+\omega_s\left[(1+\lambda)+\frac{\tau_f}{2}(\eta+\lambda)\right]+\beta_1\beta_w\bar{\varepsilon}_y\tau_f^2}{n_S+\omega_s(1+\eta+2\lambda)+2\beta_1\beta_w\bar{\varepsilon}_y\tau_f} & \geq \tau_f \\ \frac{n_S+\omega_s\left[(1+\lambda)+\frac{\tau_f}{2}(\eta+\lambda)\right]}{n_S+\omega_s(1+\eta+2\lambda)} & \leq \tau_f \end{cases}$
Curvature capacity	$\phi_{DLS} = \phi_y = \frac{\varepsilon_y}{d}\frac{1}{1-\xi_y}$
Moment capacity ($/bd^2f_c$)	$m_{DLS} = m_y = \beta_1\bar{\varepsilon}_y\frac{\xi_y^2}{1-\xi_y}\left[\frac{1}{2}\left(1+\frac{\tau_f}{2}\right)-\nu_1\xi_y\right]+$ $-\beta_1\bar{\varepsilon}_y\frac{1}{1-\xi_y}\beta_w\langle\xi_y-\tau_f\rangle^2\left[\frac{1}{2}\left(1+\frac{\tau_f}{2}\right)-\tau_f-\nu_1(\xi_y-\tau_f)\right]+$ $+\frac{2\lambda\omega_s}{2-\tau_f}(1-\xi_y)\left[\frac{1}{2}\left(1-\frac{\tau_f}{2}\right)-\frac{1}{3}(1-\xi_y)\right]+\frac{1}{2}\omega_s\left(1-\frac{\tau_f}{2}\right)$
Collapse Prevention Limit State (CLS) – Crushing of compression concrete	
Curvature capacity	$\phi_{CLS} = \phi_u = \frac{1}{\xi_u d}\min\left(\varepsilon_{cu}, \varepsilon_{su}\frac{\xi_u}{1-\xi_u}\right)$
Moment capacity ($/bd^2f_c$)	$m_{CLS} = m_u = \beta_1\xi_u\left[\frac{1}{2}\left(1+\frac{\tau_f}{2}\right)-\nu_1\xi_u\right]+\beta_1\beta_w\xi_u -$ $\tau_f\left[\frac{1}{2}\left(1+\frac{\tau_f}{2}\right)-\tau_f-\nu_1(\xi_u-\tau_f)\right]+\omega_s\frac{1}{2}\left(1-\frac{\tau_f}{2}\right)(\eta+1)+$ $\frac{4\lambda\omega_s}{2-\tau_f}\left\{\left(\xi_u-\frac{\tau_f}{2}\right)\left[\frac{1}{2}\left(\xi_u-\frac{\tau_f}{2}\right)-\frac{2}{3}\bar{\varepsilon}\,\xi_u\right]+(1-\xi_u)\left[\frac{1}{2}(1-\xi_u)-\frac{2}{3}\bar{\varepsilon}\,\xi_u\right]\right\}$

The parameters governing the bending equations were made to vary in the range of interest for bridge piers and columns [21] as follows: normalized axial force, $n_S = 0.0-0.05$, lateral-to-tensile reinforcement ratio $\lambda = 1.0-2.0$, whereas compressive-to-tensile reinforcement ratio η was taken equal to 1, as it represents the most common condition in bridge piers reinforcement configurations. Fig. 5 presents the comparisons between the proposed analytical model and the fiber analysis in terms of moment-curvature diagrams for $\lambda = 1.0, 1.25, 1.5, 2.0$, $\eta = 1$ and $n_S = 0.01$. The comparison of the yielding point for cases (a) and (b) is satisfactory, with a maximum error of about 5% on curvature and 3% on moment, while at ultimate the maximum error touches a peak of about 8% on curvature and about 5% on moment. The comparison of the yielding and ultimate point for cases (c) and (d) is still acceptable. In Fig. 6 the comparison is shown of the yield and ultimate moments and curvatures as function of n_S, whereby the moment is normalized by bd^2f_c and the axial force by bdf_c, for the case $\eta = 1$ and $\lambda = 1$. Again, a very good agreement between analytical and numerical solution is observed as far as the moments in (a) are concerned, while yield and ultimate curvatures in (b), which show slight variability with the axial load, present an acceptable error. Similar accuracy is obtained for other values of λ.

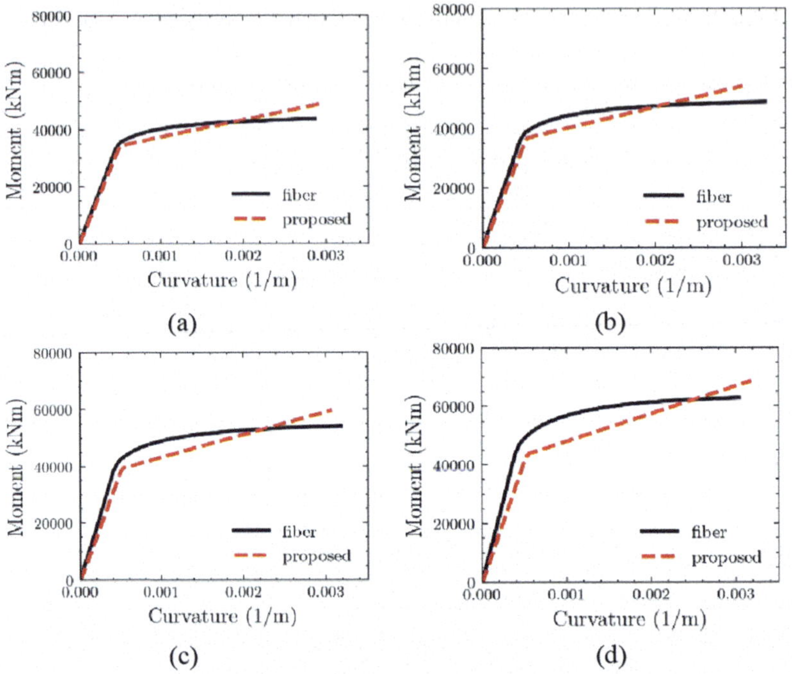

Fig. 5. Comparison of the fiber section and proposed closed-form equations in terms of moment-curvature for $\eta = 1$, $n_S = 0.01$ and for different lateral-to-tensile reinforcement ratio: a) $\lambda = 1.0$, b) $\lambda = 1.25$, c) $\lambda = 1.5$, d) $\lambda = 2.0$.

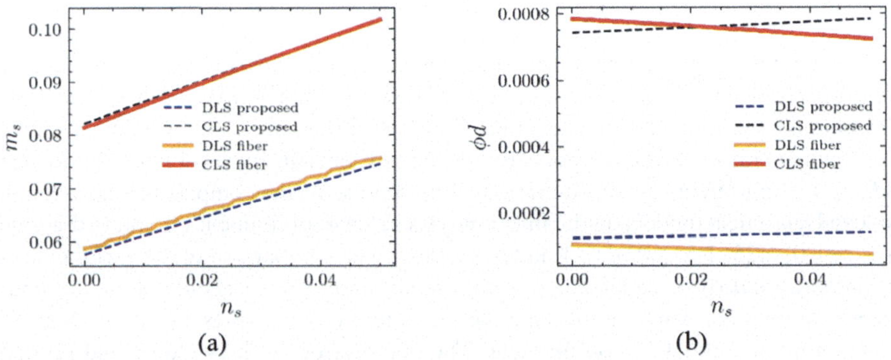

Fig. 6. Comparison of the fiber section and proposed closed-form equation for $\eta = 1$ and $\lambda = 1$. Relationship of the a) normalized axial force (n_s) and normalized moment (m_s), b) normalized axial force and normalized curvature (ϕd).

5 Conclusions

Hollow-core rectangular piers are a common feature of railway and highway bridges worldwide. For seismic performance assessments, particularly in large-scale territorial risk studies of railway or highway networks, efficient analysis tools are essential. Analytical formulations offer a practical solution to expedite these large-scale analyses. This study aimed to develop closed-form equations to describe the moment-curvature relationship of reinforced concrete (RC) hollow-core sections, reinforced on all four sides and subjected to both bending moment and axial load, including shear failure. These equations provide an effective means of determining yield and ultimate moments and curvatures, using commonly available parameters for existing bridge piers, such as material properties, cross-sectional geometry, reinforcement details, and axial forces.

The accuracy of the proposed equations has been validated through comparison with fiber-section analyses performed in OpenSeesPy, demonstrating satisfactory results.

References

1. Chopra, A.K.: Dynamics of Structures : Theory and Applications to Earthquake Engineering; Prentice Hall,: Englewood Cliffs, N.J. (1995). ISBN 0138552142
2. Fajfar, P., Gašperšič, P.: The N2 method for the seismic damage analysis of RC buildings. Earthq. Eng. Struct. Dyn. **25**, 31–46 (1996)
3. Fajfar, P.: A nonlinear analysis method for performance-based seismic design. Earthq. Spectra **16**, 573–592 (2000). https://doi.org/10.1193/1.1586128
4. Rahmat Rabi, R., Bianco, V., Monti, G.: Mechanical-analytical soil-dependent fragility curves of existing RC frames with column-driven failures. Buildings **11**, 278 (2021). https://doi.org/10.3390/BUILDINGS11070278
5. Rahmat Rabi, R., Vailati, M., Monti, G.: Simplified pushover analysis for rapid assessment of shear-type frames. Appl. Sci. **11**, 11711 (2021). https://doi.org/10.3390/APP112411711
6. Demartino, C., Vanzi, I., Monti, G., Sulpizio, C.: Precast industrial buildings in southern europe: loss of support at frictional beam-to-column connections under seismic actions. Bull. Earthq. Eng. **16**, 259–294 (2018). https://doi.org/10.1007/S10518-017-0196-5/FIGURES/19
7. Wang, X., Demartino, C., Narazaki, Y., Monti, G., Spencer, B.F.: Rapid seismic risk assessment of bridges using UAV aerial photogrammetry. Eng. Struct. **279**, 115589 (2023). https://doi.org/10.1016/J.ENGSTRUCT.2023.115589
8. Vailati, M., Monti, G.: Low-LOD fragility curves of structural units in masonry building clusters for territorial risk analysis. Eng. Struct. **286**, 116143 (2023). https://doi.org/10.1016/J.ENGSTRUCT.2023.116143
9. Demartino, C., Monti, G.: Low-LOD code-driven identification of the high seismic risk areas for industrial buildings in Italy. Bull. Earthq. Eng. **18**, 4421–4452 (2020). https://doi.org/10.1007/S10518-020-00867-3/FIGURES/20
10. Pfrang, E., Siess, C.P., Sozen, M.A.: Load-moment-curvature characteristics of reinforced concrete cross sections. J. Proc. **61**, 763–778 (1964). https://doi.org/10.14359/7807
11. Carreira, D.J., Chu, K.H.: The moment-curvature relationship of reinforced concrete. J. Proc. **83**, 191–198 (1986). https://doi.org/10.14359/10413
12. Vebo, A., Ghali, A.: Moment-curvature relation of reinforced concrete slabs. J. Struct. Div. **103**, 515–531 (1977). https://doi.org/10.1061/JSDEAG.0004579
13. Kwak, H.G., Kim, S.P.: FE analysis of RC beams based on moment-curvature relation. In: Proceedings of the 8th International Conference on Computing in Civil and Building Engineering, vol. 279, pp. 472–479 (2000). https://doi.org/10.1061/40513(279)62

14. Kwak, H.G., Kim, S.P.: Simplified monotonic moment-curvature relation considering fixed-end rotation and axial force effect. Eng. Struct. **32**, 69–79 (2010). https://doi.org/10.1016/J.ENGSTRUCT.2009.08.017
15. Kim, J.K., Lee, S.S.: The behavior of reinforced concrete columns subjected to axial force and biaxial bending. Eng. Struct. **22**, 1518–1528 (2000). https://doi.org/10.1016/S0141-0296(99)00090-5
16. Oller, S., Barbat, A.H.: Moment-curvature damage model for bridges subjected to seismic loads. Comput. Methods Appl. Mech. Eng. **195**, 4490–4511 (2006). https://doi.org/10.1016/J.CMA.2005.09.011
17. European Committee for Standardization CEN, Eurocode 2: Design of Concrete Structures (EN 1992:2004) (2004)
18. Ministry of Infrastructures and Transportation MIT, Ministerial Decree: NTC 2018 - Norme Tecniche Per Le Costruzioni (2018)
19. Monti, G., Spacone, E.: Reinforced concrete fiber beam element with bond-slip. J. Struct. Eng. **126**, 654–661 (2000). https://doi.org/10.1061/(ASCE)0733-9445(2000)126:6(654)
20. Zhu, M., McKenna, F., Scott, M.H.: OpenSeesPy: python library for the opensees finite element framework. SoftwareX **7**, 6–11 (2018). https://doi.org/10.1016/J.SOFTX.2017.10.009
21. Wehbe, N., Saiidi, M., Sanders, D., Douglas, B.: Ductililty of Rectangular Reinforced Concrete Bridge Columns with Moderate Confinement Ductility. US National Center for Earthquake Engineering Research (NCEER), 96 (1996)

Equivalent Strut Models for CLT Infills in RC Frames Based on Machine Learning

Carlotta Pia Contiguglia(✉), Cristoforo Demartino, and Camillo Nuti

Department of Architecture, Roma Tre University, Rome, Italy
Carlottapia.contiguglia@uniroma3.it

Abstract. This study introduces a novel framework for establishing the stress-strain relationship for equivalent strut models of cross-laminated timber panels infilled in moment-resisting RC frames. High-fidelity models of RC frames with CLT infills are developed using a parametric finite element model in OpenSeesPy. A comprehensive parametric study using Latin Hypercube Sampling was conducted to generate 8.000 different frame configurations, varying parameters describing both the frame, the infill and the applied loads. Calibration of the stress-strain curves was conducted using a genetic algorithm-based optimization process, aimed at minimizing the mean squared error between the global force-displacement curves of the high-fidelity model and the equivalent strut model. Optimal parameters found by the genetic algorithm for the material properties assigned to the strut were used to derive a predictive model. A comprehensive predictive model was developed using the CatBoost machine learning algorithm, incorporating 16 input variables.

Keywords: Cross-Laminated Timber (CLT) · Seismic Retrofitting · Equivalent Strut Model · Predictive Model · Machine Learning

1 Introduction

The use of infill elements in construction has a long history, evolving from ancient methods such as mud bricks, stones, and timber, to modern masonry, concrete blocks, and precast panels [1]. In the 20th century, Reinforced Concrete (RC) and steel frames revolutionized the construction industry, allowing for more versatile and structurally efficient designs. Infill panels within these frames became critical components that not only enclose space but also significantly influence the response to lateral loads from earthquakes and wind. Infill panels within RC frames have been shown to contribute to the stiffness, strength, and energy dissipation of the structure, altering load distribution and affecting the response of the frame to external forces [2, 3].

In the field of seismic retrofitting, a recent innovation involves incorporating Cross-Laminated Timber (CLT) panels as infills within RC building structures. These panels, connected using metal dowel-type fasteners, aim to enhance the in-plane response of RC frames. Despite emerging scholarly interest, comprehensive quantification of the

X. Lu et al. (Eds.): EOS 2024, LNCE 638, pp. 229–240, 2025.
https://doi.org/10.1007/978-3-031-90690-9_20

mechanical behavior of this technique and the development of simplified macro-models for structural analyses remain underexplored.

This paper is a comprehensive parametric numerical study to evaluate the seismic performance of the retrofitting solution based on CLT panels proposed by Smiroldo *et al.* [4, 5]. Specifically, the impact of different connection layouts, geometries, mechanical properties of the materials and loads on the overall stiffness, strength, and energy dissipation of the retrofitted frames is studied. The methodology includes a large dataset of non-linear static analyses, enabling the development of both simplified macro-FE models and predictive models of the seismic monotonic behavior of the system. These models are then used to assess the performance of equivalent strut models and propose a ML predictive model, providing a more accessible approach for engineers and designers to evaluate the effectiveness of CLT infills in RC frames.

2 Benchmark Retrofitting Technique

The numerical study conducted in this research refers to a benchmark retrofitting technique (Fig. 1) initially proposed by Smiroldo *et al.* [4, 5].

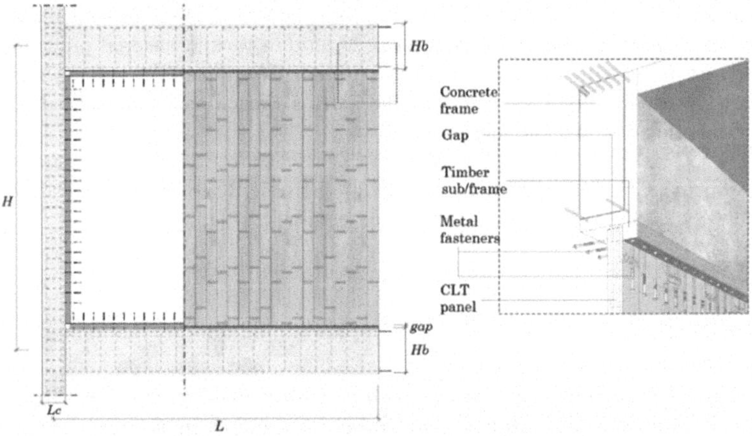

Fig. 1. RC frame infilled with CLT panel proposed by Smiroldo *et al.* (12).

This system involves replacing the existing masonry infill in a single-story, single-bay RC frame with a CLT panel. The primary goal of this retrofitting method is to improve the structural performance of the RC frame under both seismic and gravitational loads. To connect the CLT panel to the RC frame, a timber subframe is installed, which is fastened to the RC components using metal fasteners. These fasteners are designed to remain elastic while ensuring that the subframe moves with the RC frame. The connection between the CLT panel and the subframe is achieved using steel screws, which are oriented perpendicular to the panel and are primarily subjected to shear forces. The panel-to-subframe screws are intentionally designed to display dissipative behavior, meaning they are intended to yield before the frame components experience failure. This ensures that the fasteners enhance the ductility of the retrofitted frame by leveraging their post-elastic behavior.

A critical aspect of this design is the introduction of a gap between the CLT panel and the RC frame. This gap allows the fasteners to engage the forces transmitted to the panel, preventing direct contact between the panel and the frame. As a result, the ductility of the frame is increased, as the screws can deform and absorb energy before the gap closes. The CLT panel is oriented vertically in its "strong direction" to further support vertical loads during seismic events.

3 Proposed FE Models

Finite element models of a bare, infilled and simplified, one-story, one-bay RC frames were developed using OpenSeesPy, a Python library for finite element analysis [6].

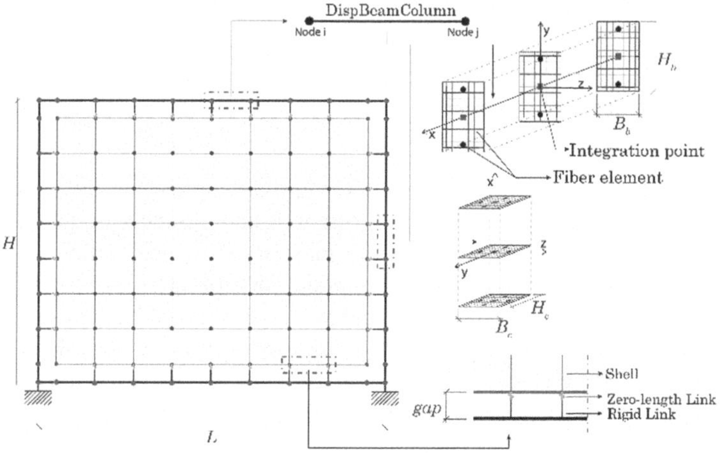

Fig. 2. FE component-level model developed in OpenSeesPy

The frame is defined by four couples of coincident points connected by rigid links representing the corners of the frame. These links ensure that both the translational and rotational degrees of freedom are constrained. Columns are fixed at the base. The beams and columns are modeled using displacement BeamColumn elements, with a P-Delta geometric transformation applied to account for large displacements. The elements are discretized to include screws, and Lobatto integration is employed with three integration points to ensure an optimal balance of accuracy and computational efficiency. Materials used include Concrete02 and Steel01 for concrete and steel respectively. Loads are assigned as nodal loads at the top of the columns to account for the presence of upper floors and as distributed loads along the length of the upper beam to account for slab load.

The high-fidelity model of the infill (Fig. 2) consists of three key elements, including shell elements to model the CLT panel, zero-length elements for the screws, and rigid links to connect zero-length elements to columns and beams and account for the semi-thickness of the concrete components. The CLT panel is assumed to remain elastic and rotate rigidly within the frame. The DowelType Timber Joint Material [7], available in the OpenSees library, is used to simulate the behavior of the screws under shear forces.

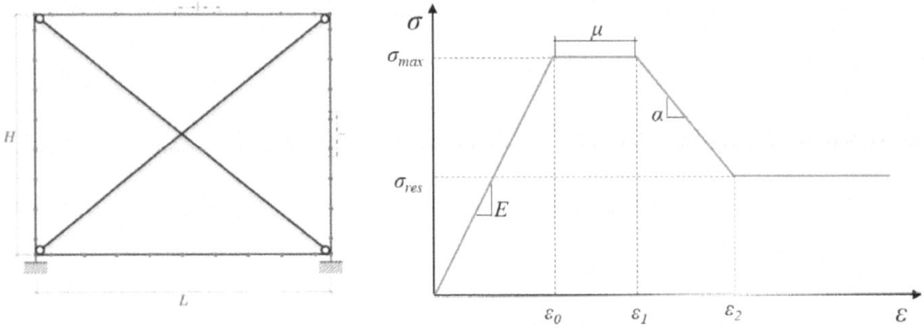

(a) (b)

Fig. 3. (a) Equivalent symmetrical struts model and (b) quadrilinear stress-strain constitutive law assigned to the equivalent struts (the curve is symmetrical for tension and compression).

The equivalent strut model (Fig. 3 (a)) is a simplified representation of the structural response of an infill often adopted in both research and engineering practice. Analogous approaches are commonly employed for masonry structures, efficiently capturing essential global behavior [8, 9]. This macro-model incorporates two symmetrical truss elements, each assigned a unit cross-sectional area, whose axial behavior replicates the in-plane seismic response of the CLT panel. The stress-strain relationship assigned in the axial direction of the elements (Fig. 3 (b)) is a quadrilinear stress-strain curve defined by key parameters such as the elastic modulus, maximum stress, ductility, and residual stress. The stress at the three interpolation points defining the curve can be evaluated as follows (Eqs. (1), (2) and (3)):

$$\varepsilon_0 = \sigma_{max}/E \tag{1}$$

$$\varepsilon_1 = \varepsilon_0 \cdot \mu \tag{2}$$

$$\varepsilon_2 = \varepsilon_1 + (\sigma_{res} - \sigma_{max}/E \cdot \alpha) \tag{3}$$

3.1 Reference Model and Dataset Generation

A reference model was defined based on the RC bare frame properties from Smiroldo [4] and a frame from a 1984 building (Fig. 4). The geometrical and material properties for the concrete and steel used in the model are provided in Table 1.

In retrofit configuration, CLT panel elastic orthotropic properties are assumed according to [10]. The mechanical behavior of the screws connecting the CLT panel to the

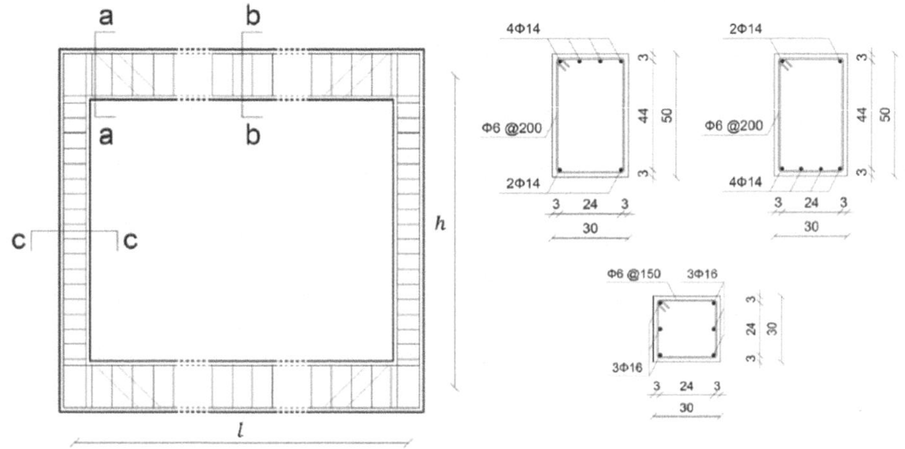

Fig. 4. Reference bare frame and cross sections of beams and columns.

Table 1. RC frame geometrical and material properties.

Frame		Beams		Columns		Concrete		Steel	
H	L	H_b	B_b	H_c	B_c	F_c	E_c	F_y	E_s
[cm]	[cm]	[cm]	[cm]	[cm]	[cm]	[MPa]	[GPa]	[MPa]	[GPa]
310	500	50	30	30	30	16.73	25.6	440	210

subframe is modeled using experimental data. The force-displacement relationship of these screws was derived from cyclic load tests [7] and applied in the numerical model to simulate their behavior under shear loads using Bézier and piece-wise linear envelopes in local axes 2 and 3, respectively.

A comprehensive dataset of infilled frames was generated, incorporating 16 variables related to frame geometry, infill properties, material characteristics, and load conditions (Table 2). This dataset, containing 8000 configurations, was generated using Latin Hypercube Sampling (LHS), a method well-suited for exploring high-dimensional spaces. LHS ensures that the entire range of variable combinations is sampled efficiently, reducing the sample size needed for accurate analysis [11].

4 Evaluation of Optimal Parameters

The optimization problem solved using the PyGAD Genetic Algorithm (GA) [12] is structured to find the optimal parameters of the stress strain curve assigned to the equivalent struts in the axial direction Fig. 5.

The quadrilinear curve models four key phases of the infill's response: 1) elastic behavior, 2) ductile response due to the screws' post-elastic deformation, 3) strain softening, and 4) residual stress. The parameters that define this curve – elastic modulus

Table 2. Variables of the parametric models.

Variable	Definition	Units	Lower bound	Upper bound	Discrete step
Frame length	L	[mm]	4000.0	7000.0	100.0
Frame height	H	[mm]	3000.0	4200.0	100.0
Column base	B_c	[mm]	300.0	700.0	50.0
Variable	Definition	Units	Lower bound	Upper bound	Discrete step
Column height	H_c	[mm]	300.0	700.0	50.0
Beam base/Column base	B_b/B_c	[–]	1	2	
Beam height	H_b	[mm]	L/10	L/15	
Column reinforcement ratio	ρ_{col}	[%]	0.005	0.04	
Beam top reinforcement ratio	ρ_{top}	[%]	0.01	0.04	
Beam bottom reinforcement ratio	ρ_{bot}/ρ_{top}	[%]	1/3	1	
Concrete compressive strength	f_c	[MPa]	–20.0	–45.0	
Steel yield strength	f_y	[MPa]	350.0	550.0	
Compressive stress	v	[%]	5	20	
Beam uniform load	q	[kN/m]	25.0	50.0	
CLT panel thickness	T_{CLT}	[mm]	60.0	150.0	10.0
Number of screws along the beam	$n_{screws,beam}$	[–]	30.0	50.0	1
Number of screws along the beam	$n_{screws,col}$	[–]	30.0	50.0	1

(E), maximum stress (σ_{max}), softening slope (α), residual stress (σ_{res}), and ductility (μ) – are contained in a design vector $x = [E, \sigma_{max}, \alpha, \sigma_{res}, \mu]^T$. . The optimization problem is constrained to avoid that the residual stress is greater than the maximum stress. The fitness function is defined as the Mean Squared Error (MSE) between the predicted force-displacement curve of the equivalent strut and the target curve obtained from the high-fidelity component-level simulations.

To provide an example, the optimization procedure was initially applied to 25 force-displacement curves obtained from the pushover analysis of high-fidelity models, where the number of screws along the beams was varied between 25 and 50. The bounds for each parameter during the optimization process are shown in Table 3.

The optimized strut model effectively replicates both trends shown by the high-fidelity model curves, Shape A (in orange) and Shape B (in green), ensuring close alignment between the simplified and high-fidelity models (Fig. 6). In this small set of analysis, where only the number of screws along the column was varied, the distribution

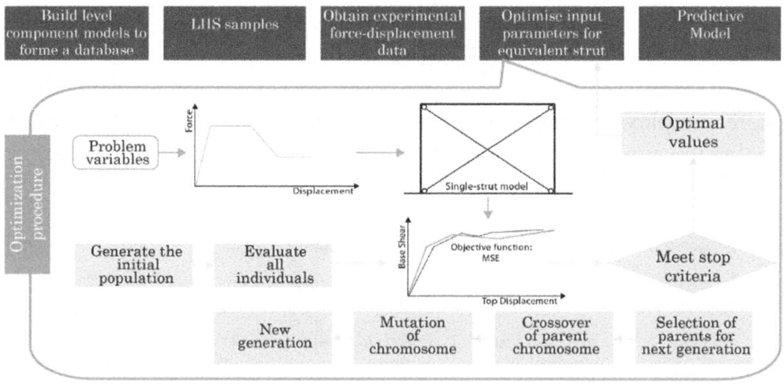

Fig. 5. Flowchart of the optimization procedure [13].

Table 3. Lower bound (x_l) and upper (x_u) bound used in the one variable model optimization problem.

E [MPa]		σ_{max} [MPa]	α [–]	σ_{res} [MPa]	μ [–]
x_l	1.00E + 07	7.00E + 04	–2.00E + 00	1.00E + 03	1.00E + 00
x_u	4.00E + 07	5.00E + 05	–1.00E-05	7.00E + 04	2.00E + 00

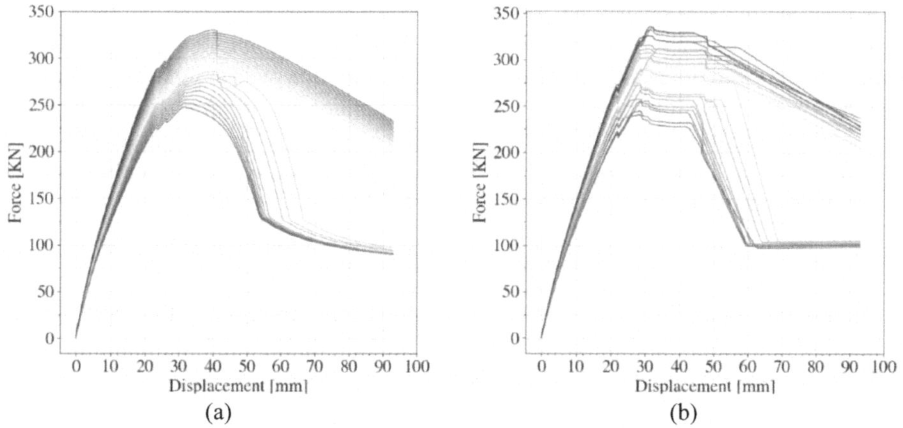

Fig. 6. Comparison of the curves obtained by the high-fidelity model (a) and the equivalent symmetrical struts model (b).

of the elastic modulus against the number of screws revealed a linear dependence with the number of screws, which aligned with expected results (Fig. 7). Similarly, the maximum stress followed a similar trend. However, the softening slope (α) and residual stress (σ_{res}) showed more complex behavior. In Shape A curves, the softening slope was inversely proportional to the number of screws, while in Shape B, it remained relatively constant.

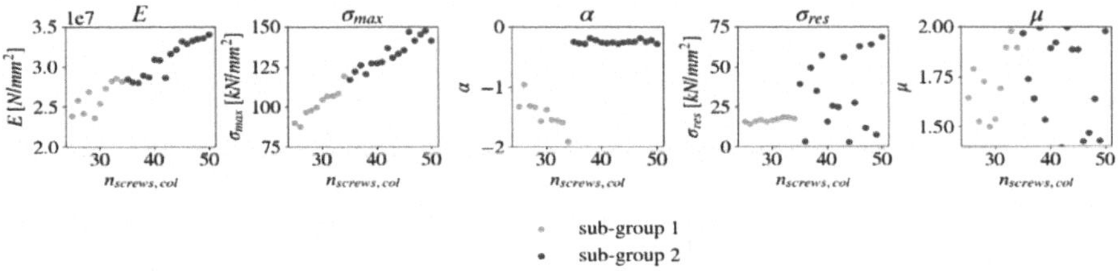

Fig. 7. Scatter plots of the values of the optimal parameters against the number of screws along the columns for the stress-strain curves of the equivalent struts.

The residual stress also exhibited a linear increase with screws in Shape A but was less predictable in Shape B, likely due to the limitations in the pushover curves at higher drift levels. The ductility parameter (μ) did not show a clear pattern in the optimization process and was found randomly by the algorithm. In this example, deriving simple equations to correlate design variables with model parameters is feasible. In contrast, the complexity introduced by a 16-variable model obstructs the definition of straightforward equations, prompting the use of a machine learning model.

A second set of 600 analyses on the reference model is defined by varying the number of screws along columns and beams. Two distinct force-displacement trends, Shape A and B, were identified using the Fréchet Distance, a measure of shape similarity computed in Python. Curves, represented as point sequences, yielded a distance matrix for clustering, and two clusters emerged: 390 curves (Shape A) and 40 curves (Shape B) (Fig. 8). Fig. 9 includes a pairplot illustrating variable relationships, revealing a linear relation that defines the cluster transition (Eq. (4)):

$$n_{screws,beam} = 1.75n_{screws,col} \tag{4}$$

For frames with an L/H ratio of 1.6, the curve shape transitions when the screw density along beams surpasses that of the columns.

This classification is critical for developing the predictive model.

5 Proposed Model

A predictive model based on CatBoost [14] machine learning algorithm was calibrated. The scatter plots (Fig. 10) illustrate the comparison between measured and predicted values of key parameters (E, σ_{max}, α, σ_{res}, and μ) for both the training and testing datasets. Ideally, every point would line exactly with the diagonal, indicating a perfect correspondence between the measured and expected values. For elastic modulus (E), the model performs well, with most data points near the diagonal line, indicating accurate

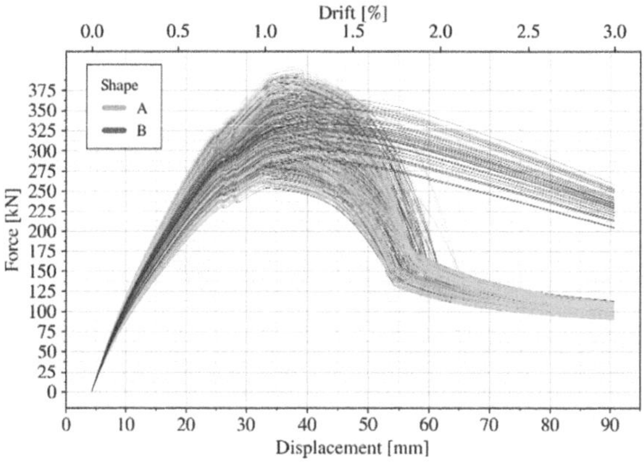

Fig. 8. Force-displacement curves clustered into two groups, Shape A (orange) and Shape B (green).

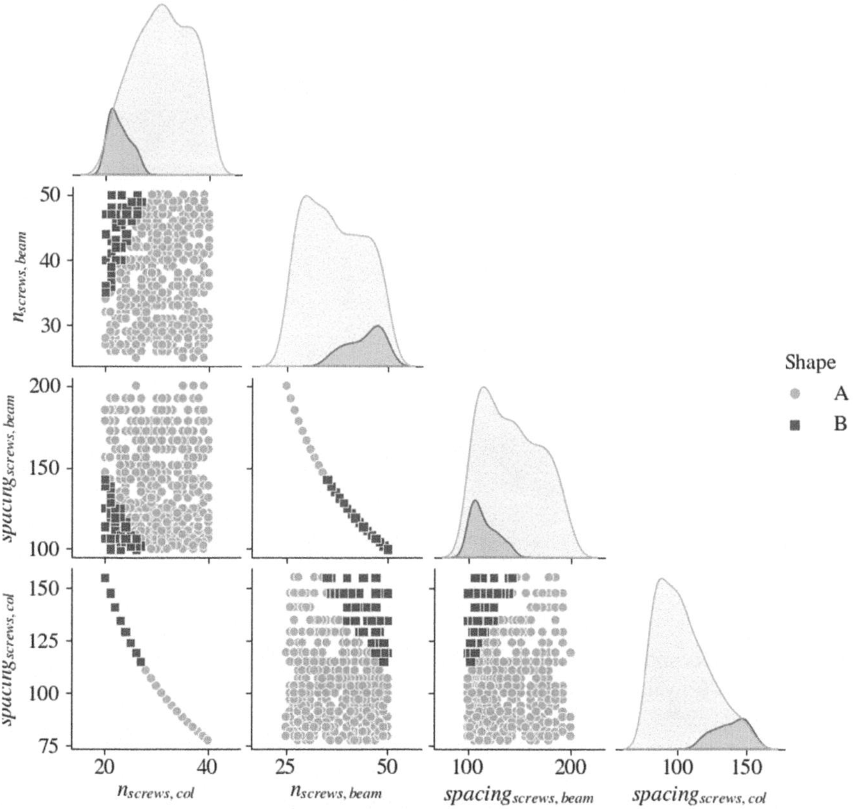

Fig. 9. Pairplot showing variable relationships for Shape A (orange) and Shape B (green), highlighting clustering and linear trends.

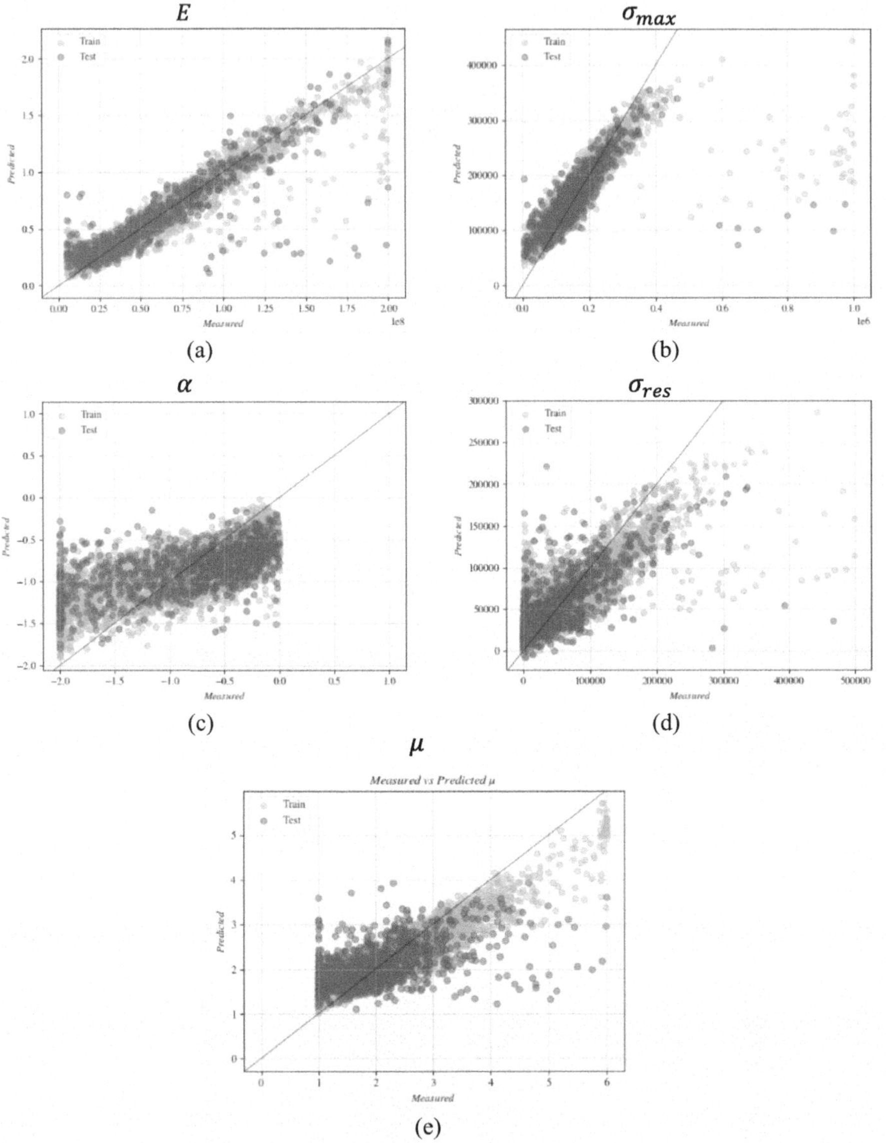

Fig. 10. Scatter plots show the correlation between measured and predicted E (a), σ_{max} (b), α (c), σ_{res} (d), and μ (e) values for training and testing datasets.

predictions. The R^2 value for the training data is 0.90, while for the test data, it is slightly lower at 0.75, suggesting good generalization without overfitting (Table 4). For maximum stress (σ_{max}), the model shows a slight discrepancy, particularly over-predicting lower values and under-predicting higher values. The R^2 values are 0.48 for training data and 0.39 for testing, suggesting the model struggles with σ_{max} predictions, especially for test data. For strain softening slope (α) and residual stress (σ_{res}), the scatter plots show that the model's predictions are relatively close to the measured values, although some overfitting is observed (R^2 of 0.54 for α in the training data and 0.33 for testing). Similarly, the model shows lower performance for σ_{res} in the test data (= 0.49), indicating that additional refinement is needed.

Finally, for ductility (μ), the model fits well for the training data (= 0.84), but its performance on the test set is lower (= 0.23), highlighting potential overfitting or sensitivity to specific test data conditions.

Table 4. Performance metrics for the train and the test set for each model.

Model	Metric	E	σ_{max}	σ_{res}	α	μ
Train	R^2	0.9	0.48	0.65	0.54	0.84
	RMSE	11004778	69760	36097	0.4	0.36
	MAE	6508498	26437	21710	0.32	0.26
	MedAE	4343638	16102	14488	0.27	0.2
Test	R^2	0.75	0.39	0.49	0.33	0.23
	RMSE	17916197	66560	40528	0.49	0.75
	MAE	9121447	27444	25987	0.39	0.53
	MedAE	5329237	17698	17643	0.34	0.39

6 Conclusions

This study introduced a framework for modeling and analyzing the seismic performance of CLT panels infilled in RC frames. High-fidelity finite element models and parametric analyses demonstrated the effectiveness of CLT infills in enhancing frame stiffness, strength, and energy dissipation, confirming their potential for seismic retrofitting.

A simplified equivalent strut model, calibrated using a genetic algorithm, successfully replicated the global force-displacement behavior. The complexity of 16 design variables required the use of a machine learning approach to develop a predictive model for the axial stress-strain behavior of the trusses. The CatBoost-based predictive model showed strong performance in predicting elastic modulus but struggled with other stress-strain parameters. Refinement of the dataset, such as outlier removal and curves clustering, is expected to improve its accuracy.

This methodology provides a valuable tool for designing retrofitted RC frames with CLT infills. Future work will focus on dataset refinement and extending the framework to dynamic seismic analyses.

References

1. Mascarenhas-Mateus, J., Paula Pires, A.: History of Construction Cultures, Volume 1
2. Moretti, M.L.: Seismic design of masonry and reinforced concrete infilled frames: a comprehensive overview. Am. J. Eng. Appl. Sci. **8**(4), 748–766 (2015). https://doi.org/10.3844/ajeassp.2015.748.766
3. Phan, L.T., Cheek, G.S., Todd, D.R., Prabhakar, A.: Strengthening methodology for lightly reinforced concrete frames: recommended design guidelines for strengthening with infill walls. Gaithersburg (1995)
4. Smiroldo, F., Paviani, I., Giongo, I., Zanon, S., Albatici, R., Piazza, M.: An integrated approach to improve seismic and energetic behaviour of RC framed buildings using timber panels. Sustainability (Switzerland) **13**(20) (2021). https://doi.org/10.3390/su132011304
5. Smiroldo, F., Giongo, I., Piazza, M.: Use of timber panels to reduce the seismic vulnerability of concrete frame structures. Eng Struct. **244** (2021). https://doi.org/10.1016/j.engstruct.2021.112797
6. Mazzoni, S., Mckenna, F., Scott, M.H., Fenves, G.L., Iii, A.: Open System for Earthquake Engineering Simulation (OpenSees) OpenSees Command Language Manual (2006)
7. Dong, H., He, M., Wang, X., Christopoulos, C., Li, Z., Shu, Z.: Development of a uniaxial hysteretic model for dowel-type timber joints in OpenSees. Constr Build Mater. **288** (2021). https://doi.org/10.1016/j.conbuildmat.2021.123112
8. Asteris, P.G., Cavaleri, L., Di Trapani, F., Sarhosis, V.: A macro-modelling approach for the analysis of infilled frame structures considering the effects of openings and vertical loads. Struct. Infrastruct. Eng. **12**(5), 551–566 (2016). https://doi.org/10.1080/15732479.2015.1030761
9. Asteris, P.G., Cavaleri, L., Di Trapani, F., Tsaris, A.K.: Numerical modelling of out-of-plane response of infilled frames: State of the art and future challenges for the equivalent strut macromodels. Elsevier Ltd. (2017). https://doi.org/10.1016/j.engstruct.2016.10.012
10. Shahnewaz, M., Alam, S., Tannert, T.: In-plane strength and stiffness of cross-laminated timber shear walls. Buildings **8**(8) (2018). https://doi.org/10.3390/buildings8080100
11. "https://docs.scipy.org/doc/scipy/reference/generated/scipy.stats.qmc.LatinHypercube.html."
12. Gad, A.F.: PyGAD: An intuitive genetic algorithm python library. Multimed Tools Appl. 1–14 (2023). http://arxiv.org/abs/2106.06158
13. Wu, J.R., Di Sarno, L., Freddi, F., D'Aniello, M.: Modelling of masonry infills in existing steel moment-resisting frames: Nonlinear force-displacement relationship. Eng. Struct. **267** (2022). https://doi.org/10.1016/j.engstruct.2022.114699
14. https://catboost.ai/en/docs/concepts/parameter-tuning

Integration of 3D Printed Fiber Concrete and OpenSees TO for Enhanced Structural Design

Junsong Wang[1], Haoran Ye[2], Yating Sun[2], Zixin Zhang[1](✉), Chenxu Qi[2], Chenzhe Zhang[2], Chengwei Song[2], Cristoforo Demartino[3], and Asif Usmani[1]

[1] Department of Building Environment and Energy Engineering, The Hong Kong Polytechnic University, Hong Kong, Hong Kong 999077, China
zixin96.zhang@polyu.edu.hk
[2] Zhejiang University - University of Illinois at Urbana Champaign Institute, Zhejiang University, 718 East Haizhou Road, Haining 314400, Zhejiang, China
[3] Department of Architecture, Roma Tre University, Largo G. B. Marzi 10, 00153 Rome, Italy

Abstract. This study examines the mechanical performance of 3D-printed fiber-reinforced concrete through comprehensive experimental testing, including compression and splitting tests conducted across three principal print orientations. Specimens were oriented to reflect actual printing processes, distinguishing between horizontal print planes and vertical layers relative to nozzle movement. The results reveal significant anisotropic behavior and varying strength properties contingent on print orientation. Specifically, the FX orientation (parallel to the printing direction) demonstrated compressive strengths comparable to traditional cast concrete, while FY (perpendicular) and FZ (vertical) orientations exhibited progressively lower strengths due to weaker interlayer bonding. Building on these findings, a customized constitutive model for 3D-printed fiber concrete was developed and integrated into the OpenSees framework, enabling precise finite element simulations. Furthermore, the integration of 3D-printed fiber concrete with OpenSees Topology Optimization (TO) facilitates automated structural model generation and innovative design processes, leveraging iterative topology optimization and isogeometric analysis to enhance structural performance and material efficiency. This integration represents a significant advancement in structural engineering and computational simulation, offering viable solutions for the future of construction technology and supporting the broader adoption of additive manufacturing in the construction industry.

Keywords: 3D-Printed Concrete · Anisotropic Behavior · Topology Optimization · OpenSees Framework

1 Introduction

The emergence of 3D-printed concrete has significantly transformed modern construction by enabling the creation of complex, customized geometries with enhanced material efficiency [1]. This additive manufacturing technique offers precise control over structural forms, reducing material waste and providing design flexibility beyond traditional

X. Lu et al. (Eds.): EOS 2024, LNCE 638, pp. 241–252, 2025.
https://doi.org/10.1007/978-3-031-90690-9_21

casting methods [2]. However, the inherent layer-by-layer deposition process introduces anisotropy, which adversely affects the material's mechanical behavior and structural performance [3]. The strength of interlayer bonding becomes crucial in determining the concrete's compressive and tensile strengths, stiffness, and failure mechanisms [4].

OpenSees, an open-source finite element analysis framework initially developed for earthquake engineering, has evolved into a versatile platform for modeling complex structural behaviors [5–7]. The integration of Isogeometric Analysis (IGA) further enhances OpenSees' capabilities for simulating advanced materials like 3D-printed fiber-reinforced concrete [8]. IGA provides precise geometric and deformation modeling, allowing for accurate representation of complex structures.

This study leverages OpenSees with extended IGA functionality to model the mechanic behavior of 3D-printed concrete. Experimental compression and tensile tests inform a calibrated constitutive model that incorporates print orientation. These models are integrated into the OpenSees TO framework to optimize structural designs and explore innovative applications. A case study involving a simply supported beam demonstrates the practical application of the developed constitutive model, highlighting the potential of 3D-printed fiber-reinforced concrete in optimized structural designs. This comprehensive approach—from experimental validation to computational modeling and case application—demonstrates the feasibility and advantages of integrating 3D-printed concrete into advanced structural systems, paving the way for future innovations in construction technology.

2 Development of Isogeometric Analyses in the OpenSees Framework

The integration of IGA within the OpenSees framework represents a significant advancement in structural modeling and simulation. This section delineates the development and implementation of isogeometric analyses in OpenSees, focusing on the incorporation of TO models and the associated computational strategies.

2.1 Optimization Models in the ITO Tool: Minimum Volume with Stress Constraint

The OpenSees-ITO (Isogeometric Topology Optimization) tool package supports two primary topology optimization models: minimum compliance and minimum volume with stress constraints [8]. These models are designed to facilitate the creation of optimized structural layouts that balance material efficiency with structural performance.

The minimum volume model aims to design lightweight structures while ensuring structural safety by maintaining the von Mises stress below a specified limit. The optimization problem is formulated as follows:

$$\begin{aligned} \min_{\rho} \quad & V(\rho) = \sum_{e} \hat{\rho}_e(\rho) V_e \\ & \sigma_{e,i}^{M}(\rho) \le \overline{\sigma}, 0 \le i \le N_{GP} \\ s.t. \quad & K(\rho)U(\rho) = F \\ & 0 \le \rho \le 1 \end{aligned} \tag{1}$$

where: $V(\rho)$ is the total volume of the structure. ρ_e represents the design variable (density) for element *e*. *Ve* is the volume of element *e*. $\sigma_{e,i}^{M}(\rho)$ denotes the von Mises stress in element *e* at integration point *i*. $\overline{\sigma}$ is the maximum allowable stress. $K(\rho)$ K(ρ) K(ρ) is the stiffness matrix dependent on the design variables. U(ρ) U(ρ) $U(\rho)$ and F are the displacement vector and external force vector, respectively.

2.2 Graphical User Interface for the OPS-ITO Framework

The OpenSees-ITO framework is facilitated through an executable program equipped with a user-friendly graphical user interface (GUI) (see Fig. 1). The GUI is designed to streamline the setup and execution of topology optimization tasks by categorizing input parameters into distinct sections:

1. *Model Setup:* Defines model dimensions, selects the type of optimization, and specifies geometric configurations. Users can choose from predefined simple models (e.g., rectangular or box-shaped) or import complex, user-defined geometries using the SIMOPackage pre-processor, which supports geometry imports from Matlab (.mat) files.
2. *Material and Boundary Conditions*: Specifies material properties, defines boundary constraints, and sets up loading scenarios pertinent to the structural analysis.
3. *Optimization Parameters*: Configures parameters governing the optimization iterations, such as convergence criteria, optimization goals, and filter settings to prevent numerical issues like checkerboarding.

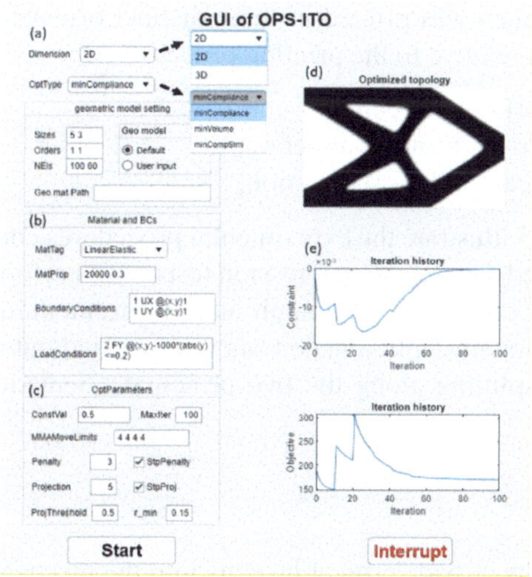

Fig. 1. Graphical user interface of OPS-ITO platform [8]

The GUI accommodates both 2D and 3D optimization problems, supporting minimum compliance and minimum volume with von Mises stress constraints. It offers real-time visualization of topology configurations, allowing users to observe the evolution of the optimized design dynamically.

3 3D Print Concrete and TO Method Modeling

This section presents the experimental program designed to validate the constitutive models for 3D-printed fiber-reinforced concrete. The program encompasses material characterization tests aimed at capturing the mechanical properties and behaviors essential for developing robust constitutive models that account for anisotropy, plasticity, and interlayer effects in 3D-printed concrete structures.

3.1 Material and Test Setup

The experimental program utilized both plain and fiber-reinforced 3D-printed concrete. PP fibers were incorporated into the concrete mix to enhance tensile strength, crack resistance, and overall durability. The concrete mix was meticulously formulated to ensure optimal printability and strength, maintaining appropriate workability for the 3D printing process.

Two distinct types of cylindrical specimens were fabricated for material characterization: specimens for compression tests, designed to evaluate compressive strength and elastic modulus, and specimens for split tensile tests, intended to assess the tensile strength of the concrete.

Each type of specimen was printed in three distinct orientations corresponding to the principal directions relative to the printing process:

- *FX:* Parallel to the printing direction.
- *FY*: Perpendicular to the printing direction.
- *FZ*: Along the vertical (building) direction.

Figures 2, 3, 4 and 5 illustrate the experimental procedures conducted on cylindrical specimens of 3D-printed concrete. Compression tests were performed using a hydraulic press to determine the compressive strength and elastic modulus in the FX, FY, and FZ orientations. Additionally, split tensile tests were carried out to evaluate the tensile strength by inducing splitting along the two principal orientations of the cylindrical specimens.

3.2 Test Results

The experimental results provide critical insights into the mechanical properties of 3D-printed fiber-reinforced concrete. The experimental results provide critical insights into the mechanical properties of 3D-printed fiber-reinforced concrete, forming the foundation for calibrating constitutive models and validating the proposed TO methods within the OpenSees framework.

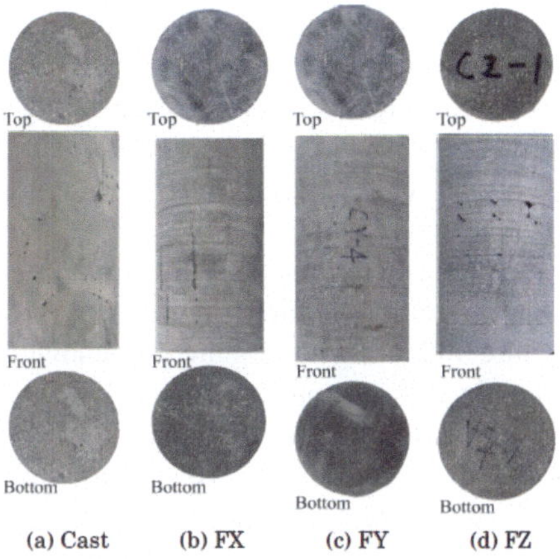

Fig. 2. Compression test specimens

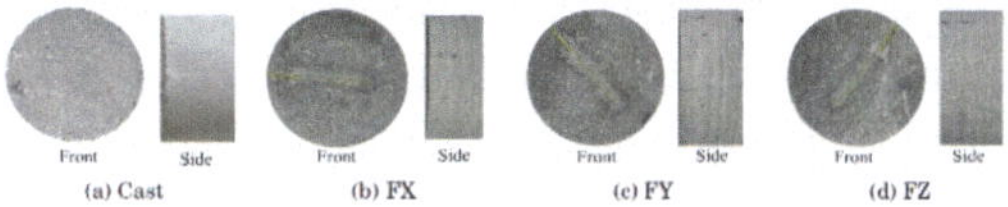

Fig. 3. Split test specimens

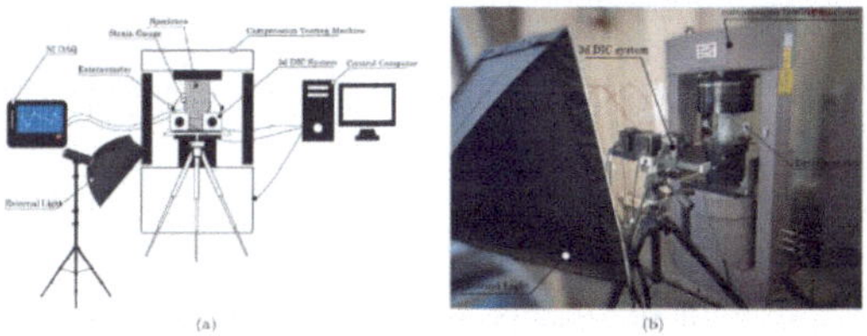

Fig. 4. Compression test

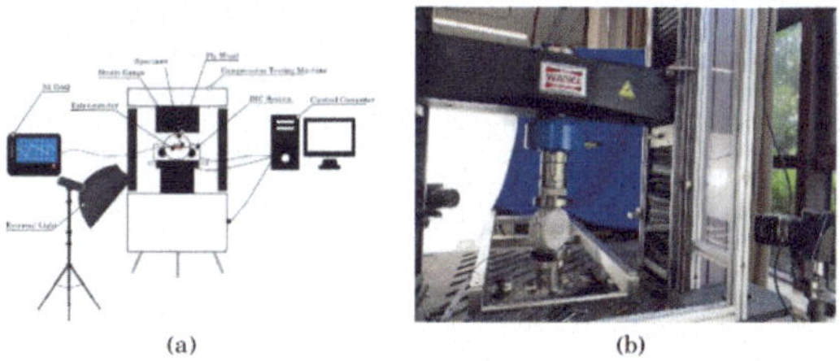

Fig. 5. Split test

Compression Test

The compression tests revealed significant anisotropic behavior in the 3D-printed fiber-reinforced concrete. As shown in Fig. 10, specimens oriented parallel to the printing direction (FX) demonstrated the highest compressive strength of 16.5 MPa, closely approaching that of traditional cast concrete (17 MPa). This high strength is attributed to effective fiber alignment and uniform layer deposition, which enhance load resistance. In contrast, specimens oriented perpendicular to the printing direction (FY) demonstrated moderate strength (12 MPa), while those oriented along the vertical direction (FZ) showed the lowest strength (8.5 MPa) due to weaker interlayer bonding, which limits vertical load-bearing capacity.

Strain propagation patterns further elucidated the failure mechanisms observed during compression testing. Cast concrete samples (Fig. 6) displayed uniform strain progression from initiation to fracture, reflecting homogeneous behavior. Similarly, FX-oriented samples (Fig. 7) showed localized strain concentrations along well-bonded layers, resulting in failure patterns akin to cast concrete. Conversely, FY-oriented samples (Fig. 8) exhibited less uniform strain propagation with weaker bonding along layer interfaces, and FZ-oriented samples (Fig. 9) presented the most non-uniform strain distribution, with significant strain concentrations at interlayer boundaries leading to early crack initiation and failure. These observations were corroborated by the stress-strain curves, where FX samples demonstrated a steep rise and gradual post-peak decline, indicative of strong interlayer bonding. FY samples displayed a less steep rise with earlier post-peak softening, reflecting weaker perpendicular bonding, while FZ samples exhibited gradual stress buildup and rapid post-peak decline, highlighting poor vertical strength and limited interlayer adhesion.

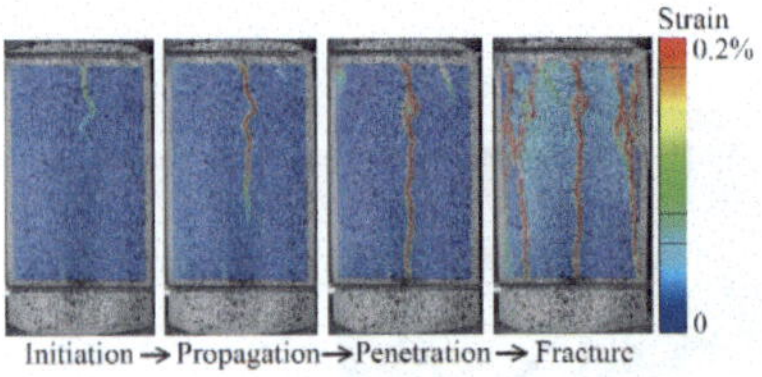

Fig. 6. Creak propagation for strain, Cast

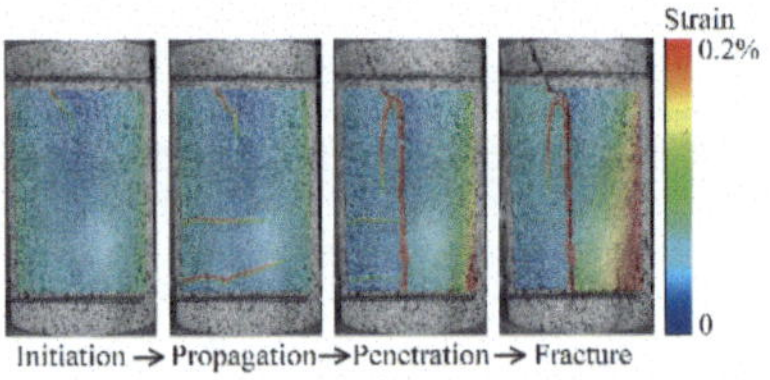

Fig. 7. Creak propagation for strain, FX

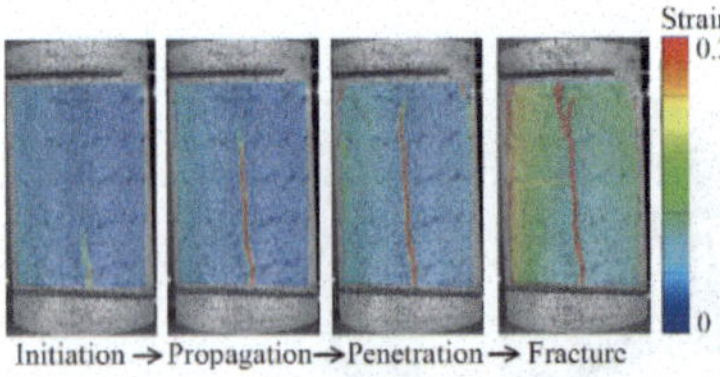

Fig. 8. Creak propagation for strain, FY

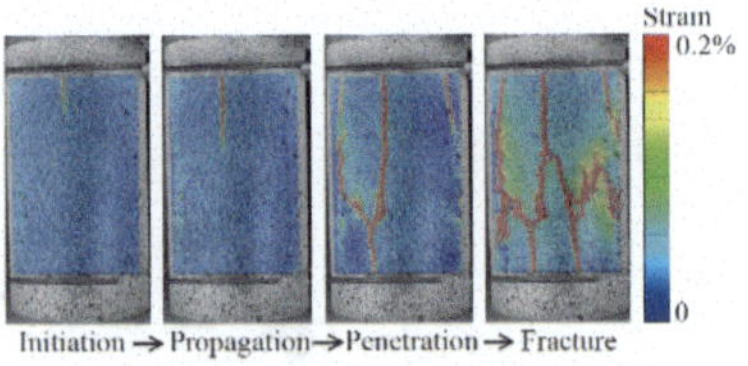

Fig. 9. Creak propagation for strain, FZ

Split Tensile Test

The split tensile test further confirmed the orientation-dependent behavior of the 3D-printed fiber-reinforced concrete. As shown in Fig. 14, specimens oriented parallel to the printing direction (FX) achieved the highest tensile strength (3 MPa), followed by perpendicular (FY) (2.5 MPa) and vertical (FZ) (2 MPa) orientations, aligning with the compression test results and underscoring the importance of print direction and interlayer bonding. Strain propagation in tensile tests revealed that FX-oriented samples (Fig. 11)

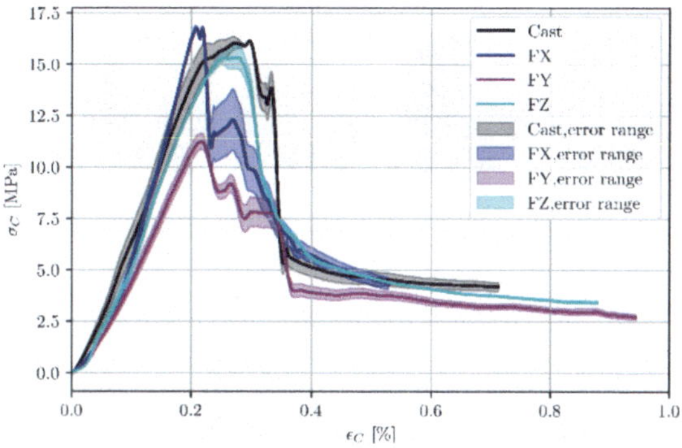

Fig. 10. Stress Vs. strain for compression test

exhibited uniform and gradual strain propagation with strong tensile resistance due to fiber alignment and effective layer bonding. The corresponding stress-strain curve (Fig. 14) revealed a steep initial rise and stable post-peak behavior. In contrast, FY-oriented samples (Fig. 12) displayed less uniform strain propagation, with cracks forming along weaker bonding interfaces perpendicular to the load direction, and the stress-strain curve showed earlier post-peak softening. FZ-oriented samples (Fig. 13) presented the lowest tensile strength and most uneven strain propagation, with cracks initiating at interlayer boundaries and propagating vertically, leading to early failure. The stress-strain curve for FZ samples indicated gradual stress buildup followed by rapid failure.

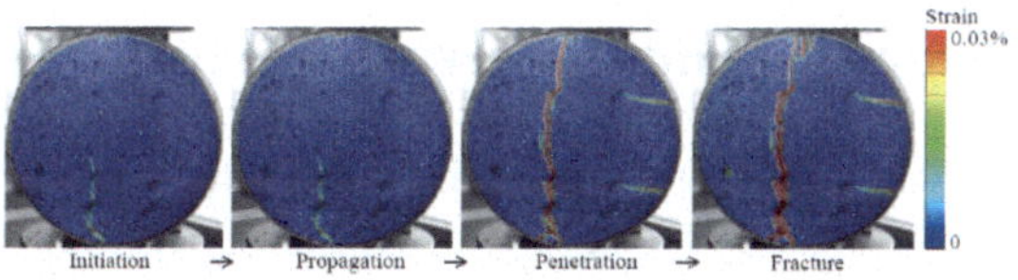

Fig. 11. Creak propagation for strain FX

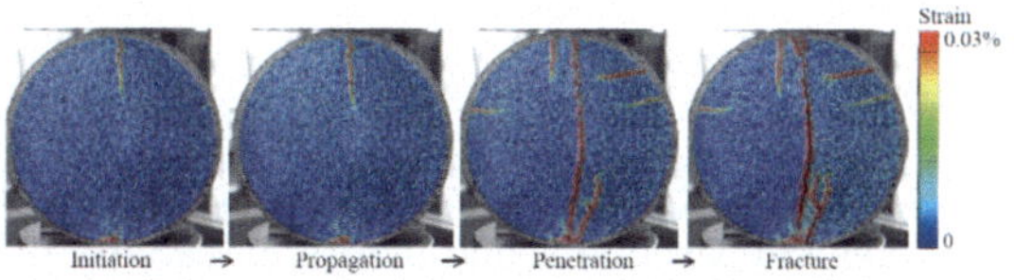

Fig. 12. Creak propagation for strain FY

These findings emphasize the critical role of print orientation and interlayer bonding in determining the mechanical performance of 3D-printed fiber-reinforced concrete. The comprehensive stress-strain data and strain propagation patterns obtained from these tests serve as key inputs for developing constitutive models that accurately simulate 3D-printed concrete behavior within the OpenSees TO framework. The experimental results highlight the potential of 3D-printed fiber-reinforced concrete as a viable alternative to

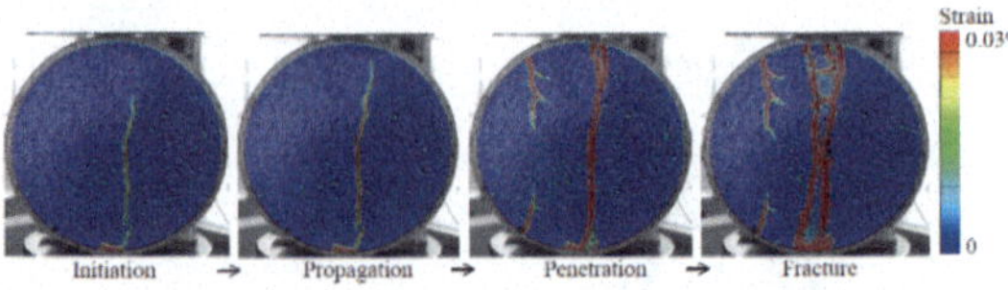

Fig. 13. Creak propagation for strain FZ

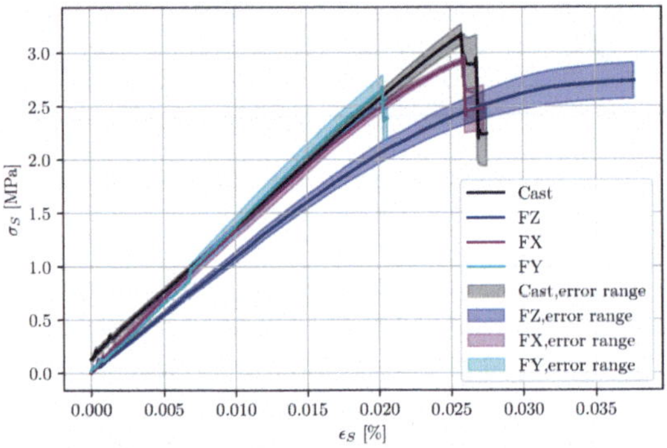

Fig. 14. Stress Vs. strain for split test

traditional cast concrete, particularly in applications requiring customized geometries and efficient material use.

4 TO Modeling for 3D Print Concrete Structure

These TO models provide a comprehensive framework for analyzing and optimizing 3D-printed concrete structures. By addressing different aspects of material behavior and reinforcement strategies, the models facilitate a deeper understanding of how topology optimization can enhance the performance and applicability of 3D-printed concrete in construction.

4.1 Modeling Selection

Two TO models (Model 1 and Model 2) were developed to evaluate the mechanical behavior of 3D-printed concrete structures and their applicability in real-world scenarios. Each model incorporates distinct reinforcement strategies and material properties to simulate different structural responses effectively. The design domain for both models was discretized into finite elements, with each element assigned a design variable representing its material density.

Model 1: Simply Supported Beam with Varying Compression-to-Tension Ratios

This model analyzes a simply supported beam with a 2-m span and 0.5-m height under different compression-to-tension strength ratios (1 and 2). For a tensile strength of 2.5 MPa, the corresponding compressive strengths were set at 2.5 MPa and 5.0 MPa, respectively. These variations assess how changes in the compression-to-tension ratio influence the optimized structural shape, demonstrating concrete's distinct behavior under varying loading conditions.

Model 2: Reinforced Concrete with Stress Constraints

The second model incorporates steel reinforcement using the von Mises yield criterion with stress constraints of 15 MPa and 10 MPa. This approach simulates the effect of embedded steel bars to enhance tensile properties. The objective is to investigate how different stress constraints affect the optimized topology, particularly in regions with high stress concentrations, thereby ensuring structural integrity while minimizing material usage.

To ensure manufacturability and eliminate numerical artifacts like checkerboard patterns, a filtering technique aligned with the 30 mm nozzle size used in the 3D printing process was employed. Figure 15 illustrates the filtering process utilized in both models, ensuring that the optimized designs are manufacturable by considering the spherical filtering range based on the nozzle size.

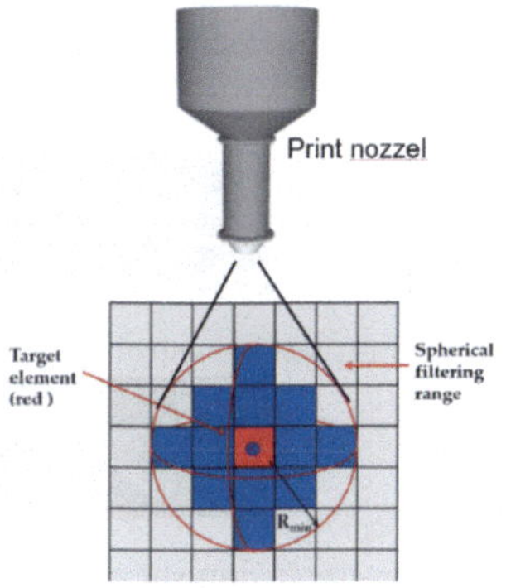

Fig. 15. Filter operator Vs. nozzel

4.2 Prototype Beam

The primary prototype consisted of a simply supported beam structure subjected to uniform loading, as depicted in Fig. 16. This experimental setup was designed to provide a clear assessment of the beam's load-bearing capacity and overall structural performance under realistic conditions.

The optimized beam geometries corresponding to compression-to-tension strength ratios of 1 and 2 are illustrated in Fig. 17. Analysis of the element density plots and stress distribution maps revealed distinct patterns based on the stress ratios. For a compression-to-tension strength ratio of 1, the optimized design exhibited a symmetric distribution of material, with load paths strategically oriented to achieve uniform stress distribution throughout the structure. In contrast, when the ratio was increased to 2, the design adapted by concentrating material in compression-dominant zones while reducing material allocation in tension zones. This redistribution effectively leveraged the higher compressive strength of the material, thereby enhancing structural efficiency.

Furthermore, the optimized designs under stress constraints of 15 MPa and 10 MPa are presented in Fig. 18. At a stress constraint of 15 MPa, the structure demonstrated a balanced material distribution, with steel reinforcement effectively bridging tension-dominant zones. The stress distribution map revealed smooth transitions between high-stress regions, indicating the reinforcement's role in enhancing tensile strength. Conversely, with a reduced stress constraint of 10 MPa, the design became more refined, allocating increased material to high-stress areas. This adaptation minimized stress concentrations, ensuring that the structure remained within the allowable stress limits.

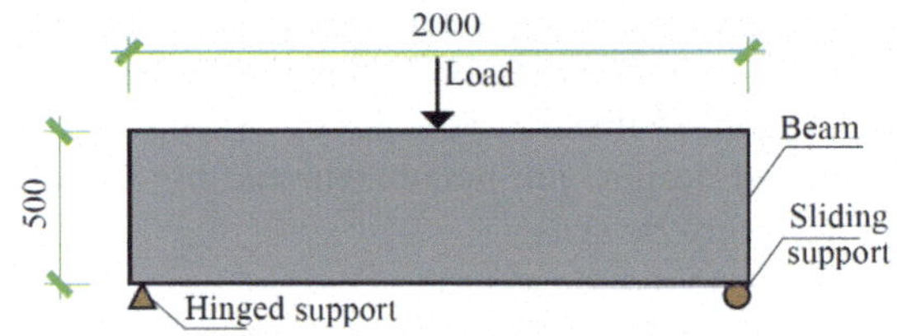

Fig. 16. TO prototype: simple support beam

(a) Compression-to-tension strength ratio=1

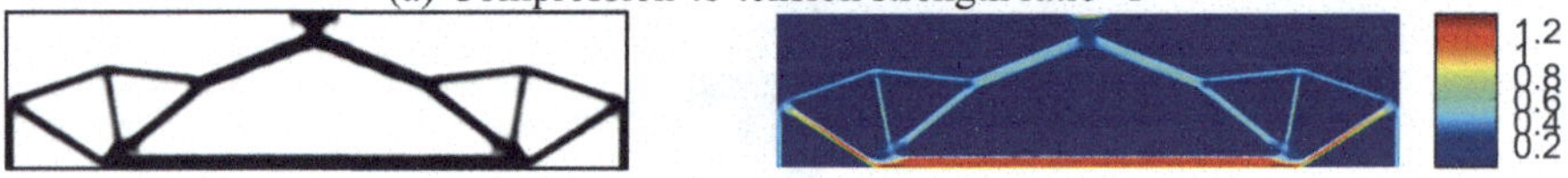

(b) Compression-to-tension strength ratio= 2

Fig. 17. Model 1: TO model of simple supported beam: Element densities, Stress distribution, Stress ratio

(a) Stress constrain, 15Mpa

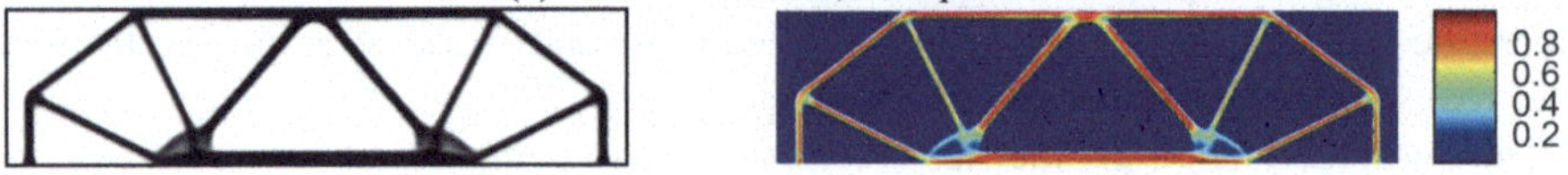

(b) Stress constrain, 10Mpa

Fig. 18. Model 2: Steel plate reinforced concrete: Element densities, Stress distribution, Stress ratio

The results from the two models demonstrate the effectiveness of topology optimization in designing 3D-printed concrete structures. Model 1 highlights the impact of varying compression-to-tension strength ratios on the optimized shape, showcasing the adaptability of concrete to different loading scenarios. Model 2 emphasizes the role of steel reinforcement in enhancing tensile properties and redistributing stresses within the structure.

By integrating the optimized designs with manufacturability considerations (e.g., nozzle size), this study bridges the gap between computational optimization and practical 3D printing applications. These findings serve as a foundation for future advancements in 3D-printed concrete technology, enabling more efficient and robust structural designs.

5 Conclusion

This study successfully evaluates the mechanical performance of 3D-printed fiber-reinforced concrete and demonstrates the effectiveness of integrating topology optimization (TO) within the OpenSees framework for enhanced structural design. Experimental tests revealed significant anisotropic behavior, with specimens aligned parallel to the printing direction (FX) exhibiting strengths comparable to traditional cast concrete, while perpendicular (FY) and vertical (FZ) orientations showed reduced performance due to weaker interlayer bonding.

The development and application of two distinct TO models underscored the adaptability of concrete structures under varying compression-to-tension ratios and the benefits of incorporating steel reinforcement. Model 1 highlighted how adjusting stress ratios can optimize material distribution for different loading scenarios, whereas Model 2 demonstrated that steel reinforcement effectively enhances tensile properties and redistributes stresses, ensuring structural integrity.

By aligning optimized designs with practical considerations such as nozzle size, this research bridges the gap between computational optimization and practical 3D printing applications. The findings lay a foundation for future advancements in 3D-printed concrete technology, promoting the development of more efficient and robust structural designs. Future work should focus on refining printing techniques to minimize anisotropy, exploring additional reinforcement methods, and expanding optimization models to further enhance the reliability and performance of 3D-printed structures.

References

1. Mechtcherine, V., Nerella, V.N., Will, F., Näther, M., Otto, J., Krause, M.: Large-scale digital concrete construction–CONPrint3D concept for on-site, monolithic 3D-printing. Autom. Constr. **107**, 102933 (2019)
2. Bedarf, P., Dutto, A., Zanini, M., Dillenburger, B.: Foam 3D printing for construction: A review of applications, materials, and processes. Autom. Constr. **130**, 103861 (2021)
3. Murcia, D.H., Genedy, M., Taha, M.R.: Examining the significance of infill printing pattern on the anisotropy of 3D printed concrete. Constr. Build. Mater. **262**, 120559 (2020)
4. Hou, S., Duan, Z., Xiao, J., Ye, J.: A review of 3D printed concrete: Performance requirements, testing measurements and mix design. Constr. Build. Mater. **273**, 121745 (2021)
5. McKenna, F.: OpenSees: a framework for earthquake engineering simulation. Computing in Science & Engineering **13**(4), 58–66 (2011)

6. Wang, J., et al.: An active-bending sheltered pathway based on bamboo strips for indoor temporary applications: Design and construction. Eng. Struct. **307**, 117863 (2024)
7. Jiang, L., Jiang, Y., Zhang, Z., Usmani, A.: Thermal analysis infrastructure in OpenSees for fire and its smart application interface towards natural fire modelling. Fire Technology, 1–26 (2021)
8. Jiang, J., Usmani, A.: Modeling of steel frame structures in fire using OpenSees. Comput. Struct. **118**, 90–99 (2013)
9. Zhang, Z., Jiang, L., Yarlagadda, T., Zheng, Y., Usmani, A.: OPS-ITO: development of isogeometric analysis and topology optimization in OpenSEES for free-form structural design. Comput. Aided Des. **160**, 103517 (2023)

OpenSees-Assisted Optimization of Sensor Placement for Moving Force Identification Using an Augmented Kalman Filter

Chenyu Zhou[1,2], Yongjia Xu[3], Cristoforo Demartino[4], and Yasutaka Narazaki[2](✉)

[1] College of Civil Engineering and Architecture, Zhejiang University, Hangzhou 310058, Zhejiang, China

[2] Zhejiang University - University of Illinois Urbana-Champaign Institute, Zhejiang University, Haining 314400, Zhejiang, China
narazaki@intl.zju.edu.cn

[3] Department of Civil Engineering, Central South University, Changsha 410017, Hunan, China

[4] Department of Architecture, Roma Tre University, Largo G. B. Marzi 10, 00153 Rome, Italy

Abstract. Identifying the forces acting on bridges is crucial for enhancing safety and guiding maintenance and design improvements. This study investigates the optimization of sensor placement for moving force identification (MFI), addressing limitations in existing research, which often simplifies sensor placement analyses. An augmented Kalman filter (AKF) is employed for force identification, using acceleration measurement of the bridge. Finite element simulations, conducted using OpenSees, are integrated to generate the dynamic responses of the bridge under moving forces, providing data for the filter and optimization. To optimize sensor placement, the Non-dominated Sorting Genetic Algorithm II (NSGA-II) is applied, enabling a multi-objective analysis of trade-offs between sensor count and identification accuracy. Results demonstrate the feasibility of integrating AKF, OpenSees simulations, and NSGA-II for sensor optimization, offering practical strategies for configuring sensors under real-world constraints. The findings highlight the impact of sensor placement on MFI accuracy, showing the balance between sensor count and error reduction. This research contributes to MFI techniques by incorporating modeling with OpenSees and offers insights for sensor configuration design in bridge structure applications.

Keywords: Optimal sensor placement · Moving Force Identification · Augmented Kalman filter · Non-dominated Sorting Genetic Algorithm II

1 Introduction

Bridges are critical components of urban infrastructure, essential for city operations. Understanding the forces on a bridge can greatly enhance personnel safety and inform improvements in bridge design and maintenance [1].

Since Moses introduced the strain-based identification algorithm for bridges [2], significant efforts have been made to develop methods for identifying moving forces

X. Lu et al. (Eds.): EOS 2024, LNCE 638, pp. 253–260, 2025.
https://doi.org/10.1007/978-3-031-90690-9_22

on bridge structures. Law proposed a time-domain transformation technique to identify moving forces on a simply supported beam using vibration responses [3]. Chen introduced the preconditioned least squares QR-factorization method for force identification [4]. Zhou developed a semi-analytical method for moving force identification (MFI) based on finite element (FE) modeling and discrete cosine transforms [5]. Research in this area has improved MFI techniques considerably.

However, optimizing sensor placement for force identification has received relatively little attention. The configuration of sensors on a bridge plays a pivotal role in ensuring the accuracy of load identification. Sensitivity analyses of sensor placement are typically conducted using methods proposed in the literature for force identification [6, 7]. Nonetheless, these studies often focus on the influence of sensor count or conduct simplified effect analyses. A more comprehensive understanding of sensor placement optimization in the context of MFI is needed.

This study analyzes the trade-offs among sensor location, sensor count, and the identification error of moving forces. The augmented Kalman filter (AKF), a proven effective approach for moving force identification [8], is employed in this research. To address the time-variant nature of applying AKF for MFI, the Non-dominated Sorting Genetic Algorithm II (NSGA-II), a genetic algorithm designed for multi-objective optimization, is applied [9]. The acceleration time history of the bridge structure is utilized as the measurement input for the Kalman filter.

The rest of this paper is structured as follows: Sect. 2 presents the modeling equations for the Kalman filter. Section 3 introduces the optimization approach for the sensor placement. Section 4 presents a numerical example. Finally, Sect. 5 provides conclusions and discussions.

2 Dynamic Modeling

This section presents the modeling of the interaction between moving forces and the bridge, as well as the equations for the Kalman filter.

2.1 Equation of Motion

Figure 1 illustrates a simply supported beam subjected to two moving forces, denoted as f_1 and f_2 with a given velocity s and a fixed axle distance b.

The equation of motion can be represented as:

$$\boldsymbol{M}\ddot{\boldsymbol{u}}(t) + \boldsymbol{C}\dot{\boldsymbol{u}}(t) + \boldsymbol{K}\boldsymbol{u}(t) = \boldsymbol{S}_f(t)\boldsymbol{f}(t) \tag{1}$$

where $\boldsymbol{u}(t) \in \mathbb{R}^{n_d}$ is the displacement vector, n_d denotes the number of degrees of freedom (DOFs) of the model. $\boldsymbol{M} \in \mathbb{R}^{n_d \times n_d}$, $\boldsymbol{K} \in \mathbb{R}^{n_d \times n_d}$, and $\boldsymbol{C} \in \mathbb{R}^{n_d \times n_d}$ are the global mass matrix, stiffness matrix, and damping matrix, respectively, $\boldsymbol{f}(t) \in \mathbb{R}^{n_f}$ represents the force vector where n_f denotes the number of the forces, $\boldsymbol{S}_f(t) \in \mathbb{R}^{n_d \times n_f}$ is the force distribution matrix indicating how moving loads apply to the structure. The Rayleigh damping is applied in this study to construct the damping matrix $\boldsymbol{C}$. OpenSees, an open-source software framework for finite element modeling, is utilized for this study [10]. OpenSeesPy, the package for OpenSees in Python, is employed in the study for running the analysis [11].

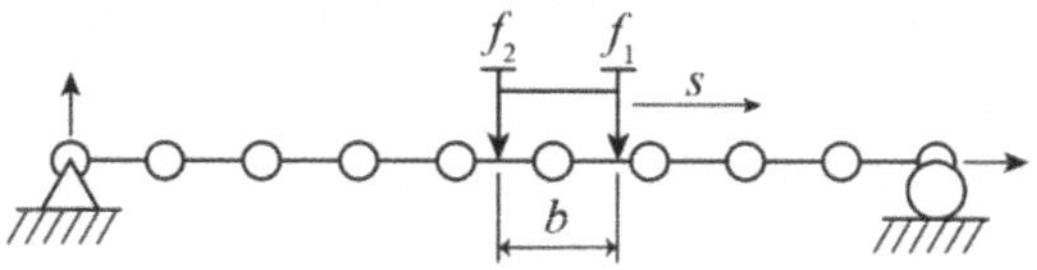

Fig. 1. Diagram of a simply supported beam subjected to a two moving forces.

2.2 Sensor Configuration Vector

The vector denoting the sensor configuration is given as follows:

$$\boldsymbol{d} = \left[d_1, d_2, \cdots, d_{n_m}\right]^{\mathrm{T}} \in \mathbb{R}^{n_m} \tag{2}$$

where n_m is the number of the measurement points and d_i represents the i-th selected sensor locations. It should be noted that in the optimization process, n_m can vary for different individuals.

Given $\boldsymbol{d}$, $\boldsymbol{M}$, $\boldsymbol{K}$, and $\boldsymbol{C}$ by the function of OpenSeesPy, the augmented output influence matrix $\boldsymbol{H}_k(\boldsymbol{d})$ can be constructed through the traversing function, the interpolation function, and the discretization function.

2.3 Augmented Kalman Filter

The augmented state-space representation can be expressed as:

$$\begin{cases} \boldsymbol{X}_k = \boldsymbol{A}_{k-1}\boldsymbol{X}_{k-1} + \boldsymbol{w}_{k-1} \\ \ddagger_k = \boldsymbol{H}_k(\boldsymbol{d})\boldsymbol{X}_k + \boldsymbol{v}_k \end{cases} \tag{3}$$

where $\boldsymbol{X}_k \in \mathbb{R}^{2n_d+n_f}$ is the augmented stat vector, $\ddagger_k \in \mathbb{R}^{n_z}$ denotes the measured vector and n_z is the number of the measured points. $\boldsymbol{w}_{k-1}$ is the process noise denoting the zero-mean noise in the state equation with the associated covariance matrix $\boldsymbol{Q} \in \mathbb{R}^{(2n_d+n_f)\times(2n_d+n_f)}$, and $\boldsymbol{v}_k$ denotes the zero-mean measurement noise with associated covariance matrix $\boldsymbol{R} \in \mathbb{R}^{n_z \times n_z}$.

The applied equations of the Kalman filter are given as:
i. Measurement update

$$\begin{aligned} \boldsymbol{L}_k &= \boldsymbol{P}_k^- \boldsymbol{H}_k(\boldsymbol{d})^{\mathrm{T}}\left(\boldsymbol{H}_k(\boldsymbol{d})\boldsymbol{P}_k^- \boldsymbol{H}_k(\boldsymbol{d})^{\mathrm{T}} + \boldsymbol{R}\right)^{-1} \\ \widehat{\boldsymbol{X}}_k &= \widehat{\boldsymbol{X}}_k^- + \boldsymbol{L}_k\left(\ddagger_k - \boldsymbol{H}_k(\boldsymbol{d})\widehat{\boldsymbol{X}}_k^-\right) \\ \boldsymbol{P}_k &= (\boldsymbol{I} - \boldsymbol{L}_k\boldsymbol{H}_k(\boldsymbol{d}))\boldsymbol{P}_k^-(\boldsymbol{I} - \boldsymbol{L}_k\boldsymbol{H}_k(\boldsymbol{d}))^{\mathrm{T}} + \boldsymbol{L}_k\boldsymbol{R}\boldsymbol{L}_k^{\mathrm{T}} \end{aligned} \tag{4}$$

ii. Time update

$$\begin{aligned} \widehat{\boldsymbol{X}}_{k+1}^- &= \boldsymbol{A}_k\widehat{\boldsymbol{X}}_k \\ \boldsymbol{P}_{k+1}^- &= \boldsymbol{A}_k\boldsymbol{P}_k\boldsymbol{A}_k^{\mathrm{T}} + \boldsymbol{Q} \end{aligned} \tag{5}$$

where $\widehat{\boldsymbol{X}}_k^-$ and $\widehat{\boldsymbol{X}}_k$ are the prior and posterior estimation of the augmented state vector at k-th step respectively, $\boldsymbol{P}_k^-$ and $\boldsymbol{P}_k$ are the prior and posterior error covariance matrix at k-th step respectively, and $\boldsymbol{L}_k$ denotes the Kalman gain.

3 Optimization Method

This section presents the optimization method applied in this study.

3.1 Error Equation

To evaluate the accuracy of the identification, the error is computed as the absolute of the difference between the true and estimated mean axle force vectors, expressed as:

$$\mathrm{E} = \frac{1}{n_f}\sum\nolimits_1^{n_f}\frac{|f_{true} - f_i|}{|f_{true}|} \times 100\% \tag{6}$$

where f_i represents the mean identified force vector over the selected period and f_{true} is the true force value.

3.2 Optimization Equation

The optimization equation is given as follows:

$$\begin{gathered} \min_{\boldsymbol{d}}\begin{cases} E(\boldsymbol{d}) \\ W(\boldsymbol{d}) \end{cases} \\ \text{subject to} \\ \boldsymbol{d} = \left[d_1, d_2, \cdots, d_{n_m}\right]^{\mathbf{T}} \in \mathbb{R}^{n_m}, n_{min} \le n_m \le n_{max} \\ 0 \le d_i \le L \end{gathered} \tag{7}$$

where E($\boldsymbol{d}$) is the first cost function calculate by Eq. 6 and W($\boldsymbol{d}$) is the second cost function denotes the sensor count in the sensor placement vector. n_{min} And n_{max} denote the selected minimum and maximum number of the sensor used in the placement. L Denotes the length of the structure.

3.3 Non-dominated Sorting Genetic Algorithm II

NSGA-II operates on four main principles [9]: Non-Dominated Sorting, Elite Preservation, Crowding Distance, and Selection. In this study, the NSGA-II process is implemented as follows: (1) Evaluate the offspring by the AKF; (2) Combine and rank the population using non-dominated sorting; (3) Select the next generation using elite preservation and crowding distance; (4) Apply genetic operators, such as crossover and mutation, to generate new offspring; (5) Repeat until the termination condition is met. The common genetic algorithm operators, such as crossover and mutation, are adapted to suit the optimization process.

4 Numerical Example

This section presents a numerical example illustrating the proposed framework.

4.1 Finite Element Model

A three-span continuous bridge model subjected to moving forces is analyzed in this study which is presented in the Fig. 2. The model, developed using OpenSees, has 36 elastic beam-column elements in total, with horizontal DOFs fixed, resulting in 70 total DOFs. The model parameters are summarized in Table 1, and the properties of the first three modes, obtained from the OpenSees analysis, are presented in Fig. 3. The time step is set to 0.005 s, corresponding to a sampling frequency of 200 Hz.

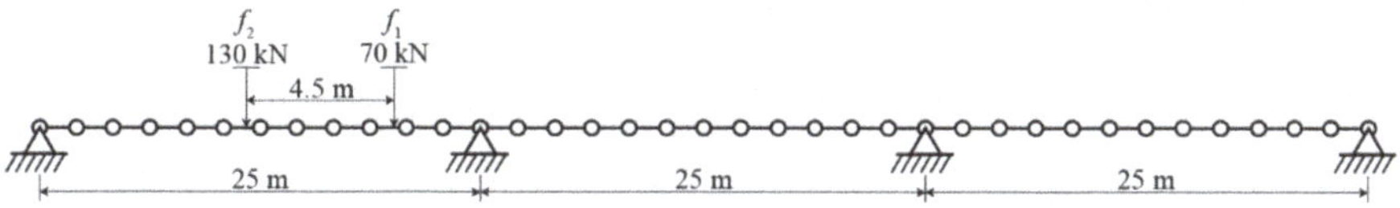

Fig. 2. Finite element model of a three-span continuous bridge and the configuration of two moving forces.

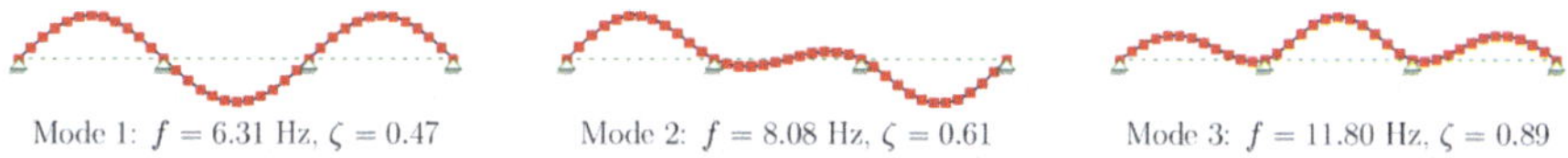

Fig. 3. First 3 mode shapes with corresponding frequencies and damping ratios for the model obtained from the OpenSees analysis.

4.2 Parameter Setting

The first span is selected as the identification period. The vehicle velocity is set at 30 m/s. Noise levels for both location noise and measurement noise are set at 3%, following the method described in [9]. Model error is introduced by applying normally distributed random perturbations, with a mean of 0 and a standard deviation of 3%, to the elastic modulus and damping coefficients of the finite element model in OpenSees. Table 2 shows the comparison between the FE model and the model with imperfections.

Table 1. Model parameter for the span of the bridge in OpenSees.

Parameter	Value	Parameter	Value
Length (m)	25	Section inertial moment (m^4)	1
Element number	12	Elastic modulus (Pa)	3.15×10^{10}
Density (kg/m^3)	2500	Damping coefficient α	1.5×10^{-1}
Section area (m^2)	2	Damping coefficient β	5×10^{-5}

Table 2. Comparison of frequencies and damping ratios between the finite element model and the model with imperfections in OpenSees.

Property	Statistic	Mode 1	Mode 2	Mode 3
Frequency	Analysis value (Hz)	6.28	8.05	11.76
	Error (%)	0.48	0.37	0.34
Damping ratio	Analysis value	0.46	0.58	0.85
	Error (%)	2.13	4.92	4.49

The bridge is assumed to be undeformed before the crossing, with the initial displacement and velocity state variances of $\boldsymbol{P}_1^-$ set to $0\,\mathrm{m}^2$ and $0\,\mathrm{m}^2/\mathrm{s}^2$, respectively. The initial force variance of $\boldsymbol{P}_1^-$ is assigned a large value of $4 \times 10^4\,\mathrm{kN}^2$ to represent a non-informative prior. The process noise covariance matrix $\boldsymbol{Q}$ is determined using a trial-and-error approach, with corresponding values for displacement, velocity, and inputs set to 10^{-16}, 10^{-16}, and 10^6, respectively. The measurement noise covariance matrix $\boldsymbol{R}$ is set based on the introduced noise in the structural response.

The available range for the number of the sensor is set from $n_{min} = 3$ to $n_{max} = 12$. The Monte Carlo method is applied, and the average of 50 identification error trials is applied to evaluate the identification error. Genetic algorithm parameters are defined as follows: a population size of 40, 20 offspring per generation, and a termination condition of 40 generations. The probabilities for crossover and mutation are set to 0.80 and 0.05, respectively. The random distribution and uniform distribution of the placements are generated as the initial sampling for the optimization.

4.3 Optimization Result

This subsection presents the results of optimizing the sensor placement using the acceleration under the presented bridge model and parameter setting.

Figure 4 shows the accumulated extreme point difference, a value for evaluating the progress of the optimization across generations. The accumulated difference illustrates the improvement from the initial sampling. A decreasing slope in the later generations indicates that the optimization process has reached a point where termination is appropriate.

Figure 5 illustrates the optimized results of the final generation using acceleration. As the number of sensors increases, the identification error for weight decreases, which is consistent with common knowledge. 5 Ranks are retained in the final generation.

Figure 6 presents the sensor layouts for Rank 1 individuals in the final generation, based on acceleration measurements. Each row represents the sensor placements of Rank 1 individuals. Notably, in most Rank 1 sensor layouts, the majority of sensors are concentrated within the first span, corresponding to the selected identification period.

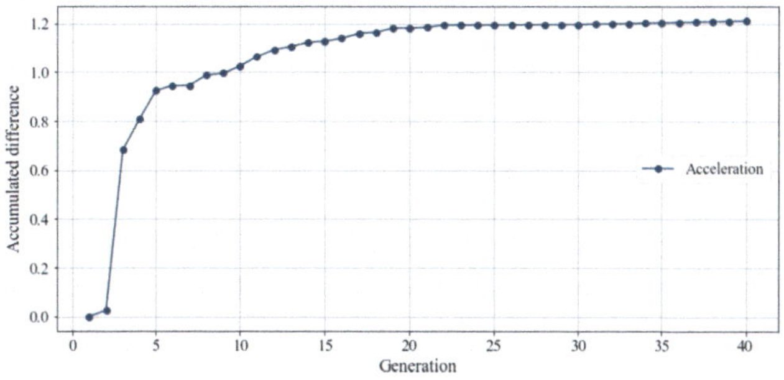

Fig. 4. Accumulated difference across optimization generations.

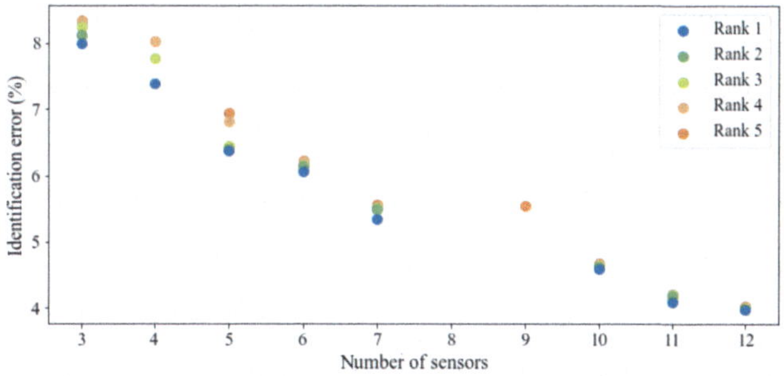

Fig. 5. Optimized result of the final generation categorized by ranks.

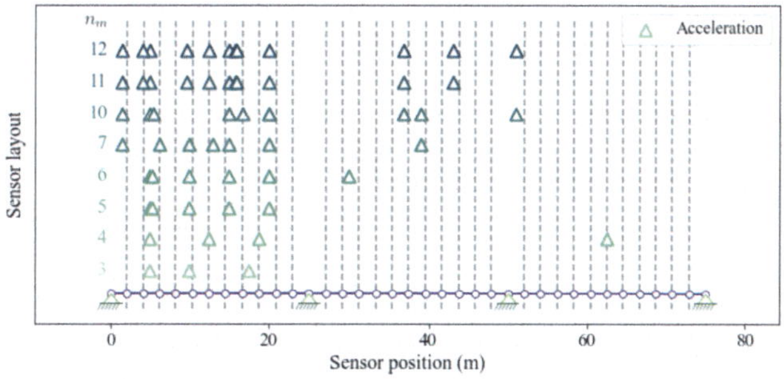

Fig. 6. Optimized sensor layout for Rank 1 individuals from Fig. 5.

5 Conclusions

This study presents an analysis of sensor placement optimization for moving force identification by integrating the AKF and NSGA-II. The trade-off between the number of sensors and the identification error has been conducted.

The following conclusions can be drawn: (1) This study confirms the feasibility of optimizing sensor placement for the AKF in identifying moving forces using NSGA-II. (2) The investigation of the trade-off between sensor counts and identification error offers valuable insights for sensor placement design.

Future work will focus on extending the proposed methodology to more complex scenarios and real-world applications.

References

1. Lydon, M., Taylor, S.E., Robinson, D., Mufti, A., Brien, E.J.O.: Recent developments in bridge weigh in motion (B-WIM). J. Civ. Struct. Heal. Monit. **6**, 69–81 (2016)
2. Moses, F.: Weigh-in-motion system using instrumented bridges. Transp. Eng. J. ASCE **105**(3), 233–249 (1979)
3. Law, S.S., Chan, T.H., Zeng, Q.H.: Moving force identification—a frequency and time domains analysis (1999)
4. Chen, Z., Chan, T.H., Nguyen, A., Yu, L.: Identification of vehicle axle loads from bridge responses using preconditioned least square QR-factorization algorithm. Mech. Syst. Signal Process. **128**, 479–496 (2019)
5. Zhou, X., He, W., Zeng, Y., Zhang, Y.: A semi-analytical method for moving force identification of bridge structures based on the discrete cosine transform and FEM. Mech. Syst. Signal Process. **180**, 109444 (2022)
6. He, W.Y., Li, Y.L., Yi, J.X., Ren, W.X.: Time-domain identification of moving load on beam type bridges considering interval uncertainty in finite element model. Mech. Syst. Signal Process. **191**, 110168 (2023)
7. Liu, H., Yu, L., Luo, Z., Pan, C.: Compressed sensing for moving force identification using redundant dictionaries. Mech. Syst. Signal Process. **138**, 106535 (2020)
8. Zhou, C., Butala, M.D., Xu, Y., Demartino, C., Spencer, B.F., Jr.: FE-based bridge weigh-in-motion based on an adaptive augmented Kalman filter. Mech. Syst. Signal Process. **218**, 111530 (2024)
9. Deb, K., Pratap, A., Agarwal, S., Meyarivan, T.A.M.T.: A fast and elitist multiobjective genetic algorithm: NSGA-II. IEEE Trans. Evol. Comput. **6**(2), 182–197 (2002)
10. McKenna, F.: OpenSees: a framework for earthquake engineering simulation. Computing in Science & Engineering **13**(4), 58–66 (2011)
11. Zhu, M., McKenna, F., Scott, M.H.: OpenSeesPy: Python library for the OpenSees finite element framework. SoftwareX **7**, 6–11 (2018)

Author Index

X. Lu et al. (Eds.): EOS 2024, LNCE 638, pp. 261–262, 2025.
https://doi.org/10.1007/978-3-031-90690-9

The manufacturer's authorised representative in the EU is Springer Nature Customer Service Centre GmbH, Europaplatz 3, 69115 Heidelberg, Germany. If you have any concerns regarding our products, please contact ProductSafety@springernature.com

Printed and bound by CPI Group (UK) Ltd, Croydon, CR0 4YY
30/09/2026
02225596-0004